AF368960

High-Performance Eco-Efficient Concrete

High-Performance Eco-Efficient Concrete

Editors

Carlos Thomas
Jorge de Brito
Valeria Corinaldesi

MDPI • Basel • Beijing • Wuhan • Barcelona • Belgrade • Manchester • Tokyo • Cluj • Tianjin

Editors
Carlos Thomas
University of Cantabria
Spain

Jorge de Brito
University of Lisbon
Portugal

Valeria Corinaldesi
Università Politecnica delle
Marche
Italy

Editorial Office
MDPI
St. Alban-Anlage 66
4052 Basel, Switzerland

This is a reprint of articles from the Special Issue published online in the open access journal *Applied Sciences* (ISSN 2076-3417) (available at: https://www.mdpi.com/journal/applsci/special_issues/Eco-Efficient_Concrete).

For citation purposes, cite each article independently as indicated on the article page online and as indicated below:

LastName, A.A.; LastName, B.B.; LastName, C.C. Article Title. *Journal Name* **Year**, *Volume Number*, Page Range.

ISBN 978-3-0365-0860-3 (Hbk)
ISBN 978-3-0365-0861-0 (PDF)

Contents

About the Editors

Carlos Thomas is an Associate Professor at the Materials Science and Engineering Laboratory in the Civil Engineering School of the University of Cantabria, Spain. He received his International Ph.D. Cum Laude and his Extraordinary Doctorate Award in 2013. His research activity has focused on the evaluation of construction, demolition and industry wastes for the manufacture of recycled mortars and concrete. In the last five years, he has participated as a Principal Investigator in 5 R+D+i projects and as a researcher in another 10 R+D+i projects, with both public and private funding, related to recycled materials. He is the author of 20 Q1 JCR papers, one of them certified as one of the 25 most downloaded in ScienceDirect, and more than 80 congress communications and conferences. He has undertaken research stays in France, Germany, Brazil and Portugal and he is one of the promoters of the "Spanish Recycled Concrete Network" that has led to the publication of this book. carlos.thomas@unican.es

Jorge de Brito is a Full Professor of Civil Engineering at Instituto Superior Técnico (IST), University of Lisbon, Portugal, where he received his Ph.D. in 1993. He is the author of over 500 research articles in peer-reviewed journals, and 8 books, including "Recycled Aggregate in Concrete: Use of Industrial, Construction and Demolition Waste" (Springer) and "Sustainable Construction Materials: Recycled aggregates" (Woodhead). He was co-editor of "Handbook of Recycled Concrete and Demolition Waste" (Woodhead) and "New Trends on eco-efficient and recycled concrete" (Woodhead). He is Editor-in-Chief of the *Journal of Building Engineering* (Elsevier) and is part of the Editorial Board of 45 peer-reviewed journals. He belongs to several CIB, CEN, RILEM and IABSE committees. He was the Head of the CERIS Research Centre (around 400 researchers) in 2017–2018 and is the Head of the Department of Civil Engineering, Architecture and Georesources, and Director of the Eco-Construction and Rehabilitation Doctoral Program from IST, University of Lisbon. jb@civil.ist.utl.pt.

Valeria Corinaldesi is a Full Professor of "Materials Science and Technology" at Università Politecnica delle Marche (Italy). She completed her M.Sc. Degree in Civil Engineering at the University of Ancona (Italy) in 1998; her Ph.D. in Materials' Engineering at the University of Bologna (Italy) in 2002. She is the co-author of more than 300 publications, mainly on peer-reviewed international journals and conference proceedings and the co-inventor of 3 national patents and 1 european patent. Her overall scientific production received more than 3000 citations with H-index = 32, Orcid code: 0000-0001-9372-5032. She contributed to more than 50 international conferences as a speaker and/or member of either scientific or organizing committees. She is a referee for many international scientific journals indexed on WoS and Scopus databases; in particular, she is a member of the Editorial Board of the 'Journal of Building Engineering'. She is a Member of the World Road Association-PIARC (Italian National Committee), Ministry of Infrastructures and Transports, of the American Concrete Institute—Italy Chapter, of RILEM TC "Structural behavior and innovation of recycled aggregate concrete". v.corinaldesi@staff.univpm.it.

applied sciences

MDPI

Editorial

Special Issue High-Performance Eco-Efficient Concrete

Carlos Thomas [1,*], Jorge de Brito [2] and Valeria Corinaldesi [3]

1   LADICIM, E.T.S. de Ingenieros de Caminos, Canales y Puertos, Universidad de Cantabria, Av./Los Castros 44, 39005 Santander, Spain
2   Instituto Superior Técnico, Universidade de Lisboa, Av. Rovisco Pais, 1049-001 Lisbon, Portugal; jb@civil.ist.utl.pt
3   Department of Materials, Università Politecnica delle Marche, Via Brecce Bianche 12, 60131 Ancona, Italy; v.corinaldesi@univpm.it
*   Correspondence: thomasc@unican.es

Citation: Thomas, C.; de Brito, J.; Corinaldesi, V. Special Issue High-Performance Eco-Efficient Concrete. *Appl. Sci.* **2021**, *11*, 1163. https://doi.org/10.3390/app11031163

Received: 1 January 2021
Accepted: 25 January 2021
Published: 28 January 2021

Publisher's Note: MDPI stays neutral with regard to jurisdictional claims in published maps and institutional affiliations.

1. Introduction

The benefits of recycling in the construction sector have been widely demonstrated and are unquestionable. The use of recycled aggregates, steel slags, and low-impact cements implies an important reduction of the environmental footprint, and eco-efficient concretes made with them must be a priority. However, these materials show, in some cases, losses of mechanical and durability behavior compared with natural materials. This is why we must invest our efforts in finding high-performance eco-efficient concretes that can compete—or even surpass—traditional concrete. To achieve this, research and dissemination of its results is essential. The objective of this Special Issue is to group the most recent and relevant research in relation to high-performance eco-efficient concrete into a single document. Subsequently, the possibility of publishing a book with the contributions of all authors will be assessed.

So far, 16 papers have been published in the Special Issue out of a total of 21 submitted. The next sections provide a brief summary of each of the papers published.

2. High-Frequency Fatigue Testing of Recycled Aggregate Concrete

Sainz-Aja et al. [1] show that concrete fatigue behavior has not been extensively studied, in part because of the difficulty and cost. Some concrete elements subjected to this type of load include the railway superstructure of sleepers or slab track, bridges for both road and rail track, and the foundations of wind turbine towers or offshore structures. In order to address fatigue problems, a methodology was proposed that reduces the lengthy testing time and high cost by increasing the test frequency up to the resonance frequency of the set formed by the specimen and the test machine. After comparing this test method with conventional frequency tests, it was found that tests performed at a high frequency (90 Hz) were more conservative than those performed at a moderate frequency (10 Hz); this effect was magnified in those concretes with recycled aggregates coming from crushed concrete (RC-S). In addition, it was found that the resonance frequency of the specimen–test machine set was a parameter capable of identifying whether the specimen was close to failure.

3. Mechanical and Durability Properties of Concrete with Coarse Recycled Aggregate Produced with Electric Arc Furnace Slag Concrete

Tamayo et al. [2] show the search for more sustainable construction materials, capable of complying with quality standards and current innovation policies, aimed at saving natural resources and reducing global pollution, is one of the greatest present societal challenges. In this study, an innovative recycled aggregate concrete (RAC) is designed and produced based on the use of a coarse recycled aggregate (CRA) crushing concrete with electric arc furnace slags as aggregate. These slags are a by-product of the steelmaking industry and their use, which avoids the use of natural aggregates, is a new trend in concrete and pavement technology. This paper has investigated the effects of incorporating this type of CRA in concrete at several replacement levels (0%, 20%, 50%, and 100% by volume), by means of the physical, mechanical, and durability characterization of the mixes. The analysis of the results has allowed the benefits and disadvantages of these new CRAs to be established, by comparing them with those of a natural aggregate concrete (NAC) mix (with 0% CRA incorporation) and with the data available in the literature for concrete made with more common CRA, based on construction and demolition waste (CDW). Compared to NAC, similar compressive strength and tensile strength values for all replacement ratios have been obtained. The modulus of elasticity, the resistance to chloride penetration, and the resistance to carbonation are less affected by these CRA than when CRA from CDW waste is used. Slight increases in bulk density over 7% were observed for total replacement. Overall, functionally good mechanical and durability properties have been obtained.

4. High Performance Self-Compacting Concrete with Electric Arc Furnace Slag Aggregate and Cupola Slag Powder

Sosa et al. [3] present the development of self-compacting concretes with electric arc furnace slags is a novelty in the field of materials and the production of high-performance concretes with these characteristics is a further achievement. To obtain these high-strength, low-permeability concretes, steel slag aggregates and cupola slag powder are used. To prove the effectiveness of these concretes, they are compared with control concretes that use diabase aggregates, fly ash, and limestone supplementary cementitious materials (SCMs, also called fillers), and intermediate mix proportions. The high density SCMs give the fresh concrete self-compacting thixotropy using high-density aggregates with no segregation. Moreover, the temporal evolution of the mechanical properties of mortars and concretes shows pozzolanic reactions for the cupola slag. The fulfillment of the demands in terms of stability, flowability, and mechanical properties required for this type of concrete, and the savings of natural resources derived from the valorization of waste, make these sustainable concretes a viable option for countless applications in civil engineering.

5. Economic and Technical Viability of Using Shotcrete with Coarse Recycled Concrete Aggregates in Deep Tunnels

This work [4] analyzes the technical and economic viability of using coarse recycled aggregates from crushed concrete in shotcrete, as a primary lining support in tunnels. Four incorporation ratios of coarse natural aggregate (CNA) with coarse recycled concrete aggregates from concrete (CRCA) were studied: 0%, 20%, 50%, and 100%. The mechanical properties of the dry-mix shotcrete were obtained in an independent experimental campaign. Initially, the technical viability of CRCA shotcrete was validated for deep rock tunnels, based on the convergence-confinement method. Two cases were studied to determine the equivalent thickness for each combination of replacement ratio using CRCA shotcrete: (i) similar stiffness and (ii) similar yield stress. Subsequently, an economic assessment was performed. The stiffness criterion increased the thickness below 10% in both the 20% and 50% replacement ratios, which shows their technical viability with very marginal cost increase (<5%). On the other hand, the maximum pressure criteria required higher increments, close to 30% in the 50% replacement ratio. A full replacement was proven impracticable in both analyses.

6. Experimental Characterization of Prefabricated Bridge Deck Panels Prepared with Prestressed and Sustainable Ultra-High Performance Concrete

Enhanced quality and reduced on-site construction time are the basic features of prefabricated bridge elements and systems [5]. Prefabricated lightweight bridge decks have already started finding their place in accelerated bridge construction (ABC). Therefore, the development of deck panels using high strength and high performance concrete has become an active area of research. Further optimization in such deck systems is possible using prestressing or replacement of raw materials with sustainable and recyclable materials. This research involves experimental evaluation of six full-depth precast prestressed high strength fiber-reinforced concrete (HSFRC) and six partial-depth sustainable ultra-high performance concrete (sUHPC) composite bridge deck panels. The composite panels comprise UHPC prepared with ground granulated blast furnace slag (GGBS) with the replacement of 30% cement content overlaid by recycled aggregate concrete made with replacement of 30% of coarse aggregates with recycled aggregates. The experimental variables for six HSFRC panels were depth, level of prestressing, and shear reinforcement. The six sUHPC panels were prepared with different shear and flexural reinforcements and sUHPC-normal/recycled aggregate concrete interface. Experimental results exhibit the promise of both systems to serve as an alternative to conventional bridge deck systems.

7. Alkali Activated Paste and Concrete Based on of Biomass Bottom Ash with Phosphogypsum

There is a growing interest in the development of new cementitious binders for building construction activities. In this study, biomass bottom ash (BBA) was used as an aluminosilicate precursor and phosphogypsum (PG) was used as a calcium source [6]. The mixtures of BBA and PG were activated with the sodium hydroxide solution or the mixture of sodium hydroxide solution and sodium silicate hydrate solution. Alkali activated binders were investigated using X-ray powder diffraction (XRD), X-ray fluorescence (XRF), and scanning electron microscopy (SEM) test methods. The compressive strength of hardened paste and fine-grained concrete was also evaluated. After 28 days, the highest compressive strength reached 30.0 MPa—when the BBA was substituted with 15% PG and activated with NaOH solution—which is 14 MPa more than control sample. In addition, BBA fine-grained concrete samples based on BBA with 15% PG substitute activated with $NaOH/Na_2SiO_3$ solution showed higher compressive strength compered to when NaOH activator was used 15.4 MPa and 12.9 MPa respectfully. The $NaOH/Na2SiO_3$ activator solution resulted reduced

8. Mechanical Properties and Freeze–Thaw Durability of Basalt Fiber Reactive Powder Concrete

Basalt fiber has a great advantage on the mechanical properties and durability of reactive powder concrete (RPC) because of its superior mechanical properties and chemical corrosion resistance. In this paper, basalt fiber was adopted to modified RPC, and plain reactive powder concrete (PRPC), basalt fiber reactive powder concrete (BFRPC) and steel fiber reactive powder concrete (SFRPC) were prepared [7]. The mechanical properties and freeze–thaw durability of BFRPC with different basalt fiber contents were tested and compared with PRPC and SFRPC to investigate the effects of basalt fiber contents and fiber type on the mechanical properties and freeze–thaw durability of RPC. Besides, the mass loss rate and compressive strength loss rate of RPC under two freeze–thaw conditions (fresh-water freeze–thaw and chloride-salt freeze–thaw) were tested to evaluate the effects of freeze–thaw conditions on the freeze–thaw durability of RPC. The experiment results showed that the mechanical properties and freeze–thaw resistance of RPC increased as the basalt fiber content increase. Compared with the fresh-water freeze–thaw cycle, the damage of the chloride-salt freeze–thaw cycle on RPC was great. Based on the freeze–thaw experiment results, it was found that SFRPC was sensitive to the corrosion of chloride salts and compared with the steel fiber, the improvement of basalt fiber on the freeze–thaw resistance of RPC was great.

9. A Low-Autogenous-Shrinkage Alkali-Activated Slag and Fly Ash Concrete

Alkali-activated slag and fly ash (AASF) materials are emerging as promising alternatives to conventional Portland cement. Despite the superior mechanical properties of AASF materials, they are known to show large autogenous shrinkage, which hinders the wide application of these eco-friendly materials in infrastructure. To mitigate the autogenous shrinkage of AASF, two innovative autogenous-shrinkage-mitigating admixtures, superabsorbent polymers (SAPs) and metakaolin (MK), are applied in this study [8]. The results show that the incorporation of SAPs and MK significantly mitigates autogenous shrinkage and cracking potential of AASF paste and concrete. Moreover, the AASF concrete with SAPs and MK shows enhanced workability and tensile strength-to-compressive strength ratios. These results indicate that SAPs and MK are promising admixtures to make AASF concrete a high-performance alternative to Portland cement concrete in structural engineering.

10. Properties of Foamed Lightweight High-Performance Phosphogypsum-Based Ternary System Binder

The potential of phosphogypsum (PG) as secondary raw material in construction industry is high if compared to other raw materials from the point of view of availability, total energy consumption, and CO_2 emissions created during material processing. This work [9] investigates a green hydraulic ternary system binder based on waste phosphogypsum (PG) for the development of sustainable high-performance construction materials. Moreover, a simple, reproducible, and low-cost manufacture is followed by reaching PG utilization up to 50 wt.% of the binder. Commercial gypsum plaster was used for comparison. High-performance binder was obtained, and on a basis of it, foamed lightweight material was developed. Low water-binder ratio mixture compositions were prepared. Binder paste, mortar, and foamed binder were used for sample preparation. Chemical, mineralogical composition, and performance of the binder were evaluated. Results indicate that the used waste may be successfully employed to produce high-performance binder pastes and even mortars with a compression strength up to 90 MPa. With the use of foaming agent, lightweight (370–700 kg/m^3) foam concrete was produced with a thermal conductivity from 0.086 to 0.153 W/mK. Water tightness (softening coefficient) of such foamed material was 0.5–0.64. The proposed approach represents a viable solution to reduce the environmental footprint associated with waste disposal.

11. Effect of Fly Ash as Cement Replacement on Chloride Diffusion, Chloride Binding Capacity, and Micro-Properties of Concrete in a Water Soaking Environment

This paper [10] experimentally studies the effects of fly ash on the diffusion, bonding, and micro-properties of chloride penetration in concrete in a water soaking environment based on the natural diffusion law. Different fly ash replacement ratio of cement in normal concrete was investigated. The effect of fly ash on chloride transportation, diffusion coefficient, free chloride content, and binding chloride content were quantified, and the concrete porosity and microstructure were reported through mercury intrusion perimetry and scanning electron microscopy, respectively. It was concluded from the test results that fly ash particles and hydration products (filling and pozzolanic effects) led to the densification of microstructures in concrete. The addition of fly ash greatly reduced the deposition of chloride ions. The chloride ion diffusion coefficient considerably decreased with increasing fly ash replacement, and fly ash benefits the binding of chloride in concrete. Additionally, a new equation is proposed to predict chloride-binding capacity based on the test results.

12. Durability Assessment of Recycled Aggregate HVFA Concrete

The possibility of producing high-volume fly ash (HVFA) recycled aggregate concrete represents an important step towards the development of sustainable building materials. In fact, there is a growing need to reduce the use of non-renewable natural resources and, at the same time, to valorize industrial by-products, such as fly ash, which would otherwise be sent to the landfill. The present experimental work [11] investigates the physical and mechanical properties of concrete by replacing natural aggregates and cement with recycled aggregates and fly ash, respectively. First, the mechanical properties of four different mixtures have been analyzed and compared. Then, the effectiveness of recycled aggregate and fly ash on reducing carbonation and chloride penetration depth has been also evaluated. Finally, the corrosion behavior of the different concrete mixtures, reinforced with either bare or galvanized steel plates, has been evaluated. The results obtained show that high-volume fly ash (HVFA) recycled aggregate concrete can be produced without significative reduction in mechanical properties. Furthermore, the addition of high-volume fly ash and the total replacement of natural aggregates with recycled ones did not modify the corrosion behavior of embedded bare and galvanized steel reinforcement.

13. Mechanical Properties and Flexural Behavior of Sustainable Bamboo Fiber-Reinforced Mortar

In this study [12], a sustainable mortar mixture is developed using renewable by-products for the enhancement of mechanical properties and fracture behavior. A high-volume of fly ash—a by-product of coal combustion—is used to replace Portland cement, while waste by-products from the production of engineered bamboo composite materials are used to obtain bamboo fibers and to improve the fracture toughness of the mixture. The bamboo process waste was ground and size-fractioned by sieving. Several mixes containing different amounts of fibers were prepared for mechanical and fracture toughness assessment, evaluated via bending tests. The addition of bamboo fibers showed insignificant losses of strength, resulting in mixtures with compressive strengths of 55 MPa and above. The bamboo fibers were able to control crack propagation and show improved crack-bridging effects with higher fiber volumes, resulting in a strain-softening behavior and mixture with higher toughness. The results of this study show that the developed bamboo fiber-reinforced mortar mixture is a promising sustainable and affordable construction material with enhanced mechanical properties and fracture toughness with the potential to be used in different structural applications, especially in developing countries.

14. Industrial Low-Clinker Precast Elements Using Recycled Aggregates

Increasing amounts of sustainable concretes are being used as society becomes more aware of the environment. This paper [13] attempts to evaluate the properties of precast concrete elements formed with recycled coarse aggregate and low clinker content cement using recycled additions. To this end, six different mix proportions were characterized: a reference concrete; two concretes with 25%wt. and 50%wt. substitution of coarse aggregate made using mixed construction and demolition wastes; and others with recycled cement with low clinker content. The compressive strength, the elastic modulus, and the durability indicator decrease with the proportions of recycled aggregate replacing aggregate, and it is accentuated with the incorporation of recycled cement. However, all of the precast elements tested show good performance with slight reduction in the mechanical properties. To confirm the appropriate behavior of New Jersey precast barriers, a test that simulated the impact that simulated the impact of a vehicle was carried out.

15. Photocatalytic Recycled Mortars: Circular Economy as a Solution for Decontamination

The circular economy is an economic model of production and consumption that involves reusing, repairing, refurbishing, and recycling materials after their service life. The use of waste as secondary raw materials is one of the actions to establish this model. Construction and demolition waste (CDW) constitute one of the most important waste streams in Europe due to its high production rate per capita. Aggregates from these recycling operations are usually used in products with low mechanical requirements in the construction sector. In addition, the incorporation of photocatalytic materials in construction has emerged as a promising technology to develop products with special properties, such as air decontamination. This research [14] aims to study the decontaminating behavior of mortars manufactured with the maximum amount of mixed recycled sand without affecting their mechanical properties or durability. For this, two families of mortars were produced, one consisting of traditional Portland cement and the other of photocatalytic cement, each with four replacement rates of natural sand by mixed recycled sand from CDW. Mechanical and durability properties, as well as decontaminating capacity, were evaluated for these mortars. The results show adequate mechanical behavior, despite the incorporation of mixed recycled sand, and improved decontaminating capacity by means of NOx reduction capacity.

16. Durability of High Volume Glass Powder Self-Compacting Concrete

The transport characteristics of waste glass powder incorporated self-compacting concrete (SCC) for a number of different durability indicators are reported in this paper [15]. SCC mixes were cast at a water to binder ratio of 0.4 using glass powders with a mean particle size of 10, 20, and 40 μm, and at cement replacement levels of 20, 30, and 40%. The oxygen permeability, electrical resistivity, porosity, and chloride diffusivity were measured at different ages from 3 to 545 days of curing. The amount and particle size of the incorporated waste glass powder was found to influence the durability properties of SCC. The glass incorporated SCC mixes showed similar or better durability characteristics compared to general purpose (GP) and fly ash mixes at similar cement replacement level. A significant improvement in the transport properties of the glass SCC mixes was observed beyond 90 days.

17. High-Durability Concrete Using Eco-Friendly Slag-Pozzolanic Cements and Recycled Aggregate

Clinker production is very energy-intensive and responsible for releasing climate-relevant carbon dioxide (CO_2) into the atmosphere, and the exploitation of aggregate for concrete results in a reduction in natural resources. This contrasts with infrastructure development, surging urbanization, and the demand for construction materials with increasing requirements in terms of durability and strength. A possible answer to this is eco-efficient, high-performance concrete. This article [16] illustrates basic material investigations to both, using eco-friendly cement and recycled aggregate from tunneling to produce structural concrete and inner shell concrete, showing high impermeability and durability. By replacing energy- and CO_2-intensive cement types by slag-pozzolanic cement (CEM V) and using recycled aggregate, a significant contribution to environmental sustainability can be provided while still meeting the material requirements to achieve a service lifetime for the tunnel structure of up to 200 years. Results of this research show that alternative cements (CEM V), as well as processed tunnel spoil, indicate good applicability in terms of their properties. Despite the substitution of conventional clinker and conventional aggregate, the concrete shows good workability and promising durability in conjunction with adequate concrete strengths.

Author Contributions: Conceptualization, C.T.; methodology, C.T. and J.d.B.; validation, C.T., J.d.B. and V.C. All authors have read and agreed to the published version of the manuscript.

Funding: This research received no external funding.

Acknowledgments: Thanks are due to all of the authors and peer reviewers for their valuable contributions to this Special Issue. The MDPI management and staff are also to be congratulated for their untiring editorial support for the success of this project.

Conflicts of Interest: The authors declare no conflict of interest.

References

1. Sainz-Aja, J.; Thomas, C.; Polanco, J.A.; Carrascal, I. High-Frequency Fatigue Testing of Recycled Aggregate Concrete. *Appl. Sci.* **2019**, *10*, 10. [CrossRef]
2. Tamayo, P.; Pacheco, J.; Thomas, C.; De Brito, J.; Rico, J. Mechanical and Durability Properties of Concrete with Coarse Recycled Aggregate Produced with Electric Arc Furnace Slag Concrete. *Appl. Sci.* **2019**, *10*, 216. [CrossRef]
3. Sosa, I.; Thomas, C.; Polanco, J.A.; Setién, J.; Tamayo, P. High Performance Self-Compacting Concrete with Electric Arc Furnace Slag Aggregate and Cupola Slag Powder. *Appl. Sci.* **2020**, *10*, 773. [CrossRef]
4. Duarte, G.; Gomes, R.; De Brito, J.; Miguel, B.; Nobre, J. Economic and Technical Viability of Using Shotcrete with Coarse Recycled Concrete Aggregates in Deep Tunnels. *Appl. Sci.* **2020**, *10*, 2697. [CrossRef]
5. Zafar, M.N.; Saleem, M.; Xia, J.; Saleem, M. Experimental Characterization of Prefabricated Bridge Deck Panels Prepared with Prestressed and Sustainable Ultra-High Performance Concrete. *Appl. Sci.* **2020**, *10*, 5132. [CrossRef]
6. Vaičiukynienė, D.; Nizevičienė, D.; Kantautas, A.; Bocullo, V.; Kielė, A. Alkali Activated Paste and Concrete Based on of Biomass Bottom Ash with Phosphogypsum. *Appl. Sci.* **2020**, *10*, 5190. [CrossRef]
7. Li, W.; Liu, H.; Zhu, B.; Lyu, X.; Gao, X.; Liang, C. Mechanical Properties and Freeze–Thaw Durability of Basalt Fiber Reactive Powder Concrete. *Appl. Sci.* **2020**, *10*, 5682. [CrossRef]

8. Li, Z.; Yao, X.; Chen, Y.; Lu, T.; Ye, G. A Low-Autogenous Shrinkage Alkali-Activated Slag and Fly Ash Concrete. *Appl. Sci.* **2020**, *10*, 6092. [CrossRef]
9. Bumanis, G.; Zorica, J.; Bajare, D. Properties of Foamed Lightweight High-Performance Phosphogypsum-Based Ternary System Binder. *Appl. Sci.* **2020**, *10*, 6222. [CrossRef]
10. Liu, J.; Liu, J.; Huang, Z.-Y.; Zhu, J.-H.; Liu, W.; Zhang, W. Effect of Fly Ash as Cement Replacement on Chloride Diffusion, Chloride Binding Capacity, and Micro-Properties of Concrete in a Water Soaking Environment. *Appl. Sci.* **2020**, *10*, 6271. [CrossRef]
11. Corinaldesi, V.; Donnini, J.; Giosuè, C.; Mobili, A.; Tittarelli, F. Durability Assessment of Recycled Aggregate HVFA Concrete. *Appl. Sci.* **2020**, *10*, 6454. [CrossRef]
12. Maier, M.; Javadian, A.; Saeidi, N.; Unluer, C.; Taylor, H.; Ostertag, C.P. Mechanical Properties and Flexural Behavior of Sustainable Bamboo Fiber-Reinforced Mortar. *Appl. Sci.* **2020**, *10*, 6587. [CrossRef]
13. Thomas, C.; Cimentada, A.I.; Cantero, B.; Medina, C.; Polanco, J.A. Industrial Low-Clinker Precast Elements Using Recycled Aggregates. *Appl. Sci.* **2020**, *10*, 6655. [CrossRef]
14. Barbudo, A.; Lozano-Lunar, A.; López-Uceda, A.; Galvin, A.P.; Ayuso, J. Photocatalytic Recycled Mortars: Circular Economy as a Solution for Decontamination. *Appl. Sci.* **2020**, *10*, 7305. [CrossRef]
15. Tariq, S.; Scott, A.N.; MacKechnie, J.R.; Shah, V. Durability of High Volume Glass Powder Self-Compacting Concrete. *Appl. Sci.* **2020**, *10*, 8058. [CrossRef]
16. Voit, K.; Zeman, O.; Janotka, I.; Adamcova, R.; Bergmeister, K. High-Durability Concrete Using Eco-Friendly Slag-Pozzolanic Cements and Recycled Aggregate. *Appl. Sci.* **2020**, *10*, 8307. [CrossRef]

 *applied sciences*

Article

High-Durability Concrete Using Eco-Friendly Slag-Pozzolanic Cements and Recycled Aggregate

Klaus Voit [1], Oliver Zeman [2,*], Ivan Janotka [3], Renata Adamcova [4] and Konrad Bergmeister [2]

[1] Institute of Applied Geology, Department of Civil Engineering and Natural Hazards, University of Natural Resources and Life Sciences, Peter Jordan-Street 82, 1190 Vienna, Austria; klaus.voit@boku.ac.at

[2] Institute of Structural Engineering, Department of Civil Engineering and Natural Hazards, University of Natural Resources and Life Sciences, Peter Jordan-Street 82, 1190 Vienna, Austria; konrad.bergmeister@boku.ac.at

[3] Department of Material Research, Building Testing and Research Institute, Studena 3, 821 04 Bratislava, Slovakia; janotka@tsus.sk

[4] Department of Engineering Geology, Faculty of Natural Sciences, Comenius University in Bratislava, Ilkovicova 6, 842 15 Bratislava, Slovakia; renata.adamcova@uniba.sk

* Correspondence: oliver.zeman@boku.ac.at

Received: 14 October 2020; Accepted: 18 November 2020; Published: 23 November 2020

Abstract: Clinker production is very energy-intensive and responsible for releasing climate-relevant carbon dioxide (CO_2) into the atmosphere, and the exploitation of aggregate for concrete results in a reduction in natural resources. This contrasts with infrastructure development, surging urbanization, and the demand for construction materials with increasing requirements in terms of durability and strength. A possible answer to this is eco-efficient, high-performance concrete. This article illustrates basic material investigations to both, using eco-friendly cement and recycled aggregate from tunneling to produce structural concrete and inner shell concrete, showing high impermeability and durability. By replacing energy- and CO_2-intensive cement types by slag-pozzolanic cement (CEM V) and using recycled aggregate, a significant contribution to environmental sustainability can be provided while still meeting the material requirements to achieve a service lifetime for the tunnel structure of up to 200 years. Results of this research show that alternative cements (CEM V), as well as processed tunnel spoil, indicate good applicability in terms of their properties. Despite the substitution of conventional clinker and conventional aggregate, the concrete shows good workability and promising durability in conjunction with adequate concrete strengths.

Keywords: green cements; slag-pozzolanic cement; CEM V; tunnel spoil recycling; high durability

---

1. Introduction

1.1. Eco-Efficient Cement Production

The demand for concrete—the world´s most used construction material—continues to remain at a high level notwithstanding recent global economic fluctuations. In doing so, the concrete industry acts as one of the main contributors to CO_2 emissions accounting for approx. 5 to 7% of global anthropogenic CO_2 emissions (see [1–4]), whereby Portland cement production accounts for approx. 90% of the quoted share [5]. The rapid increase in the recent global cement production is driven by China, which produced 2.35 gigatons of cement in 2015. This corresponds to 55% of the global amount of cement produced, contributing approximately 13 to 15% of China's total CO_2 emissions. In the future, due to population growth, cement production worldwide is projected to increase between 12% and 23% by the year 2050 [6].

Such production quantities would lead to a global emission, depending on the author, of approximately 1.3 to 1.8 billion tons of CO_2 per year [2,7]. Further, CO_2 emission during cement production is derived from two sources, having a similar share: (1) Process-related CO_2 accounting for the energy demand by the use of fuels and electric energy mainly for drying, grinding, and mostly from clinker burning; (2) the emission of embodied CO_2 (ECO_2) during $CaCO_3$ decomposition during heating, when the chemically bound CO_2 from the carbonate rock is degassing [2,5]. Summed up, the amount of generated CO_2 per kilogram of produced clinker varies between 0.65 and 1.0 kg CO_2 per kg clinker [8–11] depending on the fuel type and the basis of electricity production. The process-related global CO_2 emission of concrete has been estimated at approximately 83 kg CO_2 per ton by [12], while ECO_2 accounts for an additional amount of approximately 95 to 135 kg/ton, varying depending on the specific concrete design [5,13,14], adding up to the generation of roughly 200 kg CO_2 per ton of concrete.

In view of the importance of these issues, measures have been taken to reduce the environmental implications of cement and concrete production by using various strategies. The options range from customized concrete mixing design and plant technology possibilities toward energy saving [15] to carbon capture strategies [16,17]. As a very central point with regard to embodied CO_2, the usage of alternative raw materials with the absence of carbonates in their mineral content is effective. This means the reduction in the portion of Portland cement replacing the clinker by alternative binder compositions (e.g., fly ash, furnace slag, or natural pozzolans) producing blended or—when using more than one blending material—so-called composite cements [18–20]. A current strategy in this context is the substitution of clinker by waste-based cementing materials, i.e., [19,21]. In the present article, the approach of slag-pozzolanic cements (CEM V) according to [22] is pursued.

1.2. Aggregate Recycling

Additionally, next to or supplementary to clinker optimization, aggregate recycling has a positive effect on the environmental impact of concrete production. Aggregate production is not particularly significant concerning CO_2 emission and energy consumption compared to cement production, but due to the large quantities and the high percentage by weight (approx. 80 wt%) of conventional concrete consisting of aggregate, the impact on natural resources is considerably high. Referring to this, aggregate material in particular in close range to the demand with no other intended use can be considered a potential substitute for natural aggregate, provided all crucial technical and legal requirements are met. This applies to construction and demolition waste (CDW), as well as natural excavated rock material. The use of construction and demolition waste as aggregate is the content of current research, e.g., [21,23,24].

Likewise, excavated rock material from earthworks and tunneling is intended to be recycled as aggregate for concrete [25,26]. Due to the fact that, particularly during tunneling, large volumes of concrete are used, there are efforts to reduce the ecological impact of such construction projects at the same time. As for concrete, this concerns cement, as discussed in Section 1.1, and the aggregate used. Regarding the reuse of tunnel spoil, there are numerous related studies examining the questions of reuse possibilities and the suitability of excavated rock for aggregate production. Aspects of tunnel muck recycling and tunnel spoil application opportunities have been demonstrated [27,28]. Therefore, the type of tunnel driving method has a major influence on excavated rock characteristics; this question is evaluated by [29–31]. For high-quality concrete production, careful planning, efficient rock classification [32], and rock material management [33], as well as technical considerations focusing on material analysis and data management, are fundamental [34], not to forget the juridical considerations, as tunnel spoil can initially be considered as waste from a legal point of view, which is done by [35,36].

Tunnel structures have high construction costs and, once in operation, are counted among critical infrastructure. Therefore, for these structures, a service lifetime of some hundred years is assumed. Therefore, durability is a primary focus of the admixture designs. Besides static requirements, a reliable performance capability must be guaranteed for this time period. To meet this requirement, high-performance concrete in terms of durability and density is generally applied in tunnel structures [37]. During construction of the Swiss pioneer projects of the Base Tunnels at the Lötschberg and Gotthard massif, approximately 40% and 35%, respectively, of the tunnel spoil was recycled mainly for the production of aggregate for concrete [38–40]. A concrete quality with high resistance against various environmental influences allowed the design for a projected service lifetime of 100 years [38].

This research addresses aggregate recycling from tunneling from the example of the high-priority infrastructure project of the Brenner Base Tunnel. The Brenner Base Tunnel is an approx. 55 km-long, flat railway tunnel project acting as a connecting link between Austria and Italy and the main element of the new Brenner railway from Munich to Verona. The tunnel is currently under construction, with a service lifetime designed for a lifespan of 200 years. This increased service lifetime is achieved by a high durability of concrete, accomplished by the high quality of raw materials and good processing of concrete, whereby a high density and impermeability of concrete have the greatest impact [37]. In addition, preliminary tests combining recycled tunnel spoil with slag-pozzolanic cements were conducted.

1.3. Aims, Materials, and Methods

The overall aims of the presented research can be summarized and listed as follows:

- creating various slag-pozzolanic cement admixtures reducing the clinker component while still exhibiting suitable binder properties
- characterization of the different cementitious constituents used to produce the slag-pozzolanic CEM V cement types
- identifying the cement characteristics of the designed CEM V cement types
- developing concrete formulas and gathering the concrete properties using the manufactured CEM V cements
- further ecological improvement by replacement of the standard aggregate by recycled aggregate for concrete
- acquiring the basic characteristics of the developed concrete mixtures with a focus on concrete durability

To reach the desired objectives, binder and concrete manufacture, as well as intensive material testing, was performed including the following: (1) Various testing methods were applied to characterize at first the cement admixtures and cement properties of the produced CEM V cement types:

- X-ray diffraction analysis XRD (Philips X-ray diffractometer PW1730, Malvern Panalytical GmbH, Kassel, Germany) to examine the mineralogical composition,
- X-ray fluorescence XRF (PANalytical X´Pert Pro, Malvern Panalytical GmbH, Kassel, Germany) to record the chemical composition,
- Frattini-testing to demonstrate the pozzolanic activity of the cementitious constituents,
- fresh cement mortars testing (slump, density, air content, setting time, soundness),
- hardened cement mortars testing (recording compressive strength and bending tensile strength at 2, 7, 28, 56, and 84 days of curing time to gather strength development data as well) with an additional focus on the durability-relevant parameters, porosity and passivation ability.

Subsequently, (2) standard concrete compositions using the generated cements were tested to identify the concrete characteristics:

- fresh concrete testing (flow, density, air content)
- hardened concrete testing (compressive strength, bending tensile strength, modulus of elasticity) with a focus on the durability-relevant parameters, frost-resistance and permeability

In a third step, (3) standard aggregate for concrete was replaced by recycled aggregate from nonstandard crushed metamorphic rock arising during tunnel excavation. These concrete mixtures—now consisting of a newly developed CEM V cement admixture and processed and recycled aggregate from schist rocks—were also examined with regard to workability (fresh concrete density, flow spread, air content), strength (compressive and bending tensile strength, as well as the fracture energy), and particularly with regard to durability, as the designed concrete composition may be used in tunnel constructions, where the recycled aggregate is generated, exhibiting long service lifetimes of up to 200 years.

At all investigation levels, an additional comparison to standard cement was made.

2. Slag-Pozzolanic CEM V Characteristics

2.1. General

To produce cements with reduced CO_2 impact, the strategy of Portland cement substitution by additional cementitious constituents, in this case, fly ash (FA), granulated blast-furnace slag (S), and natural pozzolana (P), in accordance with the cement standard EN 197-1 [22], was pursued in this research to produce slag-pozzolanic cement CEM V.

In contrast to the previous edition of EN 197-1, which was valid until Spring 2018, the 27 products in the family of common cements were extended to 39 products until now, inter alia, adding the main cement type CEM VI "composite cement" (having a limestone content from 6 to 20 wt%), and the former CEM V "composite cement" had been renamed to CEM V "slag-pozzolanic cement." The latter consists of 40 to 64 wt% clinker, 18–30 wt% blast-furnace slag, and 18–30 wt% Pozzolana and siliceous fly ash in the case of CEM V/A, and of 20 to 38 wt% clinker, 31–49 wt% blast-furnace slag, and 31–49 wt% Pozzolana and siliceous fly ash for CEM V/B, whereby minor additional constituents (gypsum and limestone for instance) are possible up to 5 wt% for both types.

The increase concerning cement products expresses the research performance and interest of the cement industry in additional cement types with the substitution of clinker by different constituents. The use of these additional cementitious constituents modifies the characteristics of the produced concrete. In the case of CEM V cements, early strength is naturally reduced by the pozzolanic reaction of SiO_2 and Al_2O_3 from fly ash or natural pozzolana and the latent hydraulic reaction of blast furnace slag, and therefore, is not suitable for fast-track construction. This can be overcome by the additional application of silica fume (SF) [5,41]. Then again, improvement in the long-term performance of concrete is indicated by using blended cements due to the delayed growth of calcium silicate hydrate (CSH) and aluminate phases, growing into the pores, providing an increased density of the concrete structure, e.g., [42–44].

However, in many European countries, the application of CEM V cement is not allowed (e.g., Austria) or is restricted (e.g., Germany) by national concrete standards (ÖNORM B 4710-1 [45] in the case of Austria or DIN 1045-2 [46] in the case of Germany), mainly due to the lack of experience. Therefore, in the course of this research, material characterization of the following four manufactured CEM V cement types was carried out to examine the performance of CEM V cements and concrete produced with CEM V cement:

(1) CEM V/A (S-V) 32.5 R and
(2) CEM V/B (S-V) 32.5 N based on slag and fly ash were produced in a cement plant in Austria;
(3) CEM V/A (S-V) 32.5 R, based on slag and fly ash, and
(4) CEM V/B (S-P) 32.5 N using blast furnace slag and natural pozzolana (zeolite) as the main constituents were produced in a Slovak cement factory.

In doing so, a significant reduction in the cement clinker content to a level of 45.1 wt% (1), 26.9 wt% (2), 52.9 wt% (3), and 30.9 wt% (4) for each individual cement type was achieved [47]. The proportional composition of the cementitious constituents in the case of (1) and (2) was 40 wt% slag, 50 wt% fly ash, and 10 wt% limestone. In the case of (3) and (4), the additives were added in equal shares in both cases.

2.2. Basic Cement Qualities

2.2.1. Cementitious Constituents

Subsequent to the manufacturing of the different CEM V cement types, testing of the main cement properties was performed (see also [47]). In a first step, the main chemical properties of the cementitious constituents of (3) and (4) were compared by X-ray fluorescence to illustrate the fundamentally different chemical composition of fly ash (FA), blast furnace slag (S), and the pozzolana zeolite (P) (see Table 1). In the case of FA and the natural pozzolana zeolite (P), SiO_2 and Al_2O_3 contents are—compared to S—high, suggesting high pozzolanic activity (see also Table 2).

Table 1. Chemical properties of cementitious constituents of (3) CEM V/A (S-V) 32.5 R and (4) CEM V/B (S-P) 32.5.

Compound	Mass Percentage [wt%]		
	Fly Ash (FA, V)	**Blast Furnace Slag (S)**	**Pozzolana Zeolite (P)**
SiO_2	57.8	41.2	61.4
Al_2O_3	17.9	6.2	11.5
Fe_2O_3	7.4	0.5	1.4
CaO	4.2	37.2	3.5
MgO	6.3	10.1	4.3
SO_3	2.6	2.8	3.3
Na_2O	1.4	1.4	1.3
K_2O	1.3	0.4	1.6
Loss on ignition	1.1	0.0	12.2

Table 2. Pozzolanic activity of cementitious constituents (by Frattini-test).

Compound	Bounded CaO by Pozzolanic Additives [mmol/L and (% in Saturated $Ca(OH)_2$ Solution)] after		
	1 Day	**7 Days**	**28 Days**
Fly ash (FA)	5.6 (25)	12.7 (57)	19.5 (87)
Blast furnace slag (S)	7.5 (34)	10.4 (47)	12.8 (57)
Pozzolana Zeolite (P)	19.2 (86)	21.2 (95)	21.7 (97)

Blast furnace slag (S) shows similarly high values of SiO_2 with 41.2 wt% and a high CaO and MgO content of 37.2 and 10.1 wt%, respectively. Loss on ignition is relatively high in the case of Pozzolana, because of the chemically bound water from clinoptilolite [(Na, K, Ca)$_{2-3}$Al$_3$(Al, Si)$_2$Si$_{13}$O$_{36}$·12H$_2$O].

XRD of the cementitious constituents' analysis again provides further information on the degree of crystallinity by interpretation of the peak visualization: Broader, less distinct peaks suggest low crystallinity, while explicit and high peaks indicating the opposite. XRD diagrams of the raw materials used show the following (Figure 1): FA consists of well-crystallized SiO_2 and portions of anorthite (Calcium Feldspar), while the Ca- and Mg-aluminosilicates are diffuse and show a low crystalline grade without any distinct higher-intensity peak (Figure 1). Blast furnace slag (S) can be described as poorly crystallized, showing two major peaks indicating ferrite and belite phases, and also consisting of other, different burned phases (see Figure 1). By contrast, the natural pozzolana zeolite (P) consists of the well-crystallized tectosilicate Clinoptilite (numerous high-intensity single peaks with the main intensity at angles 2Θ of 11.3 and 22.5).

Pozzolanic activity of the cementitious constituents was tested using Frattini´s testing method according to [48] by the solution of 20 g of cement in 100 mL of water at 40 °C for varying periods (1, 7, and 28 d), measuring the calcium and hydroxide ion values, calculating the oxide-bounded Ca, and comparing the bounded CaO to the calcium oxide solubility isotherm curve (% of bounded CaO) (Table 2).

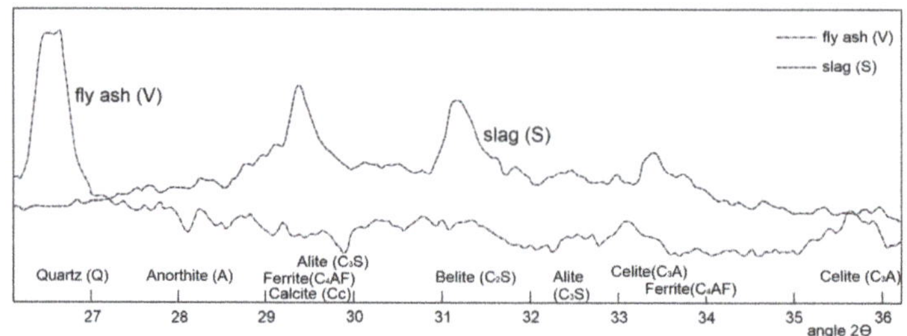

Figure 1. XRD diagrams displaying angle 2Θ-intensities of cementitious constituents fly ash (V) and furnace slag (S), whereby: Alite—$3CaO \times SiO_2$; Belite—$2CaO \times SiO_2$; Celite—$2CaO \times Al_2O_3$; Ferrite—$4CaO \times Al_2O_3 \times Fe_2O_3$.

Table 2 clearly shows that CaO is very rapidly and most strongly bound by the pozzolana zeolite, bounding approximately 97% of CaO after a period of 28 days. This is achieved by the high SiO_2 and low CaO content, cf. Table 1, resulting in a high pozzolan activity.

2.2.2. Cement Properties and Composition

Following the manufacture of the four different above-mentioned CEM V cement types (1), (2), (3), and (4) (see Section 2.1), cement testing was conducted to determine the key characteristics [49]. Fundamental characteristics are cement true density and specific surface expressed as a Blaine-value, illustrated in Table 3 for the manufactured CEM V cement types, as well as for the CEM I basic product.

Table 3. Cement true density and Blaine value.

Cement Type	True Density [kg/m^3]	Blaine-Value [cm^2/g]
CEM I 32.5 R	3088	3560
(1) CEM V/A (S-V) 32.5 R	2876	5343
(2) CEM V/B (S-V) 32.5 N	2857	4392
(3) CEM V/A (S-V) 32.5 R	2741	4484
(4) CEM V/B (S-P) 32.5 N	2790	4770

An important parameter with regard to chemical composition and environmental impact is the CaO-content of the various cement products. The lower the content of CaO, the lower the amount of ECO_2 (see Section 1.1) emitted. On the other hand, the CaO-containing components in hydrated cement are most sensitive to any aggressive attack; therefore, reducing the CaO content in the blended cement is a key condition for the expected improvement in the durability of the cement composites, see Section 2.4. Figure 2 illustrates the CaO content derived by X-ray fluorescence.

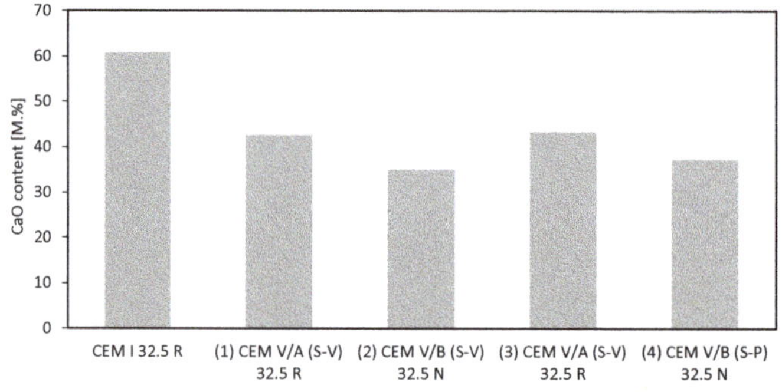

Figure 2. CaO content of the various CEM V cement types.

Figure 1 clearly shows the reduced amount of CaO in CEM V cements (replaced by SiO_2 and Al_2O_3 from cementitious constituents, see Table 1) compared to Portland clinker CEM I. Particularly, the CEM V/B with a much higher proportion of additives shows a comparable low CaO content.

Cement analysis via X-ray diffraction (Figure 3a,b) shows an amorphous content in each case, expressed by a slightly broader and diffuse appearance of the peaks, whereby the glassy, noncrystalline phases originate from blast furnace slag and fly ash, cf. Figure 1. Referring to this, cement type (1) and (2) present the highest amorphous content. Furthermore, CEM V/B (S-P) shows the lowest level of amorphous components due to the well-crystallized Clinoptilite (see also [47]).

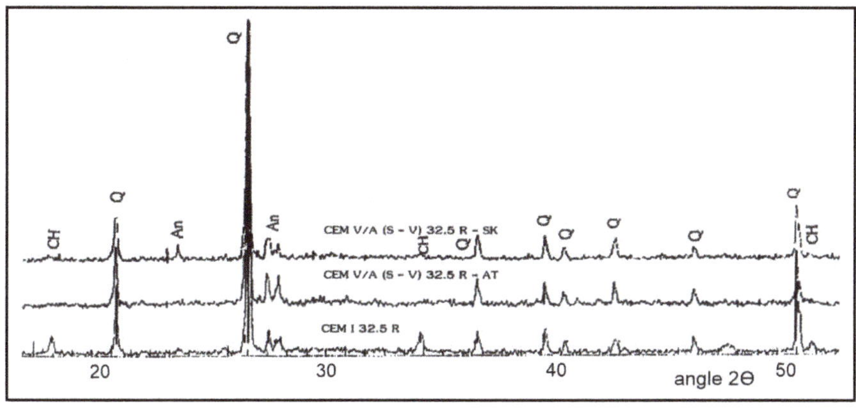

(a)

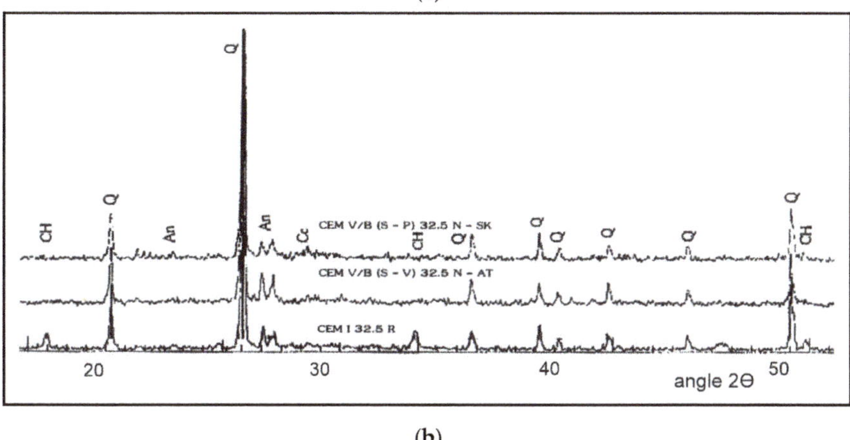

(b)

Figure 3. XRD diagrams displaying angle 2Θ-intensities of (**a**) CEM V/A types and reference cement CEM I and (**b**) CEM V/B types and reference cement CEM Iafter 365 days curing at 20 °C under water, whereby: CH—portlandite $Ca(OH)_2$; Cc—calcite, Q—quartz SiO_2, An—anorthite $(CaAl_2Si_2O_8)$.

2.3. Cement Mortars Properties

To evaluate fresh and hardened cement paste properties, fresh mortar and the derived test specimens according to [50] were produced, cf. [47,49]. Therefore, test sand was mixed with cement at a ratio 3 to 1 and the corresponding amount of water to reach a water/cement ratio of 0.5 [49]. The fresh mortar properties slump, density, and air content, as well as setting time and soundness according to [51], are illustrated in Table 4, providing information about performance and workability of the tested cement types.

Table 4. Fresh mortar properties of different CEM V cements.

Cement Type	Slump Value [mm]	Density [kg/m^3]	Air Content [%]	Init. Setting Time [min]	Final Setting Time [min]	Soundness [mm]
CEM I 32.5 R	155	2230	4.5	210	265	0.5
(1) CEM V/A (S-V) 32.5 R	175	2220	4.1	265	320	0.5
(2) CEM V/B (S-V) 32.5 N	166	2220	3.7	350	420	1.0
(3) CEM V/A (S-V) 32.5 R	135	2150	4.4	270	340	0.5
(4) CEM V/B (S-P) 32.5 N	106	2060	5.3	230	330	1.0

Cement types (3) and (4) show comparatively low slump values, indicating the need of super-plasticizers to reach a reasonable workability of the fresh mortar. By using a super-plasticizer with a dosage of approximately 0.5 wt% of the cement weight, the slump values were increased to the same level compared to (1) and (2). The fresh mortar density of CEM V/B types is lower than that of the CEM V/A types due to the lower density of the cementitious constituents compared to Portland cement clinker. The density of cement type (4) is noticeably low, due to the porous tectosilicate structure of zeolite that is probably also responsible for the comparably high air content. Setting time of the CEM V mortars is naturally delayed due to the posterior pozzolanic and latent hydraulic reactions of the additives. Soundness of all cement types falls significantly below the limit value of 10 mm according to [51], showing satisfactory volume consistency.

The compressive and bending tensile strength of hardened mortar specimens were determined for a basic mechanical characterization according to [50] with due regard to strength development with time; see Figure 4 for the concrete compressive strength and Figure 5 for the bending tensile strength.

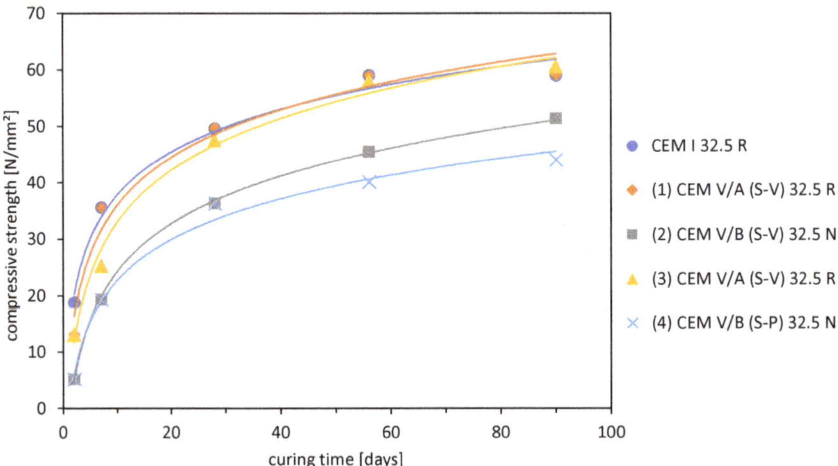

Figure 4. Compressive strength development of standard mortar specimens according to [51] (each value as a mean value of 3 individual tests), after 2, 7, 28, 56, and 90 days.

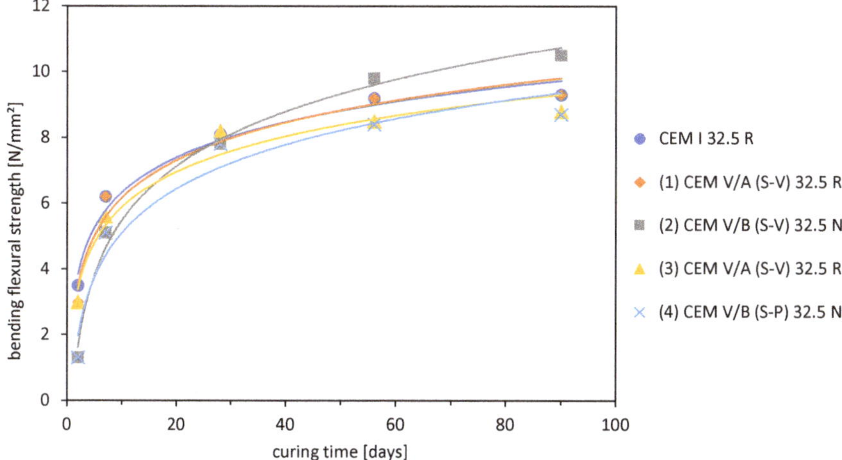

Figure 5. Bending flexural strength development of standard mortar prisms according to [51] (each value as mean value of 3 individual tests), after 2, 7, 28, 56, and 90 days.

The strength development of the considered cement types delivered the expected behavior: Standard cement CEM I 32.5 R reaches high early strength, though the CEM V/A cement types reach an equal strength level between 7 and 56 days of curing time. CEM V/B types lag behind but still show an increase in strength even after 56 days in contrast to CEM I and CEM V/A types.

Considering bending flexural strength (Figure 5), all cement types reach a similar strength level after 28 days of water curing. Subsequently, cement type (2) CEM V/B (S-V) is the only cement type still exhibiting a significant increase in flexural bending strength.

By using a super-plasticizer with a dosage of 0.5 wt% of the cement weight to improve the fresh mortar slump value of individual mixtures (cf. Table 4), the 28 day compressive strength for cement type (3) and (4) could be increased by approximately 10%, reaching a compressive strength of 52.1 and 40.3 N/mm^2, respectively.

2.4. Durability Aspects

Pore structure and the resulting permeability are key indicators to evaluate the durability of hardened cement paste and concrete. The appearance of the pore structure of cement paste and concrete is mainly influenced by the cement type. The delayed reaction of pozzolanic or hydraulic additives leads to the growth of additional CSH-phases—by consuming and reducing the Ca(OH)$_2$- content and a lower heat of hydration. As a result, the permeability and porosity of the paste or concrete is considerably reduced, as well as thermally induced cracking, which ensures an increased resistance against sulfate and chloride ions attack. Furthermore, due to the low availability of alkali in FA and S, the alkali reaction of cement with reactive components from the aggregate is inhibited, e.g., [52,53].

Pore structure was analyzed in detail for the tested cement types from hardened cement paste [47]. Therefore, the focus was laid on total pore content and the percentage of macro-pores with a diameter >50 nm, because the latter has a negative impact on concrete structure, particularly with regard to durability. Using mercury porosimetry according to [54,55], pore characterization was performed as shown in Figures 6 and 7.

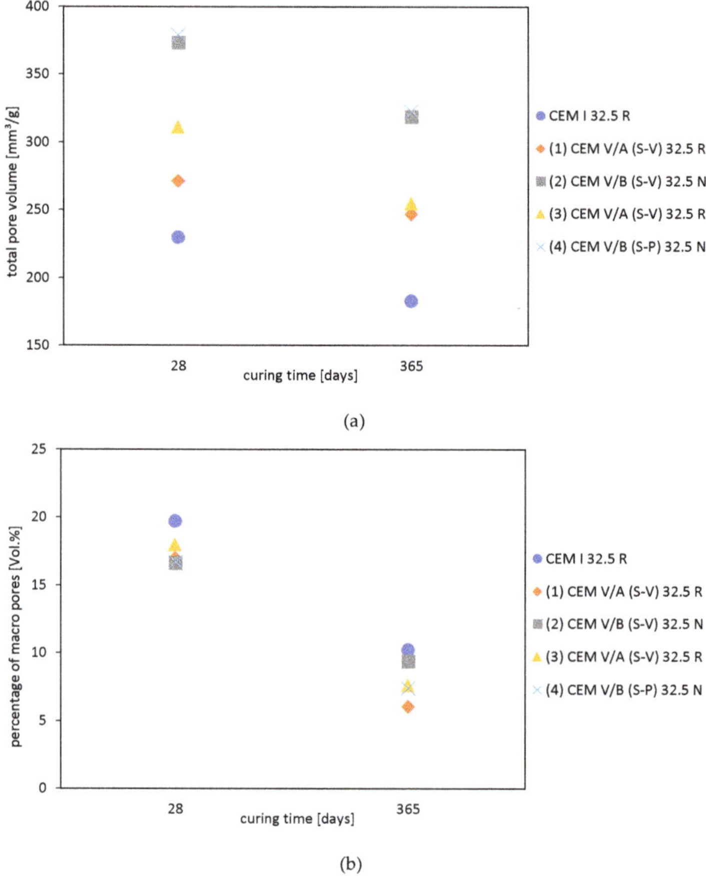

(a)

(b)

Figure 6. Development of total porosity (**a**) and percentage of macro-porosity (**b**) for a curing time of 28 and 365 days.

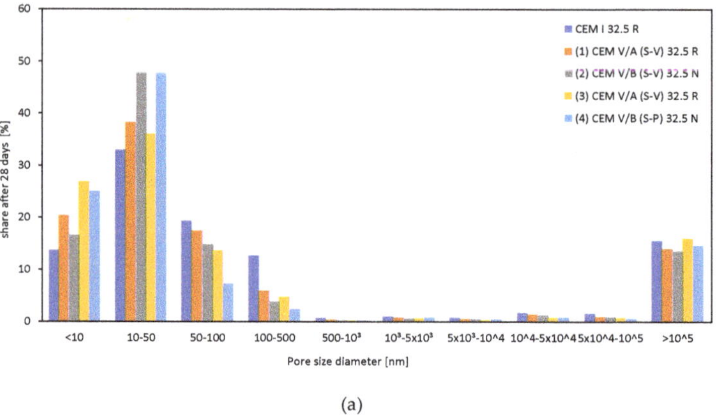

(a)

Figure 7. *Cont.*

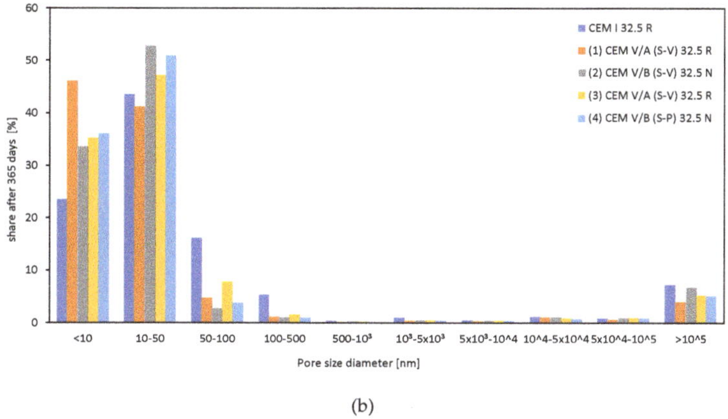

(b)

Figure 7. Comparison of pore-size distribution after 28 (**a**) and 365 days (**b**) curing time.

As illustrated in Figure 6, total porosity and percentage of macro-pores are clearly reduced during hydration of the Portland cement and—in the case of the CEM V cement types—the added cementitious constituents in the period from 28 to 365 days of curing. Therefore, CEM I 32.5 R shows clearly the lowest total porosity (Figure 7a) with CEM V/B cement types approximately having 1.5 times as much total pore volume. However, CEM I cement also has the highest amount of macro pores (Figure 6b) at 28 and 365 days of curing time, nevertheless—due to its comparably low total porosity and highest short-term reactivity (because of the highest clinker content and lack of latent hydraulic and pozzolanic additives)—showing the highest compressive strength at 28 days age compared to the CEM V cement types.

Figure 7a,b again compare the pore size distribution at 28 and 365 days of curing time, illustrating the change in pore structure. The pore structure of concretes made with CEM V-type cements shows typical symptoms of pore structure refinement in comparison with the reference CEM I concrete. During ongoing hydration, macro-pores with a diameter >50–100 nm retreat in favor of pores with smaller diameters, in the case of the CEM V-cements, more significantly compared to the CEM I cement, leaving CEM I with the highest percentage of macro-pores and demonstrating the capability of CEM V cements' pore volume densification by pozzolanic reactions of the additives. This fact indicates the increase in the volume fraction of less permeable micropores, which is reflected in the increase in the total porosity but, on the contrary, in the decrease in permeability coefficient of concrete.

In addition, carbonation is an important topic concerning durability in terms of the passivation of the reinforcement steel. Slag-pozzolanic cements are known for their lowered resistance to carbonation compared to standard Portland cement by the consumption of the strong alkaline $Ca(OH)_2$ during pozzolanic reactions as a counter-effect to that mentioned above [5,56]. The various cement types show pH values between 12.35 and 12.47 from their aqueous solution, indicating highly alkaline conditions. Additional pH-testing of concrete samples using phenolphthalein indicator solution after 56 days (7 days curing in water, followed by air storage in the laboratory at 20 °C and approx. 60% relative humidity) was performed in addition. Therefore, phenolphthalein pH-indicator solution was sprayed on cut-in-half concrete cubes. Areas with a pH value higher than 9.2 turn pink, whereas the concrete surface with a lower pH value stays colorless. Testing showed a carbonation depth <0.5 mm for all CEM V cement types, indicating sufficient passivation, inter alia, because of its dense and highly impermeable concrete structure, keeping in mind that in this test stand, the curing conditions were not aggressive and the curing time was very short. The permeability subject of concrete made of CEM V-cements is further discussed in Sections 3 and 4.

3. Concrete Properties Using Slag-Pozzolanic Cement

3.1. Concrete Formula and Fresh Concrete Properties

To evaluate and compare the performance of the different manufactured slag-pozzolanic cement types, concrete specimens were produced using a standard mixture (Table 5) with a water-binder- ratio of 0.5, only varying the type of cement. In the case of cement types (3) CEM V/A (S-V) 32.5 R and (4) CEM V/B (S-P) 32.5 N, the addition of superplasticizer was necessary to reach a good workability of the fresh concrete with a flow spread of approximately 50 mm (consistency class F3 or F4 according to [57]).

Table 5. Concrete standard mixture for comparison of produced cement types.

Compound	Amount [kg/m^3]
Cement	360
Water	176
Aggregate 0/4 mm	710
Aggregate 4/8 mm	420
Aggregate 8/16 mm	695

For all mixtures, fresh concrete properties were determined in the laboratory and are summarized in Table 6.

Table 6. Fresh concrete properties by using different slag-pozzolanic cement types.

Cement Type	Fresh Concrete Density [kg/m^3]	Flow Spread Consistency [cm] and (Consistency Class) According to [58]	Air Content [Vol.%]
CEM I 32.5 R	2283	50 (F4)	3.6
(1) CEM V/A (S-V) 32.5 R	2322	40 (F2)	2.0
(2) CEM V/B (S-V) 32.5 N	2306	45 (F3)	2.3
(3) CEM V/A (S-V) 32.5 R	2303	40 (F2)	2.6
(4) CEM V/B (S-P) 32.5 N	2190	50 (F4)	4.8

In the case of (1) CEM V/A (S-V) 32.5 R and (3) CEM V/A (S-V) 32.5 R, the consistency is plastic, not reaching a soft concrete consistency, indicating an additional demand for superplasticizers. Fresh concrete density is reasonable with the exception of (4) CEM V/B (S-P) 32.5 N, showing a low density because of the high air content.

3.2. Mechanical Properties of Concrete Using Slag-Pozzolanic Cement Types

With regard to hardened concrete, compressive strength, bending tensile strength, and modulus of elasticity were evaluated for a basic characterization of the different cement types.

Compressive strength was tested on 150 mm concrete cubes after 2, 28, 90, and 365 days of water storage at 20 ± 1 °C to present the varying strength development of the different cement or rather concrete types (Figure 8). As expected, the slag-pozzolanic cement types start at a low early strength level. CEM V/A cement types reach the strength level of the reference cement CEM I after 28 days and show a strength increase even between 90 and 365 days of concrete age. CEM V/B cements lag significantly behind, reaching a significantly lower strength level; strength development runs nearly flat after the age of 90 days.

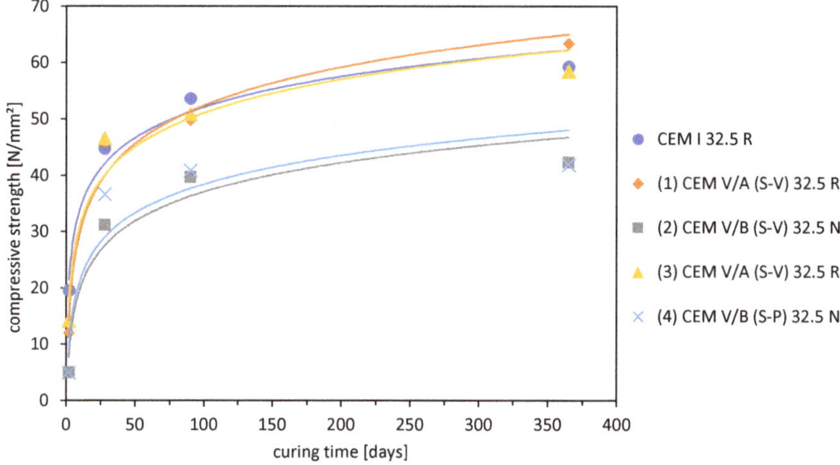

Figure 8. Concrete compressive strength using different cement types at 2, 28, and 90 days of water curing (each value as mean value of 3 individual tests).

The modulus of elasticity was determined as the secant under compression (the test specimen is loaded under axial compression, stresses and strains are recorded, and the slope of the secant to the stress–strain curve is determined) according to [58]. Test specimens (concrete prisms with dimensions of 150 × 150 × 600 mm) were water-stored at 20 ± 1 °C (the age-dependent pathway is displayed in Figure 9). The modulus of elasticity for the reference cement CEM I starts at approx. 30 GPa at the highest level and shows a stronger increase with time than the concrete types made with slag-pozzolanic cement. All CEM V cement types reach a similar level of approx. 36 GPa at a curing time of 365 days under water (see [49]).

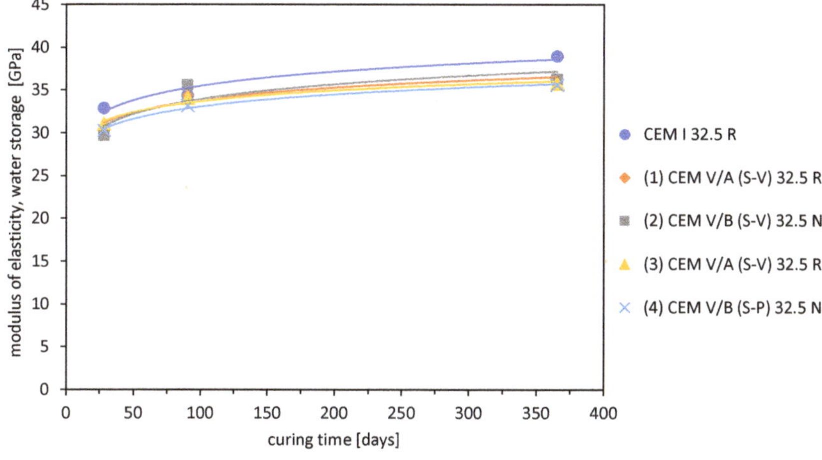

Figure 9. Modulus of elasticity (static) using different cement types at 2, 28, and 90 days of water curing (each value as mean value of 3 individual tests).

3.3. Durability Aspects: Frost and Permeability of CEM V Concretes

The negative effect of supplementary cementitious additives concerning frost resistance is well known, e.g., [59,60]. In this case, frost resistance could not be laboratory-confirmed during frost-testing, either. Testing was performed under intensified conditions without adding artificially entrained air additionally to the air content shown in Table 6. Frost resistance was evaluated indirectly via reduction in the bending flexural strength after 50 frost-thawing cycles in demineralized water. One frost-thawing cycle takes 24 h, whereby the temperature is lowered from the initial temperature of 20 °C to −20 °C within 12 h, then kept at −20 °C for 4 h and subsequently raised again to the initial temperature within 8 h. Temperature measurement is performed in the center of the concrete test specimen. Frost testing was performed after a period of 90 days of water curing at 150 x 150 x 600 mm samples. The results are illustrated in Figure 10: Bending flexural strength decreases strongly after frost impact, declining to a level between approx. 45 and 17% in the case of (1) CEM V/A (S-V) 32.5 R to less than 10%, indicating structural damage of the test specimen.

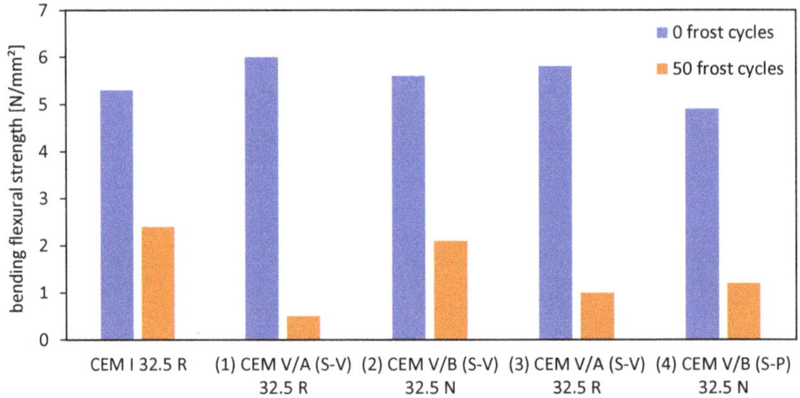

Figure 10. Bending flexural strength without and with frost impact evaluated after 90 days of water curing (each value as mean value of 3 individual tests).

Additional optimization of pore structure (porosity is reduced by the additives) and mixing design (high contents of additives increase the demand of water) will be necessary to improve frost resistance of the tested admixtures to prevent structural degradation of the concrete. An opportunity was used to build a concrete test track (washing station for trucks; Figure 11a,b) in a concrete factory in Austria in an alpine region (+815 m above the Adriatic Sea) using cement type (1) CEM V/A (S-V) 32.5 R. The concrete track itself is a horizontal structure experiencing a lot of water impact and mechanical exposure to truck tires. In winter, the input of chlorides by the truck tires is also most likely.

(**a**)   (**b**)

Figure 11. Test track construction using (1) CEM V/A (S-V) 32.5 R using a cement content of 400 kg/m^3 and a water-binder ratio of 0.45 during (**a**) concreting work and (**b**) after finishing the construction.

Considering the results from Figure 10, the mixture was adapted as the following: Cement content was increased to 400 kg/m^3, water-binder ratio was 0.45, and plastifying admixture was added to reach a flow consistency of F3 to F4 according to [57], additionally using an air-entraining admixture to reach 5% of air content. The practical realization went well and the concrete structure showed no sign of damage after winter 2019/2020, imposing at least 25 frost-thawing-cycles, implying the frost resistance of the applied concrete mixture.

Permeability of the various concrete mixtures was tested via water pressure impact. Pressure testing was conducted for concrete cubes after 90 days of water curing at 20 ± 1 °C and at atmospheric pressure according to [61]. Pressure testing was done at a water pressure of 500 kPa for the duration of 72 h. Afterward, the maximum depth of penetration was measured in millimeters. Water penetration depths for the different concrete mixtures with different CEM V-cements are illustrated in Figure 12.

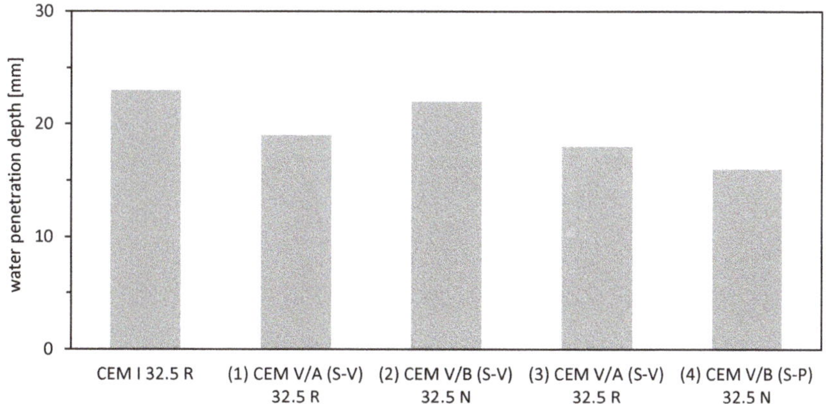

Figure 12. Water penetration depth (in mm) after water pressure test (each value as mean value of 3 individual tests).

As evident from Figure 8, water penetration lies in the range of 16 to 23 mm, whereby the water penetration was reduced by using the slag-pozzolanic cements. Especially, cement type (4) CEM V/B (S-P) 32.5 N shows a strongly reduced water penetration.

4. Concrete Applying Recycled Aggregate and Slag-Pozzolanic Cement

4.1. General

A new ecological approach was applied by merging the slag-pozzolanic cements with recycled rock aggregate from tunnel driving to maximize the environmental sustainability. For this reason, basic material characteristics were determined for a filling concrete mixture using three different rock types from excavation and a structural concrete mixture using aggregate from calcareous schist to evaluate the performance of the combination of CEM V cements and recycled rock material. Detailed information on the rock characteristics are given in [28,62,63].

4.2. Concrete Mixtures and Mechanical Characterization

4.2.1. Tentative Testing

The concrete mix for filling concrete without requirements of concrete exposure classes according to [57], using CEM V cement types as binder and recycled aggregate from tunneling as aggregate, applies a low cement level of 220 kg per m^3 of concrete, and the water-binder ratio was set at 0.55 to reach a flow spread of approx. 45 mm (consistency class F3 according to [57]). Maximum grain size was 31.5 mm, whereby for the grain fraction of 0–4 mm, standard quartz sand was used (to keep the water demand of the mixture at a manageable level concerning workability of the fresh concrete, because the high mica content of the finer fractions of recycled aggregates would increase water demand of the fresh concrete mix, see [28]), while fractions 4/8, 8/16, and 16/31.5 mm were produced from recycled rock material.

Three different types of aggregate were used from the tunnel driving of the Brenner Base Tunnel to determine the 28 day compressive strength (after 7 days of water storage and subsequent air storage at 20 ± 1°C until 28 days of concrete age; Figure 13):

- Quartz phyllite (quartz phyllite of Innsbruck), characterized by its layered appearance and high-sheet silicate layers showing low to moderate rock strength.
- Calcareous schists (Bündner Schist), fine- to medium-grained, parallel-aligned structure consisting of mica-rich schist layers and layers mainly made of quartz and calcite showing mediocre rock strength.
- Gneiss (central Gneiss), medium- to coarse-grained orthogneiss with granoblastic texture and high rock strength.

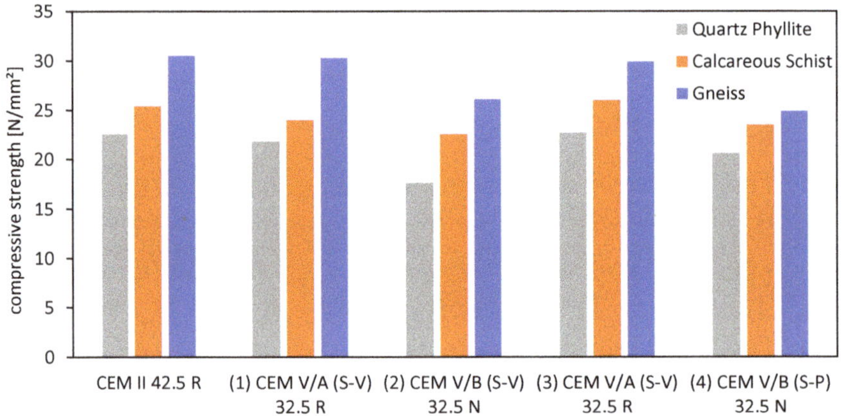

Figure 13. Compressive strength of various concrete mixtures varying cement and aggregate type (each value as mean value of 3 individual tests) after 28 days.

Figure 9 confirms the trend illustrated in Figure 6 (CEM V/B cement types show the lowest strength development after 28 days of curing time), additionally demonstrating the influence of the aggregate type: Quartz phyllite aggregate shows the lowest, calcareous schists aggregate shows mediocre, and gneiss aggregate shows the highest compressive strength due to rock characteristics, cf. [28].

4.2.2. Inner Lining Concrete

In order to use concrete with slag-pozzolanic binder and recycled aggregate, a concrete strength level of at least 50 N/mm² and a low permeability were defined as the crucial requirements. As calcareous schists are quantitatively dominant at the Brenner Base tunnel, cf. [28], concrete testing was performed using standard quartz aggregate for the grain fraction 0/4 mm and calcareous schists for 4/8, 8/16, and 16/31.5 mm. Cement content using (2) CEM V/B (S-V) 32.5 N was increased to a level of 330 kg per m³ of fresh concrete, and the water-binder-ratio was set to 0.52. Superplasticizer was used to reach a flow spread of class F4 according to [57]. Test specimens were stored for 7 days under water with subsequent air storage until 28 days of concrete age. Fresh and hardened concrete properties are illustrated in Table 7, whereby fracture energy G_F was determined by the cut-through-tensile splitting test.

Table 7. Concrete properties of (1) CEM V/A (S-V) 32.5 R cement type and calcareous schist aggregate.

Fresh Concrete Density [kg/m^3]	Flow Spread [cm]	Air Content [Vol.%]	Hardened Concrete Density [kg/m^3]	Compressive Strength f_c [N/mm^2]	Bending Flexural Strength f_{ct} [N/mm^2]	Fracture Energy G_F [N/m]
2411	55 (F4)	2.5	2376	54.7	4.6	2.5

The results from Table 7 show encouraging results concerning basic mechanical properties of the concrete composition, achieving good concrete density and workability of the fresh concrete, as well as a good strength level reaching concrete strength class C40/50 according to [57]. Figure 14 illustrates a concrete cube specimen after compressive strength testing.

Figure 14. Concrete cube test specimen after compressive strength testing according to Table 7.

Water penetration testing was performed according to [64] with an initial water pressure of 175 kPa for the first three days, subsequently increasing the water pressure to 700 kPa up to the 14th day. A water penetration depth of 19 mm could be determined, demonstrating a very low permeability comparable to the findings illustrated in Figure 8.

5. Summary and Conclusions

Cement production has a major contribution to global CO_2 emission. Composite cements with a partial substitution of CO_2-intensive clinker by additives, like slag, fly ash, or pozzolana, are a promising approach to make cement manufacture more environmentally friendly, reducing the CO_2-impact by up to two-thirds. In this research, slag-pozzolanic cements (CEM V) were produced and evaluated regarding their composition and characteristics. To evaluate their performance as binder, various concrete admixtures were produced and tested, whereby the mechanical properties regarding strength and durability were of central importance. In order to further increase the life cycle assessment of concrete using CEM V cements with slag-pozzolanic additives, conventional aggregate was partly exchanged by recycled rock aggregate originating from tunnel driving of the Brenner Base Tunnel. The results obtained in this research could encourage both the national standardization and the cement/construction industry for future applications because of its environmental, economic, and durability characteristics. In the course of this work, the following conclusions could be drawn:

-   Cementitious constituents were compared, whereby natural pozzolana zeolite (P) showed the highest SiO_2 content, and natural pozzolana zeolite (P) also demonstrated the highest pozzolanic activity compared to fly ash (FA) and blast furnace slag (S).

- The CaO content (crucial with regard to ECO_2-generation) of the manufactured CEM V cement types was considerably smaller compared to CEM I Portland cement.
- Individual CEM V types showed low slump values, indicating the need of superplasticizer to obtain good workability.
- Natural pozzolana zeolite (P) indicated a low fresh mortar density due to the porous tectosilicate structure, also containing a higher amount of air
- Compressive strength development was naturally behind CEM I cement types, whereby CEM V/B cement types with a higher proportion of additives showed lower strength (approximately 20 to 30%) after 90 days of curing time, at which there were no major differences considering bending tensile strength.
- Slag-pozzolanic cement types indicated high durability because of their lower porosity and permeability compared to Portland cement.
- Considering concrete production, different CEM V cement types showed a higher demand for superplasticizer to reach a good workability of fresh concrete; CEM V with natural pozzolana zeolite (P) again showed a high air content and low fresh concrete density.
- Concrete compressive strength and bending tensile strength development corresponded to the cement strength development.
- Water penetration of CEM V concrete admixtures was considerably smaller compared to Portland cement.
- Frost resistance of CEM V concretes via reduction in the bending tensile strength after freezing could not be verified in the laboratory, but could be verified at large scale using 400 kg of CEM V cement per m^3 fresh concrete and increasing the air content to 5%.
- Additional aggregate recycling was tested using different processed rock types from tunneling: Mechanical characteristics, especially water permeability, are promising for the usability of such concrete types.

In conclusion, it can be said that alternative cement types with reduced clinker content have their legitimacy for many application areas regarding their special characteristics discussed in this paper. Recycling of tunnel excavation material has already been proven at different construction projects and works well considering the essential framework conditions. In both cases, there is often a lack of experience impeding the practical implementation of these two alternatives.

Author Contributions: Conceptualization, K.V. and O.Z.; methodology, K.V., I.J., R.A. and K.B.; validation, K.V. and O.Z.; formal analysis, K.V. and I.J.; investigation, K.V. and I.J.; writing—original draft preparation, K.V. and O.Z.; writing—review & editing, I.J. and K.B.; visualization, K.V. and O.Z.; supervision, K.B.; project administration, K.V., I.J. and K.B.; funding acquisition, I.J. and K.B. All authors have read and agreed to the published version of the manuscript.

Funding: This research was funded by the EUROPEAN UNION European Regional Development Fund. Open-access funding provided by BOKU Vienna Open Access Publishing Fund. The authors would like to acknowledge the financial and technical support.

Conflicts of Interest: The authors declare no conflict of interest.

References

1. Shi, C.; Jiménez, A.F.; Palomo, Á. New cements for the 21st century: The pursuit of an alternative to Portland cement. *Cem. Concr. Res.* **2011**, *41*, 750–763. [CrossRef]
2. Zhang, C.-Y.; Han, R.; Yu, B.; Wei, Y.-M. Accounting process-related CO2 emissions from global cement production under Shared Socioeconomic Pathways. *J. Clean. Prod.* **2018**, *184*, 451–465. [CrossRef]
3. Chen, C.-H.; Lee, C.-M.; Tung, W.-C.; Wang, J.-H.; Hung, C.-H.; Hu, T.-H.; Wang, J.-C.; Lu, S.-N.; Changchien, C.-S. Evolution of full-length HBV sequences in chronic hepatitis B patients with sequential lamivudine and adefovir dipivoxil resistance. *J. Hepatol.* **2010**, *52*, 478–485. [CrossRef] [PubMed]
4. Uwasu, M.; Hara, K.; Yabar, H. World cement production and environmental implications. *Environ. Dev.* **2014**, *10*, 36–47. [CrossRef]

5.   Limbachiya, M.; Bostanci, S.C.; Kew, H. Suitability of BS EN 197-1 CEM II and CEM V cement for production of low carbon concrete. *Constr. Build. Mater.* **2014**, *71*, 397–405. [CrossRef]

6.   Ofosu-Adarkwa, J.; Xie, N.; Javed, S.A. Forecasting CO_2 emissions of China's cement industry using a hybrid Verhulst-GM(1,N) model and emissions' technical conversion. *Renew. Sustain. Energy Rev.* **2020**, *130*, 109945. [CrossRef]

7.   Barker, D.; Turner, S.; Napier-Moore, P.; Clark, M.; Davison, J. CO_2 Capture in the Cement Industry. *Energy Procedia* **2009**, *1*, 87–94. [CrossRef]

8.   Bosoaga, A.; Masek, O.; Oakey, J.E. CO_2 Capture Technologies for Cement Industry. *Energy Procedia* **2009**, *1*, 133–140. [CrossRef]

9.   Hills, T.; Leeson, D.; Florin, N.; Fennell, P.S. Carbon Capture in the Cement Industry: Technologies, Progress, and Retrofitting. *Environ. Sci. Technol.* **2015**, *50*, 368–377. [CrossRef]

10.   Deja, J.; Uliasz-Bochenczyk, A.; Mokrzycki, E. CO_2 emissions from Polish cement industry. *Int. J. Greenh. Gas Control.* **2010**, *4*, 583–588. [CrossRef]

11.   Anand, S.; Vrat, P.; Dahiya, R. Application of a system dynamics approach for assessment and mitigation of CO_2 emissions from the cement industry. *J. Environ. Manag.* **2006**, *79*, 383–398. [CrossRef] [PubMed]

12.   The Concrete Centre. *Embodied Carbon Dioxide (CO_2 e) of Concretes Used in Buildings*; MPA The Concrete Centre: London, UK, 2013.

13.   Flower, D.J.M.; Sanjayan, J.G. Green house gas emissions due to concrete manufacture. *Int. J. Life Cycle Assess.* **2007**, *12*, 282–288. [CrossRef]

14.   Purnell, P.; Black, L. Embodied carbon dioxide in concrete: Variation with common mix design parameters. *Cem. Concr. Res.* **2012**, *42*, 874–877. [CrossRef]

15.   Biernacki, J.J.; Bullard, J.W.; Sant, G.; Brown, K.; Glasser, F.P.; Jones, S.; Ley, T.; Livingston, R.; Nicoleau, L.; Olek, J.; et al. Cements in the 21stcentury: Challenges, perspectives, and opportunities. *J. Am. Ceram. Soc.* **2017**, *100*, 2746–2773. [CrossRef]

16.   Proaño, L.; Sarmiento, A.; Figueredo, M.; Cobo, M. Techno-economic evaluation of indirect carbonation for CO_2 emissions capture in cement industry: A system dynamics approach. *J. Clean. Prod.* **2020**, *263*, 121457. [CrossRef]

17.   Schneider, M. The cement industry on the way to a low-carbon future. *Cem. Concr. Res.* **2019**, *124*, 105792. [CrossRef]

18.   Mikulčić, H.; Vujanović, M.; Markovska, N.; Filkoski, R.V.; Ban, M.; Duić, N. CO_2 Emission Reduction in the Cement Industry. *Chem. Eng. Trans.* **2003**, *35*, 703–708. [CrossRef]

19.   De Carvalho, J.M.F.; De Melo, T.V.; Fontes, W.C.; Batista, J.O.D.S.; Brigolini, G.J.; Peixoto, R.A.F. More eco-efficient concrete: An approach on optimization in the production and use of waste-based supplementary cementing materials. *Constr. Build. Mater.* **2019**, *206*, 397–409. [CrossRef]

20.   Deolalkar, S. Composite Cements. In *Designing Green Cement Plants*; Elsevier Science: Amsterdam, The Netherlands, 2016; pp. 41–44.

21.   Salazar, R.A.R.; Valencia-Saavedra, W.; De Gutiérrez, R.M. Construction and Demolition Waste (CDW) Recycling-As Both Binder and Aggregates-In Alkali-Activated Materials: A Novel Re-Use Concept. *Sustainability* **2020**, *12*, 5775. [CrossRef]

22.   *ÖNORM EN 197-1: Cement—Part 1: Composition, Specification and Conformity Criteria for Common Cements*; Edition 2018-12; Austrian Standards International: Vienna, Austria, 2018.

23.   Tam, V.W.Y.; Soomro, M.; Evangelista, A.C.J. A review of recycled aggregate in concrete applications (2000–2017). *Constr. Build. Mater.* **2018**, *172*, 272–292. [CrossRef]

24.   Malešev, M.; Radonjanin, V.; Marinković, S. Recycled Concrete as Aggregate for Structural Concrete Production. *Sustainability* **2010**, *2*, 1204–1225. [CrossRef]

25.   *Working Group 14 & 15, ITA Report No. 21: Handling, Treatment and Disposal of Tunnel Spoil Materials. Working Groups 14 and 15—Underground Construction and the Environment and Mechanized Tunneling*; International Tunneling and Underground Space Association ITA: Châtelaine, Switzerland, 2019; ISBN 978-2-9701242-0-7.

26.   Burger, D.; Kohlböck, B.; Schoitsch, C. Geotechnics and mass balance of the earthworks for the second tube of the Tauern Tunnel. *Geomech. Tunn.* **2010**, *3*, 391–401. [CrossRef]

27.   Bellopede, R.; Brusco, F.; Oreste, P.; Pepino, M. Main Aspects of Tunnel Muck Recycling. *Am. J. Environ. Sci.* **2011**, *7*, 338–347. [CrossRef]

28. Bellopede, R.; Marini, P. Aggregates from tunnel muck treatments. Properties and uses. *Physicochem. Probl. Miner. Process.* **2011**, *47*, 259–266.

29. Voit, K.; Kuschel, E. Rock Material Recycling in Tunnel Engineering. *Appl. Sci.* **2020**, *10*, 2722. [CrossRef]

30. Thalmann, C. Characterisation and Possibilities of the Utilisation of Excavated Rock Material by Tunneling Boring Machines as Concrete and Shotcrete Aggregates. Ph.D. Thesis, Swiss Federal Institute of Technology Zurich (ETH), Zürich, Switzerland, 1996.

31. Gertsch, L.; Fjeld, A.; Nilsen, B. Use of TBM muck as construction material. *Tunn. Undergr. Space Technol.* **2000**, *15*, 379–402. [CrossRef]

32. Oggeri, C.; Fenoglio, T.M.; Vinai, R. Tunnel spoil classification and applicability of lime addition in weak formations for muck reuse. *Tunn. Undergr. Space Technol.* **2014**, *44*, 97–107. [CrossRef]

33. Ritter, S.; Einstein, H.H.; Galler, R. Planning the handling of tunnel excavation material—A process of decision making under uncertainty. *Tunn. Undergr. Space Technol.* **2013**, *33*, 193–201. [CrossRef]

34. Erben, H.; Galler, R. Tunnel spoil—New technologies on the way from waste to raw material/Tunnelausbruch—Neue Technologien für den Weg vom Abfall zum Rohstoff. *Géoméch. Tunn.* **2014**, *7*, 402–410. [CrossRef]

35. Reichel, P. Tunnel spoil: Tipping or the end of the definition as waste/Tunnelausbruch: Deponierung oder Abfallende. *Géoméch. Tunn.* **2014**, *7*, 419–427. [CrossRef]

36. Posch, H.; Murr, R.; Huber, H.; Kager, M.; Kolb, E. Tunnel excavation—The conflict between waste and recycling through the example of the Koralm Tunnel, contract KAT2/Tunnelausbruch—Das Spannungsfeld zwischen Abfall und Verwertung am Beispiel Koralmtunnel, Baulos KAT2. *Géoméch. Tunn.* **2014**, *7*, 437–450. [CrossRef]

37. Bergmeister, K. *Brenner Base Tunnel—Innovative Solutions in Tunneling*; Österreichische Ingenieur- und Architekten-Zeitschrift: Vienna, Germany, 2018.

38. *BLS AlpTransit Lötschberg: Final report Logistics in Excavation and Material Management (Schlussbericht Logistik Ausbruch- und Materialbewirtschaftung)*. IG-LBT Ingenieurgemeinschaft Lötschberg-Basistunnel; BLS AlpTransit AG: Thun, Switzerland, 2008. (In German)

39. Lieb, R.H. Materials management at the Gotthard Base Tunnel—Experience from 15 years of construction. *Géoméch. Tunn.* **2009**, *2*, 619–626. [CrossRef]

40. Resch, D.; Lassnig, K.; Galler, R.; Ebner, F. Tunnel excavation material—High value raw material. *Géoméch. Tunn.* **2009**, *2*, 612–618. [CrossRef]

41. Erdem, T.K.; Kırca, Ö. Use of binary and ternary blends in high strength concrete. *Constr. Build. Mater.* **2008**, *22*, 1477–1483. [CrossRef]

42. Khatri, R.; Sirivivatnanon, V.; Gross, W. Effect of different supplementary cementitious materials on mechanical properties of high performance concrete. *Cem. Concr. Res.* **1995**, *25*, 209–220. [CrossRef]

43. Bjegović, D.; Stirmer, N.; Serdar, M. Durability properties of concrete with blended cements. *Mater. Corros.* **2012**, *63*, 1087–1096. [CrossRef]

44. Asadollahfardi, G.; Yahyaei, B.; Salehi, A.M.; Ovesi, A. Effect of admixtures and supplementary cementitious material on mechanical properties and durability of concrete. *Civ. Eng. Des.* **2020**, *2*, 3–11. [CrossRef]

45. *ÖNORM B 4710-1: Concrete—Part 1: Specification, Performance, Production, Use and Conformity—Part 1: Rules for the Implementation of ÖNORM EN 206-1 for Normal and Heavy Concrete*; Edition 2018-01; Austrian Standards International: Vienna, Austria, 2018.

46. *DIN 1045-2: Concrete, Reinforced and Prestressed Concrete Structures—Part 2: Concrete-Specification, Proper-Ties, Production and Conformity—Application Rules for DIN EN 206-1*; Edition 2008-08; Construction committee of the DIN: Berlin, Germany, 2008.

47. Janotka, I.; Bergmeister, K.; Voit, K.; Zeller, H.; Bagel, L.; Sevcik, P. Verification of Austrian and Slovak CEM V(A,B) cement kinds for the use in structural concrete. In Proceedings of the International congress on durability of concrete, Trondheim, Norway, 18–21 June 2012.

48. *ÖNORM EN 196-5: Methods of Testing Cement—Part 5: Pozzolanicity Test for Pozzolanic Cement*; Edition 2011-06; Austrian Standards International: Vienna, Austria, 2011.

49. *ENVIZEO: Nutzung der Ökozementsorten CEM V (A, B) Gemäß EN 197-1 in Konstruktionsbeton*; Project report; Janotka, I. (Ed.) Self-Published: Bratislava, Slovakia, 2013.

50. *ÖNORM EN 196-1: Methods of Testing Cement—Part 1: Determination of Strength*; Edition 2016-10; Austrian Standards International: Vienna, Austria, 2016.

Appl. Sci. **2020**, *10*, 8307

51. *ÖNORM EN 196-3: Methods of Testing Cement—Part 3: Determination of Setting Times and Soundness;* Edition 2017-01; Austrian Standards International: Vienna, Austria, 2017.
52. Dziuk, D.; Giergiczny, Z.; Sokołowski, M.; Puz, T. Concrete resistant to aggressive media using a composite cement CEM V/A (S-V) 32.5 R—LH. In *Underground Infrastructure of Urban Areas 2*; CRC Press: Boca Raton, FL, USA, 2011; Volume 2, pp. 12–21. [CrossRef]
53. Li, Y.-X.; Chen, Y.-M.; Wei, J.-X.; He, X.-Y.; Zhang, H.-T.; Zhang, W.-S. A study on the relationship between porosity of the cement paste with mineral additives and compressive strength of mortar based on this paste. *Cem. Concr. Res.* **2006**, *36*, 1740–1743. [CrossRef]
54. *ASTM UOP578-02, Automated Pore Volume and Pore Size Distribution of Porous Substances by Mercury Porosimetry;* ASTM International: West Conshohocken, PA, USA, 2003.
55. *ISO 15901-1: Evaluation of Pore Size Distribution and Porosity of Solid Materials by Mercury Porosimetry and Gas Adsorption—Part 1: Mercury Porosimetry;* International Organization for Standardization ISO: Geneva, Switzerland, 2013.
56. Smrčková, E.; Bačuvčík, M.; Janotka, I. Basic Characteristics of Green Cements of CEM V/A and V/B Kind. *Adv. Mater. Res.* **2014**, *897*, 196–199. [CrossRef]
57. *ÖNORM EN 206-1: Concrete—Specification, Performance, Production and Conformity;* Edition 2017-05; Austrian Standards International: Vienna, Austria, 2017.
58. EN 12390-13: Testing Hardened Concrete. *Part 13: Determination of Secant Modulus of Elasticity in Compression;* Edition 2013; Austrian Standards International: Vienna, Austria, 2013.
59. Łaźniewska-Piekarczyk, B.; Piekarczyk, A. The problem of the equivalent time to start the internal frost resistance test on self-compacting concrete with supplementary cementitious materials. *J. Build. Eng.* **2020**, *32*, 101503. [CrossRef]
60. Reiterman, P.; Holčapek, O.; Zobal, O.; Keppert, M. Freeze-Thaw Resistance of Cement Screed with Various Supplementary Cementitious Materials. *Rev. Adv. Mater. Sci.* **2019**, *58*, 66–74. [CrossRef]
61. *EN 12390-8: Testing Hardened Concrete. Part 8: Depth of Penetration of Water under Pressure;* Edition 2009; Austrian Standards International: Vienna, Austria, 2009.
62. Voit, K. Application and Optimization of Tunnel Excavation Material of the Brenner Base Tunnel. Ph.D. Thesis, University of Natural Resources and Life Sciences, Vienna, Austria, 2013.
63. Voit, K.; Zimmermann, T. Characteristics of selected concrete with tunnel excavation material. *Constr. Build. Mater.* **2015**, *101*, 217–226. [CrossRef]
64. *ONR 23303: Test Methods for Concrete—National Application of Testing Standards for Concrete and Its Source Materials;* Edition 2010; Austrian Standards International: Vienna, Austria, 2010.

Publisher's Note: MDPI stays neutral with regard to jurisdictional claims in published maps and institutional affiliations.

 *applied sciences*

Article

Durability of High Volume Glass Powder Self-Compacting Concrete

Samia Tariq [1], Allan N. Scott [1,*], James R. Mackechnie [1,2] and Vineet Shah [1]

[1] Department of Civil and Natural Resources Engineering, University of Canterbury, Private Bag 4800, Christchurch 8140, New Zealand; saamiatariq@ymail.com (S.T.); james@concretenz.org.nz (J.R.M.); vineet.shah@canterbury.ac.nz (V.S.)

[2] Concrete NZ, PO Box 448, Level 4/70 The Terrace, Wellington 6140, New Zealand

* Correspondence: allan.scott@canterbury.ac.nz

Received: 9 October 2020; Accepted: 5 November 2020; Published: 13 November 2020

Abstract: The transport characteristics of waste glass powder incorporated self-compacting concrete (SCC) for a number of different durability indicators are reported in this paper. SCC mixes were cast at a water to binder ratio of 0.4 using glass powders with a mean particle size of 10, 20 and 40 μm and at cement replacement levels of 20, 30 and 40%. The oxygen permeability, electrical resistivity, porosity and chloride diffusivity were measured at different ages from 3 to 545 days of curing. The amount and particle size of the incorporated waste glass powder was found to influence the durability properties of SCC. The glass incorporated SCC mixes showed similar or better durability characteristics compared to general purpose (GP) and fly ash mixes at similar cement replacement level. A significant improvement in the transport properties of the glass SCC mixes was observed beyond 90 days.

Keywords: glass powder; self-compacting concrete; durability

1. Introduction

The durability of concrete can be defined as its ability to resist weathering action, chemical attack, abrasion and any other mechanism of deterioration while preserving its original form. The degree of deterioration is strongly related to the resistance of the cover layer to transport mechanisms, such as permeation, absorption, and diffusion of gas and liquid [1]. The use of supplementary cementitious materials (SCMs) in concrete helps in increasing resistance towards most of the durability related issues. The benefits of using silica fume, fly ash, slag and calcined clay has been widely demonstrated [2–6]. The use of waste glass powder as a potential SCM has also received significant interest in the last two decades [4,7–9]. The early age compressive strength for concrete containing glass powder is generally lower than the equivalent PC concrete. Similar strength values for glass powered concrete however have been reported at 90 days for replacement levels of up to 30% [10–12]. Further grinding of the glass powder can reduce the average particle size and help to offset the lower early age compressive strength [10,13]. Increasing the fineness and proportion of glass powder however has been shown to negatively affect the workability of the mix [14,15]. The irregular angular shape of the glass particles is responsible for the increased water demand compared to PC mixes. Numerous studies have reported the implications of using glass powder on the fresh and mechanical properties of concrete; however, limited information is available on the durability performance of such concrete.

The rapid development in the infrastructure sector around the world is expected to increase the demand for cement from current the levels of ~4 billion tonnes to ~6 billion tonnes by 2050 [16]. The large carbon footprint associated with the cement industry makes it imperative to develop sustainable solutions that promote the judicious use of materials and increased service life of concrete structures with minimal maintenance. The global supply of waste glass represents approximately ~3%

(130 million tonnes) of the total cement consumption [17]. While this a relatively small proportion it does provide a possible option for diverting a waste stream from land fill and may be of particular benefit in areas that have limited supplies of more traditional SCMs such as fly ash, slag or silica fume. Establishing the durability characteristics of glass powder therefore is imperative to ensure its usage in the concrete industry together with the potential economic and environmental benefits.

Concrete containing glass powder is often investigated for susceptibility of alkali silica reaction (ASR) due to the presence of high alkali contents associated with glass. Research has shown that the ASR could be mitigated by using finely ground glass powder in concrete typically smaller than 75 µm [8,11,15,18]. Moreover, using glass powder as a replacement for cement could help in reducing the deleterious expansion associated with ASR.

Shayan and Xu (2006) [15] measured the permeation characteristics of glass powder concrete mixes using rapid chloride penetration test (RCPT). A lower chloride penetrability was reported for glass mixes as compared to the control mix. Nassar and Soroushian (2012) [19] also reported lower total charge passing through glass concrete. This property was attributed to the refined pore structure resulting from the densification of microstructure due to the pozzolanic reaction of glass. Lee et al. (2018) [20] however, reported higher RCPT value and higher chloride diffusion coefficient measured using NT 492 for concrete containing 20% glass powder at 28 days as compared to PC. The chloride diffusivity of the glass powder mixes was found to decrease as the hydration progressed. Sales et al. (2017) [21] showed a reduction in permeability and increase in electrical resistivity of mortars with an increase in the cement replacement level for fine glass powder particles. Schwarz et al. (2008) [8] compared the moisture transport properties of concrete containing fly ash and glass powder at 10% cement replacement level. The glass powder modified concrete performed better than fly ash and control mix at both early and later ages.

Although a number of researchers have studied the durability of glass powder concrete for a wide spectrum of issues including ASR, abrasion and sulfate attack, limited information is available on the effect of different fineness and replacement levels of glass powder on the transport properties in concrete. In this study, the major durability indicators: porosity, permeability, resistivity, shrinkage and chloride migration for glass powder incorporated concrete are investigated from 3 to 545 days of curing. The durability indicators provide a practical way to identify the likely performance of glass powder containing SCC [22,23].

2. Methodology and Experiments

2.1. Materials

Crushed waste glass bottles were supplied by the Glass Packing Forum. The coarse crushed glass (G) was thoroughly cleaned and subsequently ground in a ball mill to a mean particle size of 10 µm (10G), 20 µm (20G) and 40 µm (40G). General purpose (GP) Portland cement conforming to ASTM Type II and two different fly ashes one belonging each to Class C (FAC) and Class F (FAF) were used in the study. The chemical composition of raw materials measured using X-ray fluorescence (XRF) is given in Table 1. Rounded alluvial Greywacke sandstone was used as the fine and coarse aggregate (sand fineness modulus (FM) = 2.7, coarse aggregate maximum aggregate size of 13 mm). The particle size grading of the aggregates is summarized in Table 2.

Table 1. Chemical composition of raw materials (%/100 g).

	GP	G	FAF	FAC
SiO_2	20.76	70.35	49.87	39.77
Al_2O_3	3.54	2.01	21.88	16.96
Fe_2O_3	2.06	1.59	7.78	7.29
CaO	62.46	10.95	8.91	23.06
MgO	0.88	0.56	2.54	2.80
SO_3	4.00	-	1.50	2.50
K_2O	0.49	0.58	1.20	0.49
Na_2O	0.24	12.88	0.50	1.91
TiO_2	0.19	0.09	1.04	1.21
P_2O_5	0.16	0.03	0.20	0.29
MnO	0.07	0.08	0.06	0.05
LOI	4.54	0.90	3.31	4.02

Table 2. Particle size grading of fine and coarse aggregates.

Sieve Size (mm)	19.0	13.2	9.5	4.75	2.36	1.18	0.60	0.30	0.15	0.075
Fine Aggregate (% Passing)	-	-	-	96	75	60	51	33	8	2
Coarse Aggregate (% Passing)	100	93	48	1	-	-	-	-	-	-

2.2. Sample Preparation

A total of 8 different self-compacting concrete (SCC) mixes were investigated in this study at a water to binder ratio of 0.4. A polycarboxylic ether-based superplasticizer was used to achieve the target flow of 700 +/− 50 mm for the SCC mixes. The raw materials were dry mixed in a 90 L orbital pan mixer for two minutes. Approximately 80% of the required water was added over a period of one minute while mixing. The remaining 20% of water, which was premixed with super plasticizer, was then added to the mixer. After the addition of all the water and superplasticizer, the material was mixed for an additional 3 to 5 min. The high binder content helped to prevent segregation of the control mix, while the use of glass powder and fly ash further aided in the stability of the mix such that viscosity modifiers or stabilizing admixtures were not required for any of the mixes. The superplasticizer dosage was varied to ensure a consistent spread for each mix. Cylindrical specimens of 100 mm diameter and 200 mm height were cast for durability testing and prisms of $75 \times 75 \times 280$ mm were cast to measure the drying shrinkage of concrete. After casting, the samples were placed in a temperature control room maintained at 20 °C and 65% relative humidity for 24 h. Thereafter, the specimens were demolded and cured in lime saturated water until the age of testing. Table 3 shows the mix design details. The GP, FAF30% and FAC30% were the control mixes for comparison with the glass powder (G) SCC mixes.

Table 3. Details of mix proportion and concrete mix design.

Type	Proportion			Binder (kg/m^3)			Aggregate (kg/m^3)		Water(kg/m^3)
	GP	FAF/FAC	G	GP	FA	G	Fine	Coarse	
GP	1.00	-	-	450	-	-	900	850	180
FAF30%	0.70	0.30	-	315	135	-	900	850	180
FAC30%	0.70	0.30	-	315	135	-	900	850	180
10G30%	0.70	-	0.30	315	-	135	900	850	180
20G20%	0.80	-	0.20	360	-	90	900	850	180
20G30%	0.70	-	0.30	315	-	135	900	850	180
20G40%	0.60	-	0.40	270	-	180	900	850	180
40G30%	0.70	-	0.30	315	-	135	900	850	180

2.3. Experiments

2.3.1. Drying Shrinkage Test

Drying shrinkage of concrete was determined in accordance with ASTM C157 (2010) [24], which consists of measuring the drying shrinkage of prismatic specimens, having dimensions of 75 mm × 75 mm × 280 mm, subject to a controlled drying environment. The specimens were removed from the curing tank at the age of 7-days after casting. Immediately after removing the specimens and wiping their surfaces dry, they were placed in the comparator. Afterward the initial measurement all the specimens were placed in the drying room at a temperature of 23 °C and relative humidity of 50%. Drying shrinkage measurements of three replicate specimens were taken at 7, 14, 28, 56, 90 and 180 days of air-drying and an average value of drying shrinkage in micro-strain was reported for each drying age. The shrinkage data provides an indication of long-term cracking risk, which could directly influence the transport characteristics through concrete.

2.3.2. Oxygen Permeability Test

The ease of movement of fluids through a porous structure under an externally applied pressure can be determined from the permeability test. The oxygen permeability test in this study was carried out as described in [25]. This test method measures the pressure decay of oxygen passing through a core of concrete placed in a falling head permeameter. Two 25 mm thick cores were cut from the centre section of two SCC cylinders to obtain a total or 4 sample cores for testing. These slices were kept in an oven at 50 °C for drying until their weight became constant, as suggested in ASTM C642 (2008) [26]. The oven-dried specimens were subjected initially to oxygen at a pressure of 100 kPa and the pressure decay with time was monitored. The test was automated by using pressure transducers attached to the data logger and was continued for either 8 h or until pressure dropped to 50 kPa, whichever approached first. The coefficient of permeability was calculated by conducting a linear regression analysis on the best-fit line achieved by plotting values of ln(Po/Pt) against t, where Po is initial pressure reading at the start of the test, and Pt are subsequent pressure readings at times 't' measured from the time of reading of initial pressure. Four specimens were used to measure coefficients of oxygen permeability at the curing ages of 3, 7, 28, 90, 180, 365 and 545 days and an average value for each curing age was reported.

2.3.3. Porosity

In this study, same specimens used for the oxygen permeability test were also used for porosity measurement. After the end of the permeability test, the specimens were vacuum saturated in tap water and the volume of permeable voids was found according to the procedure described in ASTM C642 (2008) [26].

2.3.4. Electrical Resistivity Test

Resistivity test provides a rapid indication of the likely resistance of concrete to the penetration of chloride ions and the likely subsequent rate of corrosion. This test method consists of measuring the electrical current passed through 25 mm thick slices extracted from concrete cylinders. The same specimens used for oxygen permeability and porosity tests were used to perform the resistivity measurements. The disc specimens were placed between two parallel stainless steel plates. Sponges saturated with 5M NaCl was used to make electrical contact between concrete disks and steel plates. An alternating current was applied across the specimen and the voltage was measured.

2.3.5. Bulk Diffusion Test

The apparent chloride diffusion coefficient was determined by bulk diffusion conforming to ASTM C1556. Two samples with a thickness of 75 mm were sliced from all cast cylinders at ages of

28, 90, 180, 365 and 545 days. The sides and bottom of the test specimens were sealed with an epoxy coating leaving one concrete end face exposed. The sealed specimens were vacuum-saturated in a calcium hydroxide solution, rinsed with tap water, and placed in a sodium chloride solution (165 g/L of NaCl) for at least 35-days. When the exposure time was over, the test specimens were removed from the sodium chloride solution and ten thin layers from 0 mm to 20 mm (2 mm each) were ground off parallel to the exposed face of the specimens. Then, the acid-soluble chloride content of 4 gm sample obtained from each layer was determined. The apparent chloride diffusion coefficients was calculated using Fick's law as described in ASTM C1556.

3. Results

3.1. Compressive Strength

The compressive strength results for different mixes from 28 to 545 days, along with the T_{500} times, are summarized in Table 4. The early age compressive strength for all the fly ash and glass powder mixes were lower than the GP control. The 20G30% and 20G40% replacement levels together with the 40G30% replacement had the three lowest 28 day strengths with values between 43 and 47 MPa. As hydration continued the gap between the GP control and FA and glass powder mixes narrowed. The glass powder mixes shows the slowest strength development taking more than one year time to attain similar strength values. In particular, the glass powder mixes with the largest particle size were the slowest to gain strength. Figure 1 shows the compressive strength of mixes normalized with respect to GP mix at the respective age. The continued strength development is attributed to the formation of C-S-H arising from the pozzolanic activity of glass powder. The observations show the importance of long-term investigation for such binder systems that have slower pozzolanic activity. Similar, compressive strength characteristics have been reported for glass powder concrete mixes in literature [11,27,28].

Table 4. Compressive strength (MPa) at different ages and flow time (Sec).

	Compressive Strength at Age (Days)					T_{500}
	28	90	180	365	545	
GP	65.2	74.8	84.6	89.9	94.5	5.5
FAF30%	49.6	72.2	75.8	81.6	92.8	3.1
FAC30%	68.8	72.7	80.4	92.3	100.6	4.0
10G30%	59.5	66.4	71.1	75.0	91.8	4.0
20G20%	62.5	66.6	71.8	76.3	96.3	3.8
20G30%	43.5	55.5	62.7	72.5	83.3	4.2
20G40%	46.4	51.8	53.3	58.9	79.5	4.9
40G30%	45.8	51.3	55.2	62.0	75.9	4.8

The fresh properties of concrete can have a significant impact up on the long-term durability of a structure. The presence of voids, as a result of compaction difficulties for instance, can be just as important as either the type of binder or w/c ratio. The primary goal of this investigation was to examine the effects of different glass powder replacement levels on the durability properties of the concrete. As such, the target flow for all the mixes was set at 700 mm and the amount of superplasticizer was allowed to vary to maintain the flow and minimize differences in the workability. It can be seen from the T_{500} times reported in Table 4, that there was relatively little variation with the times ranging from 3.1 s (FAF30%) to 5.5 s (GP) with most of the glass powder mixes between 4 and 5 s. The differences between either the yield shear stress as indicated by slump flow (700 mm) or plastic viscosity as indicated by T_{500} for the different mixes therefore were relatively small.

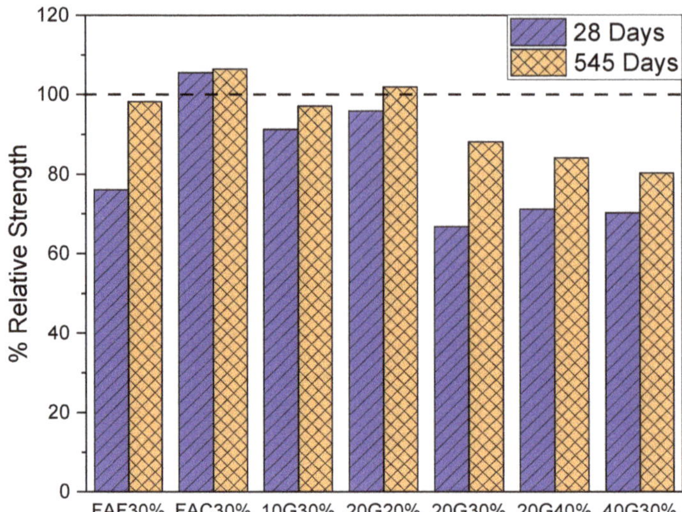

Figure 1. Relative strength of mixes normalized with respect to general purpose (GP) mix at the respective age.

3.2. Drying Shrinkage

Drying shrinkage of concrete is characterized by the time-dependent volume decrease due to moisture migration and transfer when exposed to a relatively lower humidity environment. Figure 2 shows the drying shrinkage strain of the control mixes and glass powder mixes at 30% replacement level for different particle size. GP concrete mix showed higher shrinkage measurements compared to the fly ash and glass powder SCC mixes. The finer pore structure and consumption of larger quantity of water in the pozzolanic reaction could reduce the evaporation of water due to drying. Mehdipour et al. (2018) [29] reported a larger transformation of free water to bound water in cement binders containing SCMs and hence a subsequent reduction in drying shrinkage due to the lower availability of water for evaporation. The drying shrinkage of glass mixes decreased as the glass particle size became finer, which can be related to the denser microstructure of concrete due to the presence of finer glass, which suppressed drying shrinkage. The 40G30% mix showed higher drying shrinkage compared to fly ash or the finer particle size glass powder mixes. The higher dry shrinkage could be due to overall higher porosity (Section 3.3) of the concrete because of the slower pozzolanic reaction of the coarser glass particles leading to a greater evaporation of water.

Figure 3 shows the drying shrinkage strain of the mixes at different glass powder replacement level. The shrinkage strain was found to increase with an increase in the replacement level. However, the drying shrinkage of 20G40% was similar to GP at 180 days. The higher permeable void volume and lower consumption of water in the hydration process at the higher replacement level could be the reason for higher shrinkage. Shayan and Xu [15] also reported an increase in drying shrinkage strain with increase in glass replacement level. Overall the results indicate the glass powder SCC mixes produced acceptable drying shrinkage values, below 0.075% after 56-days drying and met the requirements of the Australian Standard AS 3600 [30].

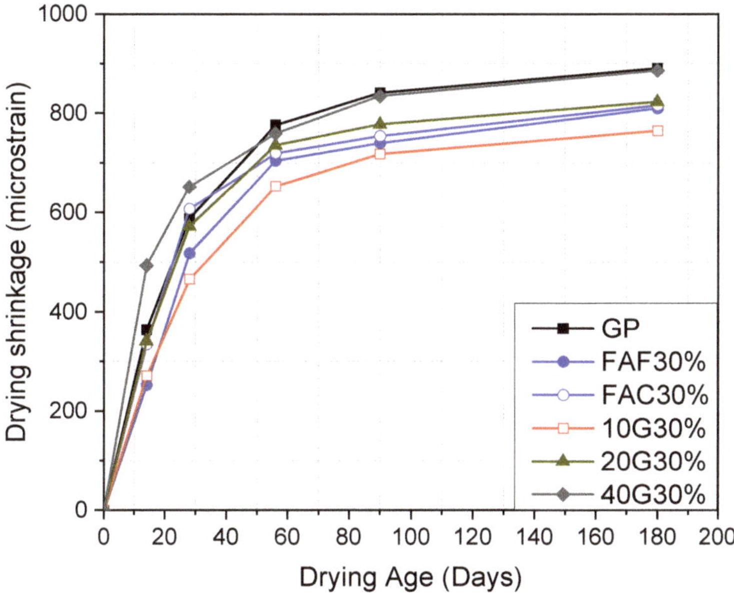

Figure 2. Drying shrinkage of self-compacting concrete (SCC) mixes at 30% cement replacement (these figures show drying time not age and the initial point needs to start at zero days).

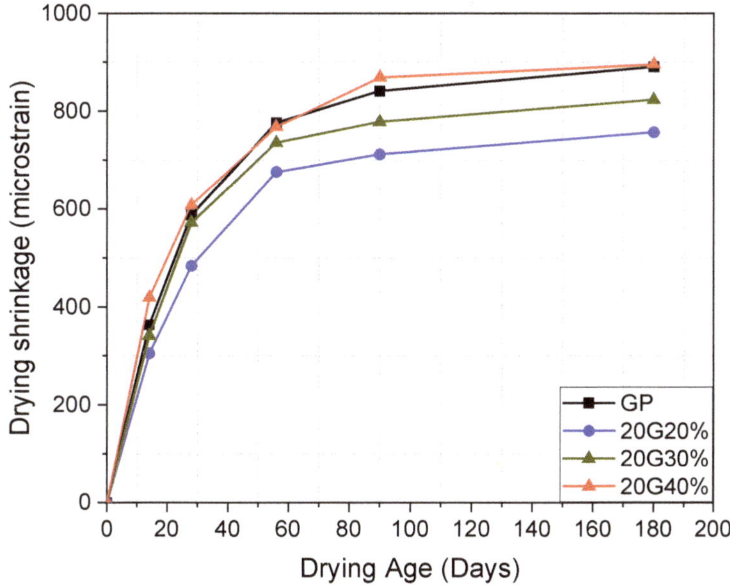

Figure 3. Drying shrinkage of SCC mixes containing glass powder with 20 µm particle size at different replacement levels.

3.3. Oxygen Permeability

Figure 4 shows the oxygen permeability coefficient for the SCC mixes at different curing ages. The GP concrete showed the highest permeability coefficient at all the ages. The permeability coefficient of fly ash concrete mixes were similar to GP concrete at 3 and 7 days, however, thereafter the permeability coefficient continued to reduce for the fly ash mixes as hydration progressed whereas the change in GP concrete was minimal. The lower permeability for fly ash mixes is associated with refinement in pore structure due to the pozzolanic reaction, which in turn increases the tortuosity of the network and thus hinders the free movement of the fluid [31,32]. The coefficient of permeability for glass powder mixes at the same replacement level was seen to be affected by their particle size. A lower resistance to gaseous permeation was observed for the coarser glass powder mixes. Similar to fly ash mixes, the permeability coefficient continued to reduce with time. Moreover, due to the long-term nature of the pozzolanic reaction, the advantages associated with glass replacement were more obvious after a long period of curing. The 10G30% and 20G30% both showed lower permeability coefficients compared to FAF30%. The finer particle size of glass powder helps in substantial pore refinement and hence creating an impermeable and denser microstructure [33]. In addition, the impervious nature of glass particle could also help in reducing the permeability. The 40G30% had a higher permeability as compared to GP at early ages (3, 7 and 28 days), however, the permeability coefficient reduced with progress in hydration. This shows that for coarser particle size, glass powder takes longer for the pozzolanic reaction to progress, which is also evident from the compressive strength data. Sales et al. (2017) [21] also reported similar observations of a reduction in oxygen permeability for mortar samples comprising of glass powder. Chaid et al. (2014) [33] attributed the reason for lower permeability in glass powder concrete to the formation of a denser microstructure. In line with the above results, Figure 5 shows the effect of glass replacement level on the permeability coefficient. The lower the replacement level the greater was the permeation resistance. The 20G40% mix exhibited a higher permeability than FAF30% until 365 days, however it showed similar value at 545 days, implying glass powder mixes even at higher replacement level could achieve similar or better transport characteristics in the long run.

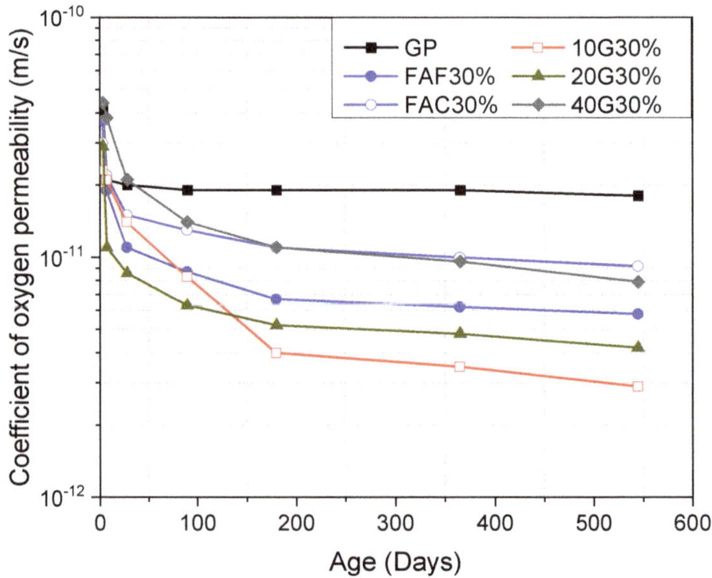

Figure 4. Coefficient of permeability SCC mixes at 30% cement replacement.

The results of the oxygen permeability against compressive strength for the different mixes is provided in Figure 6. For same compressive strength concrete, lower permeability values were observed for glass mixes irrespective of their fineness and replacement level as compared to either the GP or FAC30% mixes. It is well known that the compressive strength is not a good predictor of durability performance particularly when comparing different binder types [34,35]. The compressive strength however can provide an indication of likely relatively performance when comparing concrete with the same general mix design. A correlation of 0.82 was obtained for power based relationship between compressive strength and permeability for glass mixes ($K = 2.18 \times 10^{-9} f_c^{-1.33}$).

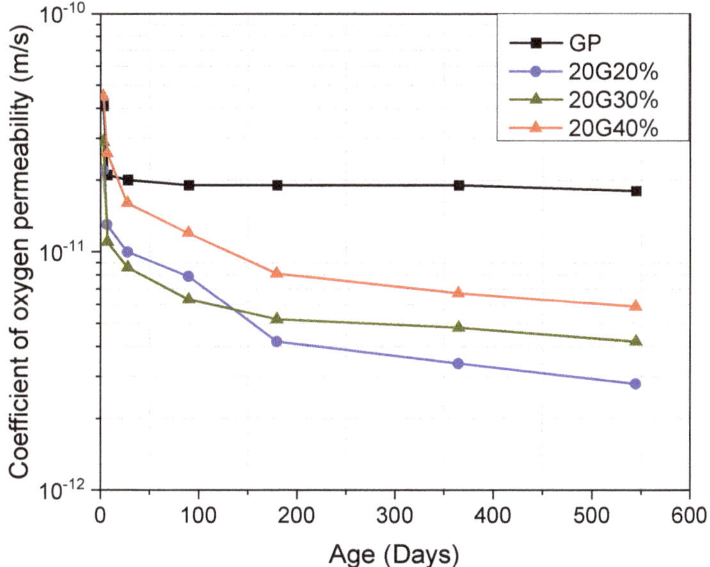

Figure 5. Coefficient of permeability of SCC mixes containing glass powder with 20 μm particle size at different replacement levels.

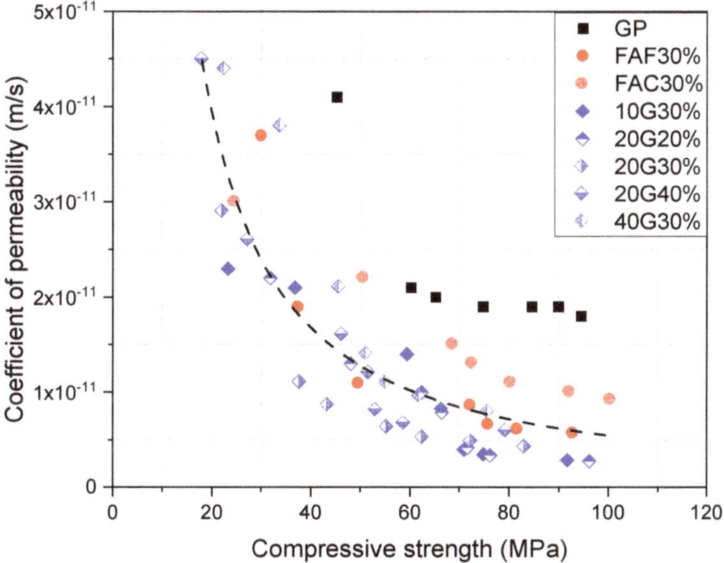

Figure 6. Relationship between compressive strength and coefficient of permeability of all the SCC mixes.

3.4. Porosity

The void content of concrete measured using water absorption test is one of the most common indicators used for assessing the durability of concrete. Figures 7 and 8 show the measured porosity for concrete mixes at different ages. The porosity reduced with continued hydration for in all the concrete mixes. The 10G30% SCC mix demonstrated a lower porosity as compared to other glass mixes at similar replacement level as well as the control mixes. Along with the pozzolanic reaction that assisted in the refinement of pore structure, the finer particle size of glass powder could help to improve particle packing in the system, which resulted in a denser and less porous concrete [19]. The findings signify the importance of particle size distribution as the presence of finer particles in the spaces between larger particles also acts to 'refine' the porosity. The porosity of 40G30% concrete mix was similar to GP despite having lower compressive strength. This implies that the pozzolanic reaction lead to precipitation of hydrates in the capillary pores which restricted the free movement of water. The volume of permeable voids increased with an increase in the glass content. The similar porosity values of 20G40% and 40G30% across ages show that the properties of concrete mixes with at higher replacement level of cement with glass powder could be maintained by varying the fineness of the glass powder. Shayan and Xu (2006) [15] reported similar results of an increase in volume of permeable voids with an increase in the glass powder content. Conversely, Du and Tan (2017) [36] showed a minimal change in porosity for up to 30% cement replacement level with glass powder, however beyond this replacement level an increase in porosity was reported as observed in 20G40% mix. A nonlinear inverse relationship was observed between coefficient of permeability and porosity as shown in Figure 9.

As the porosity of the concrete increased the permeability of the concrete also increased. As with the use of compressive strength as an indicator for durability, comparisons of overall porosity can sometimes be misleading. A single total porosity number does not provide the pore size distribution or tortuosity of the pore structure. As seen in Figure 9 a porosity of 9% would results in a coefficient of permeability of approximately 2×10^{-11} for the GP binder system, while the glass mixes were closer to 1×10^{-11}.

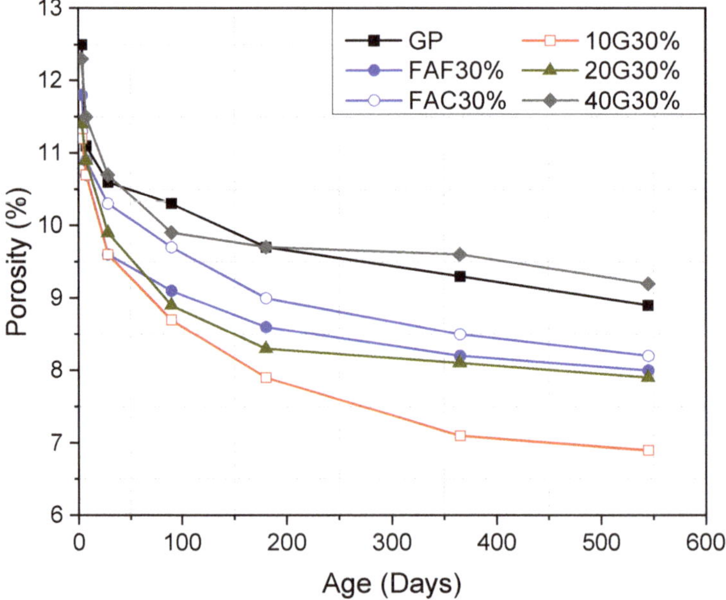

Figure 7. Porosity of SCC mixes at 30% cement replacement at different ages.

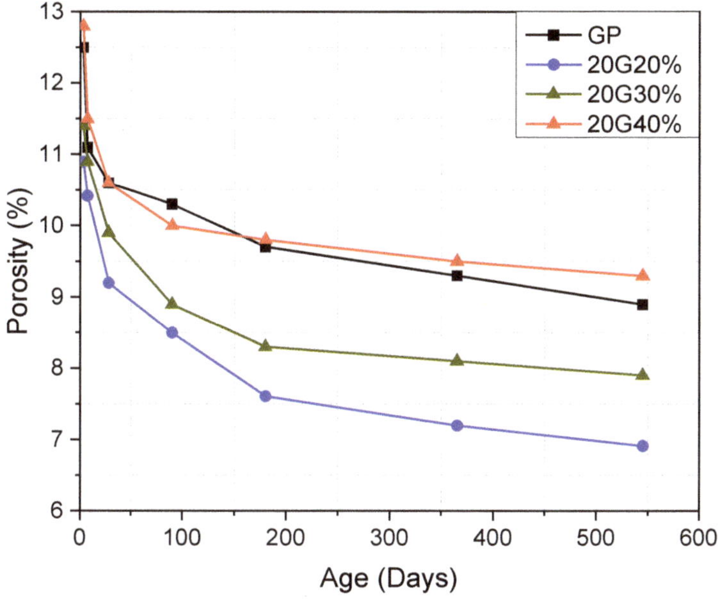

Figure 8. Porosity of SCC mixes containing glass powder with 20 µm particle size at different replacement levels.

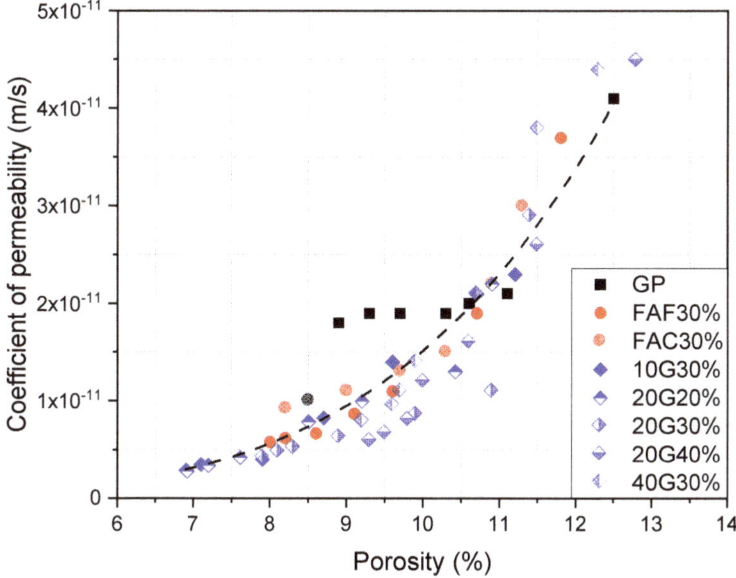

Figure 9. Relationship between coefficient of permeability and porosity.

3.5. Resistivity

The determination of concrete electrical resistivity has become an important technique in the evaluation of risk of corrosion associated with concrete structures [37,38]. The resistivity of concrete is affected by numerous factors such as pore solution composition, degree of hydration, pore structure, moisture content and temperature [39]. Figure 10 compares the resistivity of control mixes and glass powder mixes at 30% replacement. In agreement with the permeability and porosity results, the electrical resistivity of all the SCC mixes increased with curing age. The increase in resistivity is due to the continuous evolution of pore structure parameters due to hydration. The permeable void space reduced due to the precipitation of the hydration products in the capillary space, which hinder the mobility of ions, thereby resulting in an increase in resistivity of concrete. Carsana et al. (2014) [27] reported six times lower resistivity for glass powder mortar as compared to GP.

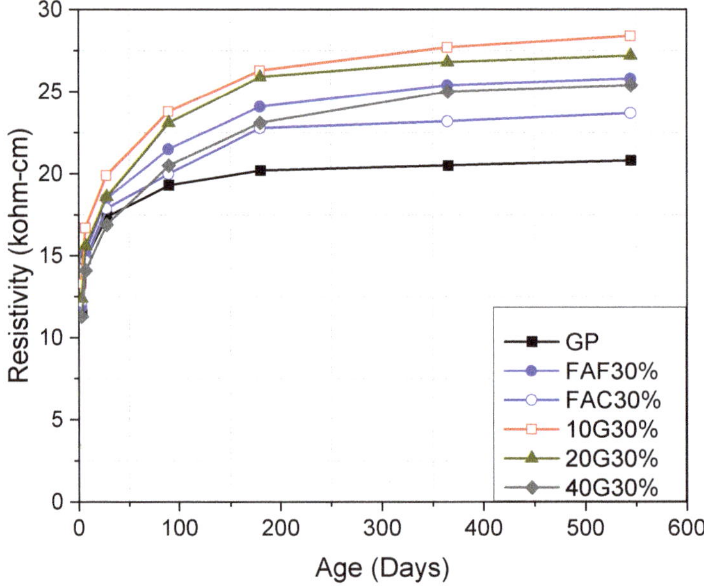

Figure 10. Electrical resistivity of SCC mixes at 30% cement replacement at different ages.

The replacement of cement with fly ash and glass powder resulted in an increase in electrical resistivity. The electrical resistivity of SCC containing glass powder increased as the glass particle size became finer. A comparison of FAF30% and 20G30%, with both having a similar particle size, showed higher electrical resistivity in glass powder containing mixes. Figure 11 shows a reduction in electrical resistivity of with increase in glass replacement level. However, even at 40% replacement 20G40% showed better electrical resistance as compared to PC.

The higher resistance towards flow of ions in SCM incorporated concrete is generally attributed to a finer pore structure due to their pozzolanic reaction. A smaller threshold pore diameter has been reporter for glass powder mixes containing up to 45% cement replacement compared to GP cement using MIP measurement techniques [36]. The smaller the threshold pore diameter, the more tortuous is the microstructure, which directly hinders the movement of ions. It should be noted, however, that the resistivity of the concrete is also affected by the composition of the pore solution and the pore solution composition of the concrete varies with w/c ratio, the degree of hydration, and use of SCMs [38–40]. Nevertheless, changes in the pore structure of the concrete are generally considered to have a greater effect on the measured electrical resistivity than changes in the pore solution composition and concentration [40]. The permeability measurements, which are not dependent on the pore solution, showed a similar behavior, implying the higher resistance offered in glass powder SCC is primarily due to pore structure modification. Evaluation of the pore solution at the various replacement levels would need to be conducted to accurately separate the contribution of the refinement in the pore structure from any modification to the pore solution composition.

The electrical resistivity and porosity values of SCC mixes at different ages are compared in Figure 12. An inverse relationship was observed between the two indicators. As expected the higher the porosity the lower was the electrical resistivity offer by the concrete. It is interesting to note that for the electrical resistivity vs. porosity all the mixes seemed to show reasonably good agreement unlike the permeability vs. strength comparison or the permeability vs. porosity comparison where the GP samples clearly showed a difference in behavior.

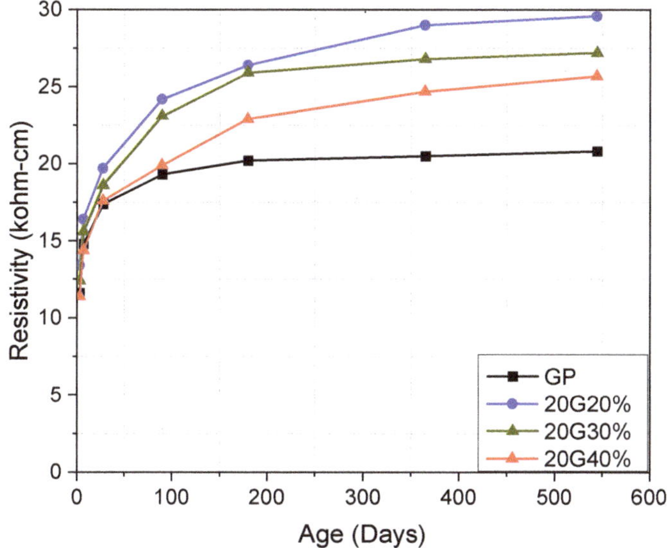

Figure 11. Electrical resistivity of SCC mixes containing glass powder with 20 μm particle size at different replacement levels.

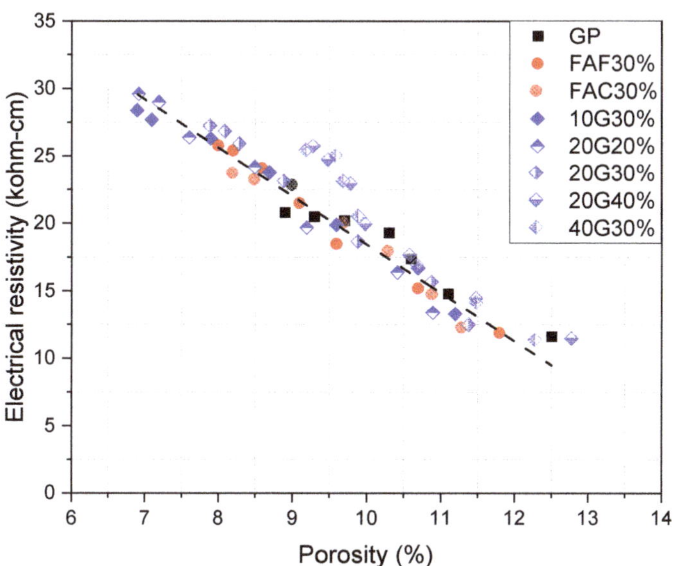

Figure 12. Relation between electrical resistivity and porosity.

3.6. Chloride Diffusion Coefficient

The apparent chloride diffusion measured at the curing ages of 28, 90, 180, 365 and 545 days for control mixes and 30% glass powder SCC mixes are shown in Figure 13, while the results for the different levels of glass replacement are provide in Figure 14. The GP concrete showed the highest chloride diffusion coefficient among all mixes. The fly ash and glass powder mixes showed lower diffusion coefficient than GP at 28 days and continued to decrease with curing age. Du and Tan (2017) [36] reported a reduction of 90% in the chloride diffusion coefficient for 30% glass powder incorporated concrete as compared to GP control. The denser microstructure accompanied with refined pores and lower connectivity contribute to the reduce diffusivity. Kamali and Ghahremaninezhad (2015) [28] also reported a reduction in chloride penetrability with increase in glass powder content in concrete. The time-dependent changes in diffusion coefficient due to continued hydration is often represented by diffusion decay index (m) or ageing factor. Taking 28 days diffusion coefficient as reference, the diffusion decay coefficient was calculated at the ages 90, 180, 365 and 545 using the following equation:

$$\frac{D_t}{D_{ref}} = \left(\frac{t_{ref}}{t}\right)^m$$

where D_t (m^2/s) is the diffusion coefficient at time t (days), D_{ref} (m^2/s) is the diffusion coefficient at t_{ref} 28 days.

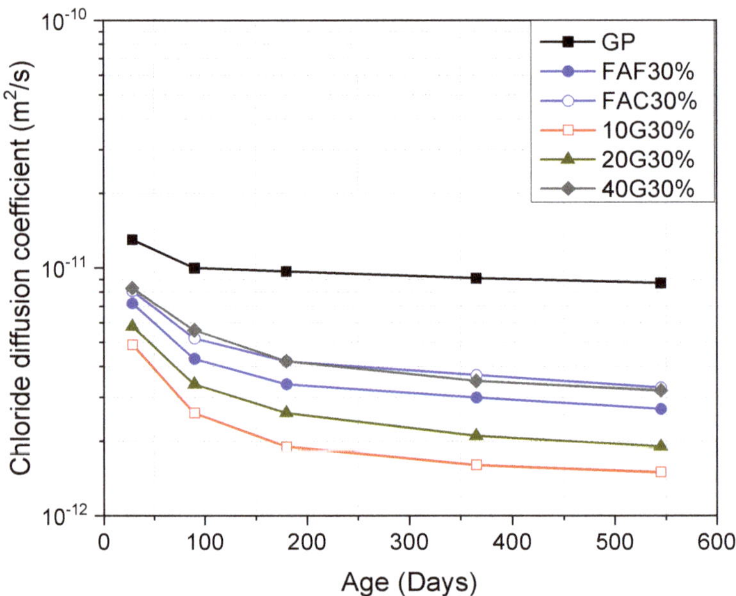

Figure 13. Chloride diffusion coefficient of SCC mixes at 30% cement replacement at different ages.

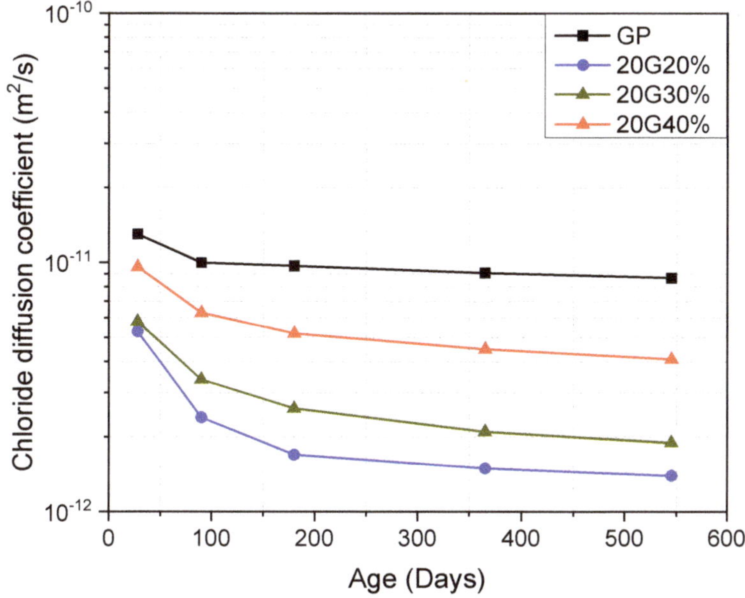

Figure 14. Chloride diffusion coefficient of SCC mixes containing glass powder with 20μm particle size at different replacement levels.

The average diffusion decay index across all ages for different mixes are given in Table 5. The diffusion decay index values of glass powder mixes are larger than those for the fly ash mixes. The decay index of 40G30% and 20G40% are similar to FAF30% indicating glass powder could bring similar changes in the pore structure that restricts the movement of chloride ions even with coarser particle size and higher replacement level. The higher ageing factor values for glass SCC mixes could improve the service life of structure subjected to chloride-induced corrosion.

Table 5. Diffusion decay index for different SCC mixes.

Mix	Diffusion Decay Index m
GP	0.16
FAF30%	0.38
FAC30%	0.34
10G30%	0.47
20G20%	0.56
20G30%	0.42
20G40%	0.32
40G30%	0.34

A good correlation was observed between chloride diffusion coefficient and electrical resistivity (Figure 15). From the results, it can be inferred that electrical resistivity provides a reasonable estimate of the diffusion coefficient value, which is easier to perform and much faster. As with the strength vs. permeability comparison, the chloride diffusion coefficient vs. resistivity relationship is most accurate when evaluating similar mix deign compositions. Where prequalification of mixes for use in marine construction applications is conducted based on bulk chloride diffusion testing, the resistivity testing may provide a good indication of individual batch performance for SCC containing glass powder.

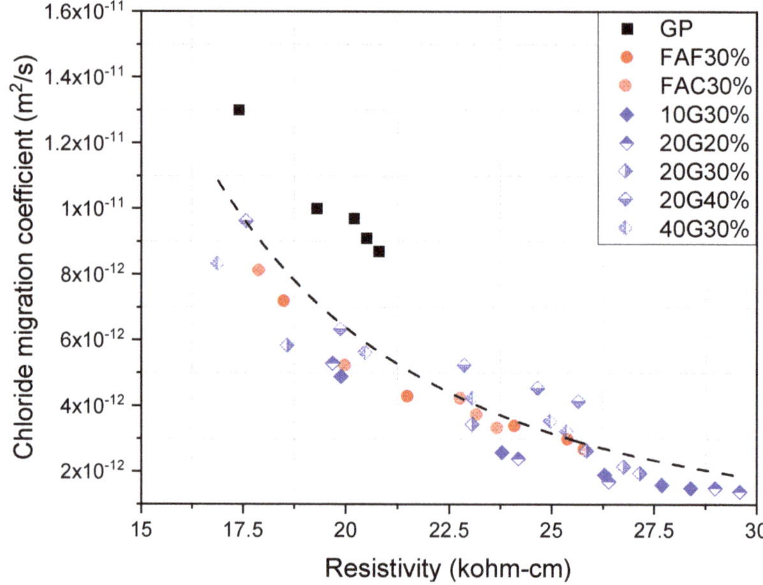

Figure 15. Relation between electrical resistivity and chloride diffusion coefficient.

The results from chloride diffusion coefficient and the time dependent decay coefficient *"m"* were in good agreement with the previous durability index tests of porosity, permeability and resistivity, such that there was a decrease in measured performance with an increase in the glass particle size and replacement level. The replacement of 20% glass powder resulted in the lowest chloride diffusion coefficient even below that of the finer glass powder at 30% replacement levels. Some percentage of glass powder is clearly necessary to refine the pore structure of the concrete, but additions beyond 20% appear to be less effective. The beneficial effect of having a more finely ground glass powder was observed at the 30% replacement level, but no data was available in this study beyond this level for the most finely ground powder. It is likely that there is improved performance below a 30% replacement, with a more finely ground glass powder, however further work is needed to identify the optimal binder replacement level and powder size.

Comparison of Service-Life in Chloride Exposure Condition

In order to understand the effect of the higher diffusion decay index of glass powder concrete mixes, the service-life in terms of corrosion initiation time was predicted using Life 365® software [41]. A concrete cover depth of 50 mm, a threshold concentration of 0.05% (by mass of concrete) and a marine splash zone environment were used to estimate the time until de-passivation of the steel. The diffusion coefficient (D_{28}) and diffusion decay coefficient (m) obtained from the reported experiments were also used in the model. Figure 16 shows the predicted time for the initiation of corrosion for the different concrete mixes. A distinct advantage in terms of increased time until corrosion initiation was observed for glass powder containing mixes. The predicted service-life for glass powder concrete mixes is greater than GP concrete irrespective of the particle size of the glass powder or the replacement levels. At a similar replacement level of 30%, an increase of approximately 2 to 3.5 times the service-life is observed for glass powder concrete mixes (10G30% and 20G30%) compared to FAF30%. A further comparison of the effect of different binder replacement levels and types on the time to initiation is provided in Table 6 for chloride thresholds levels of 0.1% and reinforcement covers of 75 mm.

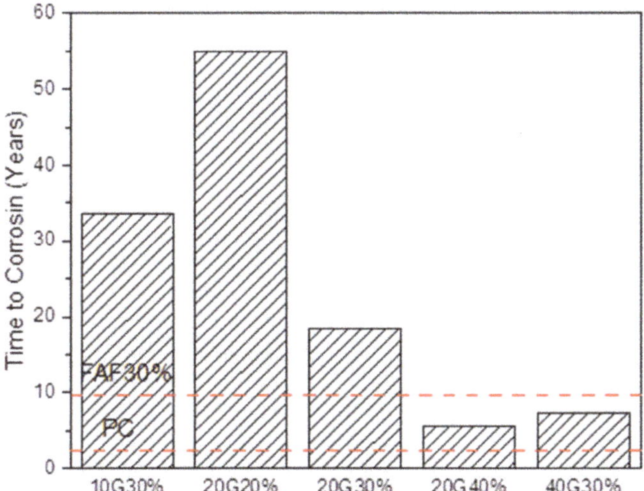

Figure 16. Comparison of corrosion initiation time for different concrete mixes.

Table 6. Comparison of time to corrosion initiation for different binder replacement levels, covers and threshold concentrations.

	OPC	FAF30%	10G30%	20G20%	20G30%	20G40%	40G30%
Cover (75), Ct (0.1)	8	47.3	143.7	150+	87.8	25.6	34.9
Cover (75), Ct (0.05)	5.4	28.5	91.8	136.3	54.9	15	20.6
Cover (50), Ct (0.1)	3.7	16.6	57.7	86.3	33.4	9.3	12.3
Cover (50), Ct (0.05)	2.37	9.9	33.6	55	18.5	5.7	7.4

The service life of real structures is affected by numerous variables. As previously noted the workability of the concrete in the fresh state can influence the degree of compaction and thereby the permeability of the concrete. The binder type, w/c ratio and level of curing are additional parameters that can all dramatically affect the quality of the cover concrete and the overall service life. In addition to the material and construction components, the site specific exposure condition is another crucial component. While a chloride threshold concentration of 0.05% and 0.1% by mass of concrete were used in this example, the actual chloride threshold concentration (Ct) can vary significantly even in the same structure as the use of a single chloride threshold value may not be appropriate for estimate actual performance. The example was intended for illustrative purposed to show the potential impact of glass powder on the durability of reinforced concrete structures assuming all other variables remained constant.

The wide range of tests carried out to corroborate the usage of glass powder as cement replacement shows promising results. The durability indicators (porosity, resistivity and oxygen permeability) of glass SCC mixes all indicate excellent durability according to the classification developed by Alexander et al. [22,23]. The results show that by optimizing the glass powder properties and replacement level durable concrete with mechanical and durable performance better than GP and fly ash mixes can be obtained. The usage of waste glass powder could bring positive economic and environmental benefits for concrete construction applications.

4. Conclusions

This study presented an overview of the influence of glass powder particle size and replacement level on the durability indicators for self-compacting concrete. Oxygen permeability, porosity, electrical resistivity, chloride diffusivity and drying shrinkage of glass modified SCC were compared with GP and fly ash mixes. The main findings from the study can be summarized as follows:

- The amount and particle size of incorporated waste glass has a significant influence on the mechanical and durability properties of SCC. Glass powder replacement level of 20–30% showed better performance as compared to fly ash and GP mixes. Moreover, the filling effect of small-sized glass particles results in improved particle packing, forming a denser and less permeable microstructure.
- The lower shrinkage strains for glass powder concrete mixes will help reduce the risk of cracking associated with long-term drying. The drying shrinkage values of all waste glass incorporating mixes, irrespective of their fineness and content, were below 0.075% after 56-days drying and met the requirements of the Australian Standard AS 3600.
- The lower oxygen permeability and chloride diffusion coefficient along with high electrical resistivity of glass incorporated SCC mixes confirms a refinement of pore structure due to pozzolanic reaction leading to increase in tortuosity and thereby inhibiting free movement of ions.
- The higher ageing coefficients for the glass powder mixes as compared to GP and fly ash mixes indicate continuous evolution and refinement of the pore structure, which results in better long-term transport performance of the binder system and considerable extension in terms of service-life for concrete exposed to marine environment.

Author Contributions: S.T. performed the experimental testing, analysed the results and prepared the initial draft of the paper under the guidance of her supervisors A.N.S. and J.R.M.; A.N.S. and J.R.M. both provided advice and guidance on research direction and assisted with the interpretation of the data in addition to writing of the manuscript. V.S. helped with analysis of the data and in the preparation and revision of the manuscript. All authors have read and agreed to the published version of the manuscript.

Funding: This research was partially funded by the Building Research Association of New Zealand (BRANZ).

Acknowledgments: The authors would like to acknowledge the contributions of BRANZ, the Glass Packaging Forum and Golden Bay Cement for their support.

Conflicts of Interest: The authors declare no conflict of interest.

References

1. Richardson, M.G. *Fundamentals of Durable Reinforced Concrete*, 1st ed.; Taylor & Francis: London, UK; New York, NY, USA, 2002.
2. Bouikni, A.; Swamy, R.N.; Bali, A. Durability properties of concrete containing 50% and 65% slag. *Constr. Build. Mater.* **2009**, *23*, 2836–2845. [CrossRef]
3. Ayub, T.; Shafiq, N.; Khan, S.U. Durability of concrete with different mineral admixtures: A review. *Sci. World J.* **2013**, *7*, 265–276.
4. Ramakrishnan, K.; Pugazhmani, G.; Sripragadeesh, R.; Muthu, D.; Venkatasubramanian, C. Experimental study on the mechanical and durability properties of concrete with waste glass powder and ground granulated blast furnace slag as supplementary cementitious materials. *Constr. Build. Mater.* **2017**, *156*, 739–749. [CrossRef]
5. Chindaprasirt, P.; Chotithanorm, C.; Cao, H.T.; Sirivivatnanon, V. Influence of fly ash fineness on the chloride penetration of concrete. *Constr. Build. Mater.* **2007**, *21*, 356–361. [CrossRef]
6. Thomas, M. The effect of supplementary cementing materials on alkali-silica reaction: A review. *Cem. Concr. Res.* **2011**, *41*, 1224–1231. [CrossRef]
7. Aliabdo, A.A.; Abd Elmoaty, A.E.M.; Aboshama, A.Y. Utilization of waste glass powder in the production of cement and concrete. *Constr. Build. Mater.* **2016**, *124*, 866–877. [CrossRef]
8. Schwarz, N.; Neithalath, N. Influence of a fine glass powder on cement hydration: Comparison to fly ash and modeling the degree of hydration. *Cem. Concr. Res.* **2008**, *38*, 429–436. [CrossRef]
9. Chen, Z.; Wang, Y.; Liao, S.; Huang, Y. Grinding kinetics of waste glass powder and its composite effect as pozzolanic admixture in cement concrete. *Constr. Build. Mater.* **2020**, *239*, 117876. [CrossRef]

10. Shao, Y.; Lefort, T.; Moras, S.; Rodriguez, D. Studies on concrete containing ground waste glass. *Cem. Concr. Res.* **2000**, *30*, 91–100. [CrossRef]

11. Matos, A.M.; Sousa-Coutinho, J. Durability of mortar using waste glass powder as cement replacement. *Constr. Build. Mater.* **2012**, *36*, 205–215. [CrossRef]

12. Metwally, I.M. Investigations on the performance of concrete made with blended finely milled waste glass. *Adv. Struct. Eng.* **2007**, *10*, 47–53. [CrossRef]

13. Shi, C.; Wu, Y.; Riefler, C.; Wang, H. Characteristics and pozzolanic reactivity of glass powders. *Cem. Concr. Res.* **2005**, *35*, 987–993. [CrossRef]

14. Rahma, A.; El Naber, N.; Issa Ismail, S. Effect of glass powder on the compression strength and the workability of concrete. *Cogent Eng.* **2017**, *4*. [CrossRef]

15. Shayan, A.; Xu, A. Performance of glass powder as a pozzolanic material in concrete: A field trial on concrete slabs. *Cem. Concr. Res.* **2006**, *36*, 457–468. [CrossRef]

16. UN Environment; Scrivener, K.L.; John, V.M.; Gartner, E.M. Eco-efficient cements: Potential, economically viable solutions for a low-CO2 cement based industry. *Cem. Concr. Res.* **2018**, *114*, 64.

17. Islam, G.M.S.; Rahman, M.H.; Kazi, N. Waste glass powder as partial replacement of cement for sustainable concrete practice. *Int. J. Sustain. Built Environ.* **2017**, *6*, 37–44. [CrossRef]

18. Rajabipour, F.; Maraghechi, H.; Fischer, G. Investigating the Alkali-Silica reaction of recycled glass aggregates in concrete materials. *J. Mater. Civ. Eng.* **2010**, *22*, 1201–1208. [CrossRef]

19. Nassar, R.U.D.; Soroushian, P. Strength and durability of recycled aggregate concrete containing milled glass as partial replacement for cement. *Constr. Build. Mater.* **2012**, *29*, 368–377. [CrossRef]

20. Lee, H.; Hanif, A.; Usman, M.; Sim, J.; Oh, H. Performance evaluation of concrete incorporating glass powder and glass sludge wastes as supplementary cementing material. *J. Clean. Prod.* **2018**, *170*, 683–693. [CrossRef]

21. Sales, B.R.C.; Sales, F.A.; Figueiredo, E.P.; José, W.; Della, N.; Mohallem, S.; Teresa, M.; Aguilar, P. Durability of mortar made with fine glass powdered particles. *Adv. Mater. Sci. Eng.* **2017**, *2017*. [CrossRef]

22. Mackechnie, J.R.; Alexander, M.G. Durability predictions using early age durability index testing. In Proceedings of the 9th Durability and Building Materials Conference, Australian Corrosion Association, Brisbane, Australia, 17–21 March 2002; pp. 1–9.

23. Beushausen, H.; Alexander, M.G. The South African durability index tests in an international comparison. *J. S. Afr. Inst. Civ. Eng.* **2008**, *50*, 25–31.

24. ASTM C157. Standard test method for length change of hardened hydraulic-cement mortar and concrete. *Am. Soc. Test. Mater.* **2010**, *215*, 1–7. [CrossRef]

25. Salvoldi, B.G.; Beushausen, H.; Alexander, M.G. Oxygen permeability of concrete and its relation to carbonation. *Constr. Build. Mater.* **2015**, *85*, 30–37. [CrossRef]

26. ASTM C642-06. Standard test method for density, absorption, and voids in hardened concrete. *Am. Soc. Test. Mater.* **2008**, 11–13. [CrossRef]

27. Carsana, M.; Frassoni, M.; Bertolini, L. Comparison of ground waste glass with other supplementary cementitious materials. *Cem. Concr. Compos.* **2014**, *45*, 39–45. [CrossRef]

28. Kamali, M.; Ghahremaninezhad, A. Effect of glass powders on the mechanical and durability properties of cementitious materials. *Constr. Build. Mater.* **2015**, *98*, 407–416. [CrossRef]

29. Mehdipour, I.; Ph, D.; Khayat, K.H. Elucidating the role of supplementary cementitious materials on shrinkage and restrained-shrinkage cracking of flowable eco-concrete. *J. Mater. Civ. Eng.* **2018**, *30*, 1–12. [CrossRef]

30. AS 3600. *Concrete Structures*; Standards Australia: Sydney, Australia, 2018.

31. Shah, V.; Bishnoi, S. Analysis of transport properties of carbonated low clinker cements. *Transp. Porous Media* **2018**, *124*, 861–881. [CrossRef]

32. Yu, Z.; Ye, G. The Pore Structure of cement paste blended with fly ash. *Constr. Build. Mater.* **2013**, *45*, 30–35. [CrossRef]

33. Chaïd, R.; Kenaï, S.; Zeroub, H.; Jauberthie, R. Microstructure and permeability of concrete with glass powder addition conserved in the sulphatic environment. *Eur. J. Environ. Civ. Eng.* **2015**, *8189*, 1–19. [CrossRef]

34. Alexander, M.; Beushausen, H. Durability, service life prediction, and modelling for reinforced concrete structures—Review and critique. *Cem. Concr. Res.* **2019**, *122*, 17–29. [CrossRef]

35. Shah, V.; Bishnoi, S. Carbonation resistance of cements containing SCMs and its realtion to various parameters of concrete. *Constr. Build. Mater.* **2018**, *178*, 219–232. [CrossRef]

Appl. Sci. **2020**, *10*, 8058

36. Du, H.; Tan, K.H. Properties of high volume glass powder concrete. *Cem. Concr. Compos.* **2017**, *75*, 22–29. [CrossRef]

37. Alonso, C.; Andrade, C.; González, J. Relation between resistivity and corrosion rate of reinforcements in carbonated mortar made with several cement types. *Cem. Concr. Res.* **1988**, *18*, 687–698. [CrossRef]

38. Hornbostel, K.; Larsen, C.K.; Geiker, M.R. Relationship between concrete resistivity and corrosion rate—A literature review. *Cem. Concr. Compos.* **2013**, *39*, 60–72. [CrossRef]

39. Malhotra, V.M.; Carino, N.J. *Handbook on Nondestructive Testing of Concrete*; CRC Press: West Conshohocken, PA, USA, 2004.

40. Mccarter, W.J.; Starrs, G.; Chrisp, T.M. Electrical conductivity, diffusion and permeability of Portland cement-based mortars. *Cem. Concr. Res.* **2000**, *30*, 1395–1400. [CrossRef]

41. Ehlen, B.M.; Thomas, M.D.A.; Bentz, E.C. Life-365 service life prediction model^{TM}. *Concr. Int.* **2009**, *31*, 41–46.

Publisher's Note: MDPI stays neutral with regard to jurisdictional claims in published maps and institutional affiliations.

Article

Photocatalytic Recycled Mortars: Circular Economy as a Solution for Decontamination

Auxi Barbudo [1,*,†], **Angélica Lozano-Lunar** [1,†], **Antonio López-Uceda** [2], **Adela P. Galvín** [1] and **Jesús Ayuso** [1]

[1]   Construction Engineering Area, Universidad de Cordoba, 14007 Córdoba, Spain;
      angelica.lozano@uco.es (A.L.-L.); apgalvin@uco.es (A.P.G.); ir1ayuje@uco.es (J.A.)
[2]   Mechanics Department, Universidad de Cordoba, 14007 Córdoba, Spain; p62louca@uco.es
*   Correspondence: abarbudo@uco.es; Tel.: +34-957218547
†   Both authors equally contributed to the paper.

Received: 2 September 2020; Accepted: 14 October 2020; Published: 19 October 2020

Abstract: The circular economy is an economic model of production and consumption that involves reusing, repairing, refurbishing, and recycling materials after their service life. The use of waste as secondary raw materials is one of the actions to establish this model. Construction and demolition waste (CDW) constitute one of the most important waste streams in Europe due to its high production rate per capita. Aggregates from these recycling operations are usually used in products with low mechanical requirements in the construction sector. In addition, the incorporation of photocatalytic materials in construction has emerged as a promising technology to develop products with special properties such as air decontamination. This research aims to study the decontaminating behavior of mortars manufactured with the maximum amount of mixed recycled sand without affecting their mechanical properties or durability. For this, two families of mortars were produced, one consisting of traditional Portland cement and the other of photocatalytic cement, each with four replacement rates of natural sand by mixed recycled sand from CDW. Mechanical and durability properties, as well as decontaminating capacity, were evaluated for these mortars. The results show adequate mechanical behavior, despite the incorporation of mixed recycled sand, and improved decontaminating capacity by means of NO_x reduction capacity.

Keywords: recycled aggregates; recycled mortar; construction and demolition waste; decontaminating; photocatalysis

---

1. Introduction

A current environmental challenge is to optimize natural resources. For this, minimizing and recovering waste materials are essential from a productive point of view of the main economic sectors. The circular economy model is a sustainable production strategy where the reuse of waste as secondary raw materials is underlined, achieving comprehensive management of waste materials [1]. So, the circular economy concept describes a cyclical system in which economic and environmental aspects are integrated [2]. Reducing the resources used and the waste generated can save resources and help reduce environmental pollution.

At the European level, this model is implemented through the Action Plan for the Implementation of the Circular Economy of the European Parliamentary Commission [3]. This proposed transition toward a more circular economy brings great opportunities. It is important to make efforts to modernize and transform the economy, shifting it toward a more sustainable direction, which will enable companies to make substantial economic gains and become more competitive in the market. This not only offers important energy savings and environmental benefits, but also creates local jobs and opportunities for social integration. The wider benefits of the circular economy also include reduced energy consumption and carbon dioxide emission levels [3].

The model highlights the important role played by the construction sector, which is still associated with strong negative environmental effects due to the high natural resource consumption for manufacturing and the large production of waste [4]. Waste from construction activities, including excavation or land formation, earthworks, civil construction and building, road works, and building renovation, are considered as construction and demolition waste (CDW) [5,6]. These represent a relevant part of the total generation of solid waste worldwide [7]. Due to this, European Directive 2008/98/EC [8] established that EU member states must reach a recycling rate of 70% for CDW in 2020 to avoid sending it to landfills. However, the potential for reuse and recycling of this waste stream is not being fully exploited. One obstacle is a lack of confidence in the quality of recycled construction and demolition materials. For this reason, on 9 November 2016, the European Commission proposed an industry-wide voluntary protocol on the management of construction and demolition waste [3]. The aim of the protocol was to improve the identification, source separation, and collection of waste, as well as logistics, processing, and quality management. The protocol could thus increase trust in the quality of recycled materials and encourage their use in the construction sector.

After CDW is properly treated, its use as recycled aggregates (RAs) in the construction sector has been widely addressed to mitigate environmental problems such as the consumption of raw materials and waste landfilling. This leads to an increase the recycling rate [9–15], becoming a practical reality that has allowed the development of specific regulations [16].

RAs have different physical, mechanical, and chemical properties from natural aggregates (NAs). They have lower density, higher water absorption, lower resistance to fragmentation, and a higher content of sulfur compounds and soluble salts [17,18]. Depending on the original waste material, recycled aggregates could be concrete, ceramic, or a mixture (recycled mixed aggregate, RMA). RMAs come from building demolitions and contain a wide range of materials, such as concrete waste, pavement material, ceramic products, and, in smaller quantities, other materials such as gypsum, glass, wood, etc. [19]. On some occasions, they are used in works with lower requirements, such as roads with low traffic intensity [20,21], bike lanes [22], urban and pedestrian roads, and unpaved rural roads [23–25], where RAs provide similar functional and structural characteristics as NAs.

Another widely studied application is in the manufacture of concrete. There are studies that support its use, even for structural concrete [26,27]. The use of recycled aggregates in the production of concretes and mortars has the following competitive advantages: (i) decreased extraction of aggregates from rivers, coasts, and quarries; (ii) exclusion of aggregates from landfills, reducing the volume of waste to be treated; and (iii) implementation of the circular economy model and approach to set recycling targets [14].

In most of these studies, only the coarse fraction was used [7,11,12,28]. It was concluded that concrete strength decreased when recycled concrete was used and the reduction could be as low as 40% [29]. No decrease in strength was reported for concrete containing up to 20% fine or 30% coarse recycled aggregates, but beyond these levels, there was a systematic decrease in strength as the content of recycled aggregates increased [29].

Of the different types of RAs, recycled crushed concrete (RCA) is the most widely used in the manufacture of new concrete [9,26,27,30–32]. Even the Spanish concrete standard (EHE-08) [16] allows the use of up to 20% RCA, but only referring to fractions greater than 4 mm. Because of the lower density of RMA, concrete made with it has lower density and higher water absorption than reference concrete [19].

Regarding the influence of recycled coarse and fine fractions, studies on recycled concrete incorporating fine recycled aggregate (FRA) from CDW did not obtain satisfactory results [19,29,30,32,33]. Kathib [29] studied the incorporation of FRA, and showed that properties such as density, dynamic modulus of elasticity, and compressive strength were reduced, the latter resulting in a decrease of 10% and more than 15% with a 50% and 100% incorporation ratio, respectively, at

90 days, in agreement with the results by Kou et al. [33]. Because of these differences, the use of fine fractions in concrete should not be dismissed, but more research on it is needed.

For this reason, the main application of FRA is in recycled mortar, for which the requirements are more tolerant. Within this group, there are different uses depending on where it will be placed: masonry mortar, interior or exterior rendering mortar, or mortar for the manufacture of paving blocks. Most of the investigations carried out to date were aimed at incorporating FRA in mortar for masonry [31,34–36]. However, this research in particular is intended for the use of recycled mortar for any application in contact with the atmosphere, such as wall cladding or paving blocks. With the results obtained, it is hoped that its use can be further expanded.

As in the case of recycled concrete, the use of mixed FRA (MFRA) in mortar has been found to be harmful to the properties of mortar [14,36]. However, this research aims to prove its use as suitable if it meets a series of quality requirements, normally linked to correct CDW treatment.

In addition to the circular economy, the mortar manufactured in this research is intended to provide a second benefit for environmental sustainability: atmospheric decontamination and reduced carbon footprint. The atmospheric pollution produced by accelerated population growth and industrialization can cause serious damage to the health of both people and ecosystems, and even to infrastructures and historical heritage. Among the main harmful emissions are nitrogen oxide gases (NO and NO_2, commonly known as NO_x) generated by transportation and various industries. These gases have high toxicity and can cause serious problems in human health, as well as environmental problems (acid rain, photochemical smog, destruction of the ozone layer, etc.) [37].

It is essential to reach the highest air quality that does not create a risk to people's health and does not cause deterioration or permanent damage to ecosystems. The current measures for reducing air pollution in cities associated with CO_2 emissions and other pollutant gases fall within two lines of action: citizen awareness policies, which advocate avoiding the use of private vehicles in favor of alternative means of transport, such as bicycles or public transport; and policies restricting the circulation of these vehicles, either with speed reductions or by prohibiting movement in downtown areas, as is carried out in cities such as Paris, London, and Madrid [38].

However, these measures are not fully effective in eliminating or reducing air pollution in urban environments, which is why complementary measures are needed in addition to conventional pollution control methods. Within these new measures, it is interesting to consider the advantages of photocatalysis.

The photocatalysis process starts from the natural principle of decontamination of nature itself. It is a similar technology to that of photovoltaic solar panels [39]. Like photosynthesis, which, thanks to sunlight, can remove CO_2 to generate organic matter, photocatalysis removes other usual pollutants in the atmosphere such as NO_x and SO_x (inorganic compounds) and volatile organic compounds (VOCs) through an oxidation process activated by solar energy. Through photocatalysis, most of the pollutants present in urban areas can be reduced, such as NO_x, SO_x, VOCs, CO, methyl mercaptan, formaldehyde, chlorinated organic compounds, poly-aromatic compounds, etc., which are aggressive in terms of both the properties of the material and the environment.

Construction materials treated with photocatalysts reduce, above all, NO_x particles that are produced by vehicles, industry, and energy production. During photocatalysis, the photocatalyst agent absorbs light energy, transfers it to a reactive compound, and triggers a chemical reaction through the formation of radicals. Titanium dioxide (TiO_2) and the products derived from it are the most widely used photocatalysts, and it is these that trigger the transformation of NO_x (nitrogen oxides) into nitrate through the action of sunlight. This can then be used to increase the shelf life of cement-based materials, while it can also be used to substantially decrease the concentration of some air pollutants, especially in semi-enclosed places such as important urban avenues, tunnels, or heavily polluted places like gas stations and some specific industries [39].

In the process of decontamination by photocatalysis, the contaminant is absorbed on the surface of the material to be later oxidized, in two stages, to an inert nitrate compound (NO_3). Finally, the inert compound is removed from the surface of the material by rain.

The incorporation of photocatalytic substances in construction materials has emerged as a promising technology to develop products with special characteristics/properties [40]. Among the multiple advantages of photocatalysis is that it is a clean technology that does not need any maintenance, and once applied, its effect is permanent, and it "cleans" the contaminated air. It also saves on costs, since areas where this mortar has been used remain clean for many years, and it destroys the dirt that is deposited on it, which favors the growth of microorganisms [41–45]. Even the possibility to eliminate pollen [46] or deposited soot [47] from the air has been studied.

A building's façade is one of its most important parts since it gives it a distinct personality. On the other hand, pavements in urban areas, steels, parking lots, etc., are among the most important parts of civil construction. In all these cases, it is important to assess the useful life of the chosen material and hence there is a need to investigate new solutions that extend any maintenance operation over time and directly contribute to improving environmental sustainability [38].

With an awareness of the needs of today's society in terms of waste reduction, specifically CDW, through the use of RA and actions on pollution in urban environments, this research aims to contribute to the development of solutions to both problems by developing recycled mortars with photocatalytic capacity that contribute to the preservation of the environment with sustainable initiatives, and are also technically viable. This study investigates the effect of photocatalytic mortars on reducing air pollution produced by traffic emissions of CO_2 and NO_x, with the added value of being made with FRA. The intention is to serve to advance the research work and thus achieve the necessary objectives for sustainable development in accordance with European and Spanish regulations. It contributes to the circular economy by using recycled materials to reduce the waste generated and the need to obtain new raw materials, and improves the quality of life of citizens who live in population centers that have severe pollution problems, which is increasingly present in urban environments.

For this, two families of mortar were manufactured with different types of cement using different replacement rates. The aim was to obtain a recycled mortar that provides the highest decontamination capacity using the highest possible percentage of recycled sand without significantly affecting its mechanical and durability properties.

The rest of the article is structured as follows: the "Materials" section details the materials for the production of the mortars; the "Experimental Program and Methods" section specifies how the research has been carried out; in the "Results and Discussion" section, the data obtained is analyzed; and finally, in the "Conclusions" section, the main advances obtained are highlighted, as well as future lines of research.

2. Materials

The specific materials selected for the research are described below.

2.1. Cements

Two types of cement were used in this research. Cement for the conventional mortar was CEM I 52'5 N (Cement without additions, high strength (52.5 MPa at 28 days), and normal initial strength). The photocatalytic cement used was i.tech ULTRA, hereinafter called Ph. CEM I. This is a Portland cement similar to CEM I, but with an addition of titanium oxide (TiO_2). Both cements come from the same manufacturer.

2.2. Aggregates

This research used two aggregates with a granule size of 0–4 mm:

1. Natural sand (NS): siliceous sand from the Gravera Dehesilla, located in Alcolea (Córdoba, Spain), containing a content of fines (particles < 0.063 mm) equal to 2.26%.
2. Recycled sand (RS): from the Gecorsa CDW treatment plant in Córdoba, Spain, corresponding to the fine fraction of a mixed recycled aggregate; 12.4% of particles were <0.063 mm and it had a solubility in acid equal to 0.87% SO_3. It presented a continuous particle size distribution curve and directly influenced mortar properties such as mechanical strength, workability, compaction, and durability [34].

The composition test carried out on the coarse fraction of said aggregate (>4 mm), according to EN 933-11, showed 18% ceramic particles (Rb), 34% concrete and mortar (Rc), 47% natural aggregate (Ru), and 1% other particles (X), among which plaster stands out. The water absorption and dry specific density of NS and RS, according to EN 1097-6, are shown in Table 1.

Table 1. Specific gravity and water absorption of aggregates. NS, natural sand; RS, recycled sand.

		NS	RS
Dry specific density	(g/cm^3)	2.645	2.573
Water absorption (WA)	(%)	0.32	3.15

The results indicate that RS had a lower specific gravity and higher percentage of water absorption compared with the corresponding properties of natural aggregates, agreeing with what was indicated by other authors [31,35]. This may be due to a higher percentage of mortar and ceramic particles. However, the percentage of water absorption after 24 h of immersion was lower than the 6–9% obtained by other authors [36,48–50]. This may be due to better treatment of the recycled aggregate, or because the original waste contained fewer porous particles.

3. Experimental Program and Methods

3.1. Experimental Program

A total of 8 mortars were manufactured and divided into 2 families, one produced with CEM I (conventional mortar family) that would be used as a reference, and one with Ph. CEM I (photocatalytic mortar family) was used to determine the decontaminating power of mortars made with TiO_2.

In order to increase to the sustainability of the sector, as much RS as possible is expected to be used. For it, each family was produced according to 4 replacement rates of NS by RS (0%, 20%, 40%, and 100%) by weight. Table 2 shows the nomenclature of mortar families produced in this research.

Table 2. Nomenclature of mortar families.

% RS	CEM I	Ph. CEM I
0	M0	PM0
20	M20	PM20
40	M40	PM40
100	M100	PM100

The mortar dosage was calculated based on EN 196-1 and is shown in Table 3. The weights of aggregates shown in this table refer to dry weight.

The amount of mixing water shown in Table 3 was constant, resulting in a water/cement (w/c) ratio equal to 0.58. Due to the low water absorption of RS compared to other recycled aggregates, an increase in the water content as a percentage of increased incorporated RS was not considered, since a high w/c ratio could produce a weaker and more porous mortar [51].

The mixing procedure was in accordance with EN 196-1. A total of 12 prismatic specimens were produced with dimensions of 40 × 40 × 160 mm. These specimens were cured in a climatic chamber at 20 ± 1 °C and 65% relative humidity until the age test.

Table 3. Mortar dosage (per m^3).

	M0/PM0	M20/PM20	M40/PM40	M100/PM100
CEM I/Ph. CEM I (kg)	323	323	323	323
NS (kg)	874	700	525	0
RS (kg)	0	174	349	873
Water (L)	187	187	187	187

3.2. Mortar Characterization

The consistency of the mortar families in the fresh state was measured in accordance with EN 1015-3. The hardened state properties analyzed in the mortar families were compressive and flexural strength (EN 1015-11), water absorption by capillarity (EN 1015-18), water absorption capacity, bulk and skeletal density and open porosity for water (Spanish Standard UNE 8398), carbonation depth (EN 13295), and photocatalytic activity (Spanish Standard UNE 83321 EX). These properties were evaluated after 28 days of curing time.

In addition, X-ray diffraction (XRD) analysis was carried out to identify the main crystalline mineral components. For that purpose, a piece of the central part of each mortar specimen was crushed and sieved through a 0.125 mm sieve. The machine used for this technique was a Bruker D8 Discover A25 with Cu-Kα radiation, and the goniometric exploration used was swept from 5° to 80° (2θ°) at a speed of 0.0142° min^{-1}. The Joint Committee on Powder Diffraction Standards database was used to identify the phases formed in the mortars [52]. A JEOL JMS-7800 scanning electron microscope (SEM) was used to determine the mortar's chemical composition.

For the carbonation depth test, the mortar specimens were introduced into a carbonation chamber under conditions of relative humidity of 55–65%, temperature of 23 ± 3 °C, and CO$_2$ concentration of 5% ± 0.1%. After 56 days of CO$_2$ exposure, a phenolphthalein pH indicator spray was used on the mortar fracture surface. The noncarbonated mortar surface showed a purple color due to its high alkaline pH. The carbonation depth was measured on the mortar fracture surface from the edge of the specimen to the purple area.

Despite the successful application of TiO$_2$ photocatalysis to cement-based materials, an ideal method to determine the photocatalytic activity is still not available. The experimental conditions and data treatment differ in many aspects (light source, UV intensity, temperature, humidity, flow rate, characteristics of test samples, contaminant analyzed), even leading to noncomparable results [42]. The experimental method proposed by Spanish standard UNE 83321 EX was used in this research. This was aimed at evaluating the degradation of nitrogen oxide, in the gas phase, of inorganic photocatalytic materials contained in cement concretes by a continuous flow test method. For the measurements and calculations required in this test, the concentration of nitrogen oxides (NO$_x$) was defined as the stoichiometric sum of nitrogen oxide (NO) and nitrogen dioxide (NO$_2$).

Likewise, a sample was extracted from the center of each prismatic specimen mortar to analyze its photocatalytic power according to the standardized methodology through the reduction capacity of NOx. The test was carried out on the photocatalytic mortar family in addition to the reference mortar (M0).

4. Results and Discussion

4.1. Consistency

The average consistency results obtained after two perpendicular measurements are shown in Table 4. It can be seen that as the percentage of RS increases, the consistency of the mortar decreases, with a minimum value corresponding to 100% RS, mainly due to its higher water absorption. Silva et al. [38] stated that this loss in consistency can also be attributed to the greater angularity of the recycled particles, avoiding effective slippage between them. To compensate for this, an amount of water corresponding to the absorption could be added [31,36,49,53,54], although the mechanical properties could be affected [35]. For this reason, the most suitable solution would be to add plasticizers to the mortar mix, thus increasing its consistency/workability, as advised by Ledesma et al. [14].

Table 4. Consistency values of mortar families (mm).

	Consistency		Consistency
M0	134	PM0	129
M20	131	PM20	125
M40	130	PM40	113
M100	111	PM100	102

It was also observed that the photocatalytic mortar family showed less consistency in all cases compared to the conventional family. The lowest consistency, registered by PM100, may be responsible for its porous appearance (Figure 1) and difficult compaction.

Figure 1. Porous appearance of PM100 versus smooth appearance of other mixes.

4.2. Mechanical Strengths

The values registered for mechanical strength at 28 days are shown in Figure 2, which illustrates the comparison of results between the two families.

Figure 2a shows that for the family made with conventional cement and 40% of NS replaced by RS (M40), the compressive strength is equal to that corresponding to 20% replacement, approximately 46 MPa, which is only 10% less than the reference mortar (M0) at 52 MPa. For full replacement (M100), the drop in compressive strength increases to 16%, with an average value of 44 MPa. This agrees with Silva et al. [51], suggesting that as the RS content increases, the compressive strength remains similar to or larger than that of the control mortar. This may be due to the reduction of effective water as the percentage of RS is increased, as explained by López Gayarre et al. [35], or because of the greater number of fine particles that can fill the gaps [49]. In this investigation, the content of fine RS was higher than NS (12.4% vs. 2.26%), favoring the filling of voids in the mortar matrix and diluting the loss of mechanical strength. However, in most mortar mixes, there is a loss of compressive strength as the replacement percentage increases [34]. In all mortar mixes, the compressive strength is greater than the value recommended by the GB 28635-2012 standard [55] (average strength ≥ 30 MPa; any individual strength ≥ 25 MPa) and values obtained in other studies [19,56].

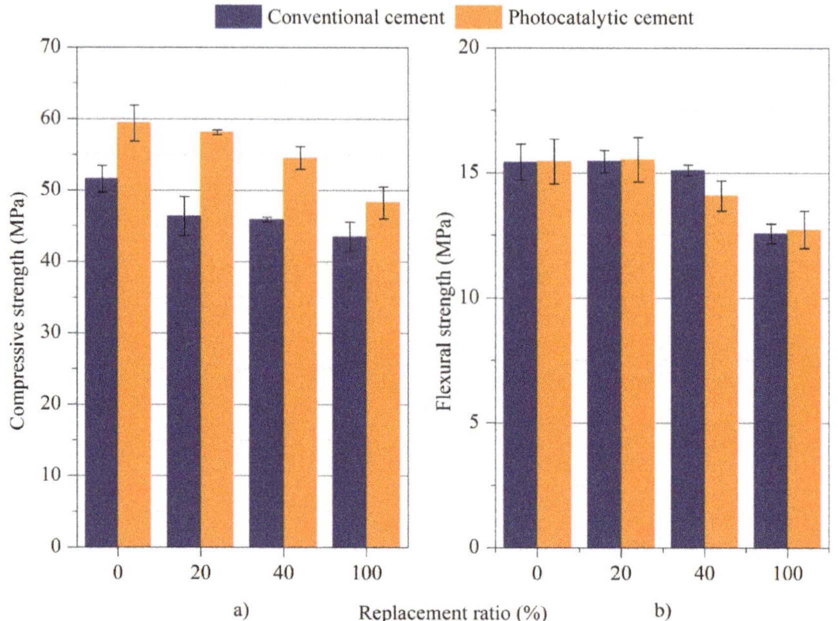

Figure 2. Mechanical strength comparison: (**a**) compressive strength; (**b**) flexural strength.

In all cases, the compressive strength obtained in the photocatalytic family is greater than that of their counterparts made with conventional cement. In this way, they start from an average resistance of 59 MPa for PM0, reaching 48 MPa for PM100.

Flexural strength can be correlated with other characteristics such as susceptibility to cracking and adhesive strength of mortar [51]. In this test (Figure 2b), a similar value of flexural strength was obtained for the 0%, 20%, and 40% replacement, 15 MPa, except for the M100 and PM100 families, for which a value of 13 MPa was found. This agrees with values obtained by Silva et al. [51], whose research used the same type and dosage of cement. Therefore, no improved behavior of the photocatalytic family was observed in this test. All values registered in the flexural strength test show very good performance of these mortars for their possible use in pavement, with values greater than 12 MPa in all mortars, exceeding the values obtained by other authors [56,57].

The differences between the strength of mixtures with 0% and 40% RS are practically insignificant, which confirms that increasing the percentage to 40% does not just mean decreased mechanical properties. This agrees with Ledesma et al. [14], who established a maximum replacement ratio of up to 50% of natural sand by mixed recycled sand without significantly affecting the hardened mortar properties.

Generally, a decrease in strength is observed as the percentage of recycled aggregate is increased, as reported in previous works. This decrease is more gradual and more noticeable in the photocatalytic family. However, the increase in RS up to 40% showed only a 7% decrease in strength with respect to 20% replacement, and up to 9% with respect to the reference mortar.

4.3. Mineralogical Analysis

The mineral phases formed in conventional and photocatalytic mortar families are shown in Figures 3 and 4, respectively. In all specimens, the main detected phase corresponds to quartz (SiO_2; 33-1161) [52]. The intensity decreased in all specimen patterns because the NS was replaced with RS, which contains less silica [58]. The main phase in M100 was sanidine ((Na, K)(Si$_3$Al)O$_8$; 10-0357) [52].

The Portlandite phase (Ca(OH)$_2$; 04-0733) [52] and ettringite (Ca$_6$Al$_2$(SO$_4$)$_3$(OH)$_{12}$·26H$_2$O; 41-1451) [52] were also observed in both families. The presence of these phases is an indicator of the Portland cement reaction [57], as shown by the mechanical performance of both mortar families (Figure 2). For this reason, the incorporation of RS into the mortar is compatible with the common Portland cement reaction.

The other detected phases in the mortars corresponded to silicates such as illite (KAl$_2$Si$_3$AlO$_{10}$(OH)$_2$; 02-0056) [52] and albite (Na(Si$_3$Al)O$_8$; 10-0393) [52], which is in agreement with authors such as Jiménez et al. [36] and Ledesma et al. [14]. Regarding carbonates, all mortars showed calcite (CaCO$_3$; 05-0586) [52]; additionally, dolomite (CaMg(CO$_3$)$_2$; 36-0426) [52] was detected in M100 and all specimens of the photocatalytic mortar family. In both mortar families, it was observed that calcite (CaCO$_3$; 05-0586) [52] had a greater presence as the replacement of NS by RS increased. This behavior has also been recorded by authors such as Gonçalves et al. [58], who studied the replacement of natural siliceous sand with recycled aggregate. Gypsum (CaSO$_4$·2H$_2$O; 33-0311) [52] was also detected in all specimens in both families, which could correspond to the RA or cement composition [36].

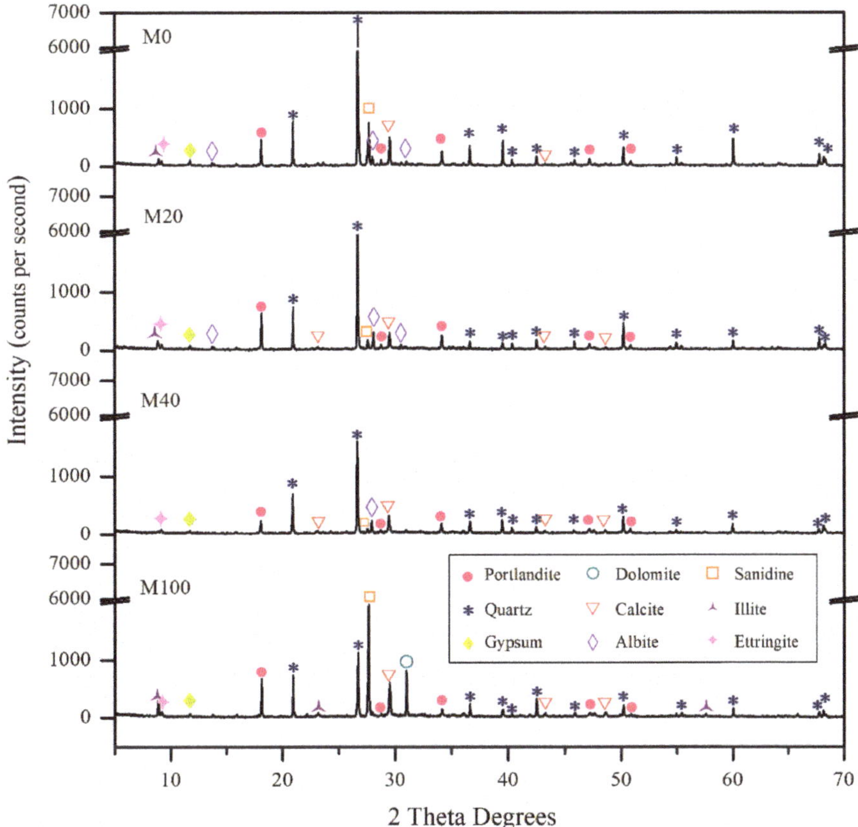

Figure 3. X-ray diffraction patterns of conventional mortar family.

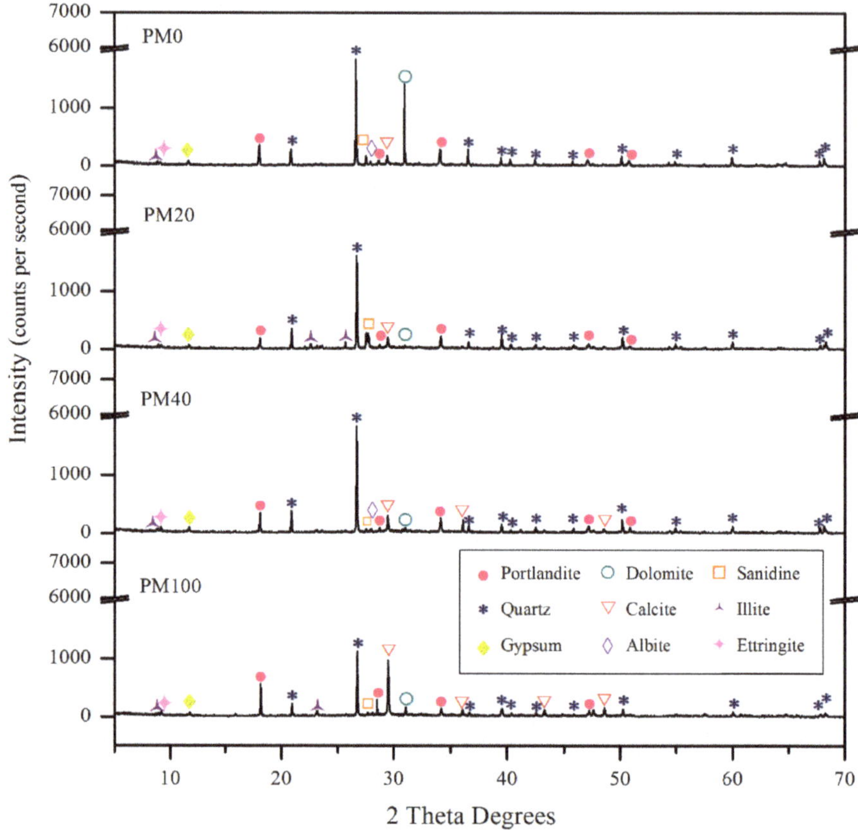

Figure 4. X-ray diffraction patterns of photocatalytic mortar family.

Additionally, an SEM study was carried out, which supported the results obtained in the mineralogical study, as shown in Table 5. In the photocatalytic mortar family, no mineralogical phase attributed to Ti was detected; however, SEM confirmed the presence of TiO_2 in all mortars with Ph. CEM I.

Table 5. Average percentage, by weight, of elements by scanning electron microscopy (SEM).

Element	M0	M20	M40	M100	PM0	PM20	PM40	PM100
C	10.30	8.53	9.31	9.40	7.79	12.15	10.42	12.16
O	49.65	48.59	50.10	45.34	48.79	46.67	45.84	49.39
Mg	0.50	0.87	0.52	0.56	0.31	0.37	0.30	0.51
Al	2.30	2.95	2.36	3.89	1.78	3.00	3.74	1.96
Si	8.01	9.72	8.27	10.61	9.46	13.78	10.61	12.18
S	0.75	0.95	1.14	0.92	0.62	0.66	0.75	0.90
K	0.34	0.39	0.41	1.06	0.71	0.32	1.64	0.39
Ca	26.34	25.58	26.51	25.14	28.96	20.40	23.69	20.48
Ti	0.00	0.00	0.00	0.00	0.52	0.55	0.63	0.63
Total:	100	100	100	100	100	100	100	100

4.4. Water Absorption by Capillarity

The capacity to absorb water indicates the ability of an unsaturated porous material to absorb and drain water by capillary action, thus making it a suitable property to indirectly assess the durability of cementitious materials. Normally, higher water absorption by capillarity contributes to worse performance since it impairs the protection against external agents [51].

The results shown in Figure 5 indicate an upward trend in water absorption by capillarity as the percentage of RS increased in both families. However, for 20% of RS (M20), the values achieved in the conventional mortar family were even lower than those of the reference mortar (M0). The higher water absorption by capillarity of recycled mortars can be due to the high absorption of RS. Similar tendencies have been observed in studies by other authors [14,31,34,36].

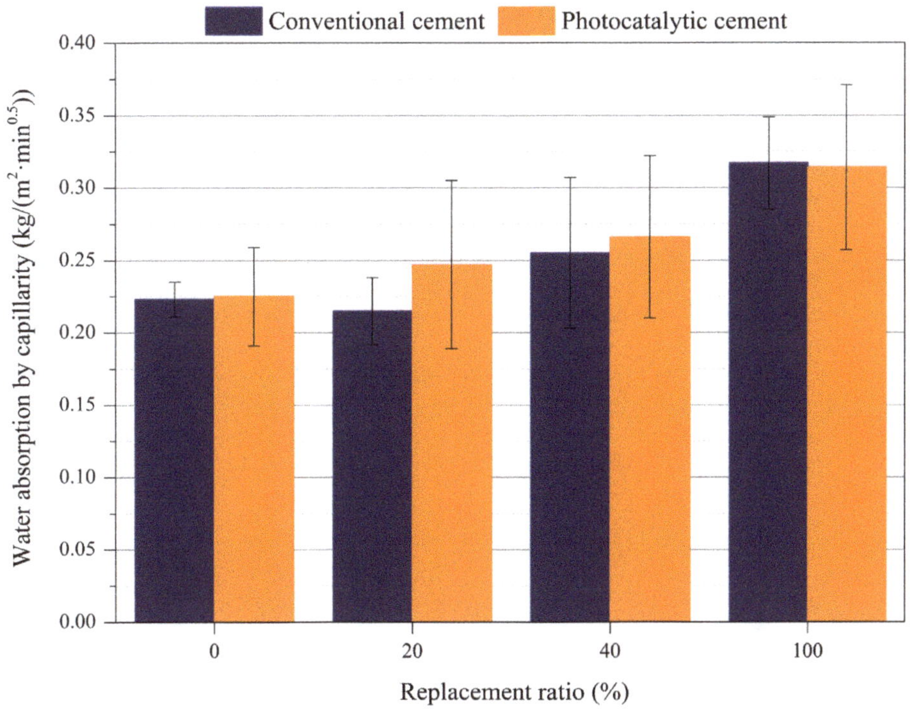

Figure 5. Water absorption by capillarity comparison.

According to López Gayarre et al. [35], the reduction in the amount of effective water as the percentage of substitution is increased (greater water absorption) reduces the porosity of fresh mortar, and for this reason, recycled mortar could present this slight increase of water absorption by capillarity.

4.5. Water Absorption Capacity, Bulk and Skeletal Density, and Open Porosity

The water absorption capacity (Figure 6a) was almost the same for all mortars produced with photocatalytic CEM I. However, for the conventional mortar family, there was a slight increase with 40% of RS, and a more appreciable increase for 100% of RS of around 25%.

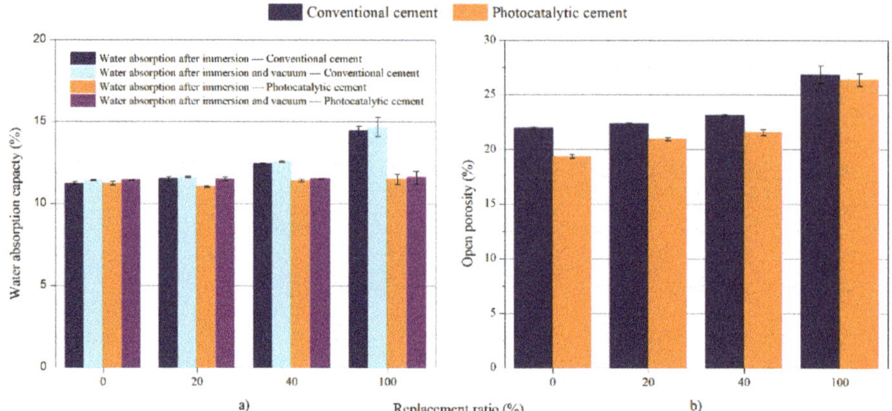

Figure 6. Comparison of results for (**a**) water absorption capacity and (**b**) open porosity.

Open porosity (Figure 6b) was higher as the RS replacement ratio increased. Poon and Cheung [59] explained this, affirming that materials with lower density lead to higher porosity of mortar blocks.

However, the density values (Figure 7a,b) were approximately the same in all cases, regardless of the type of cement or the amount of recycled aggregate. This differs from the values obtained in some studies [31,34], in which the density decreased as the replacement ratio increased. Other studies [31,49,60,61] found no significant differences in replacement ratios below 20–25%, while for higher replacement ratios, the lower dry density of FRA decreased the dry density. This result was attributed to the fact that a higher fine content (<0.063 mm) in RS allows for filling of voids at replacement ratios up to 10% of the hardened mortar.

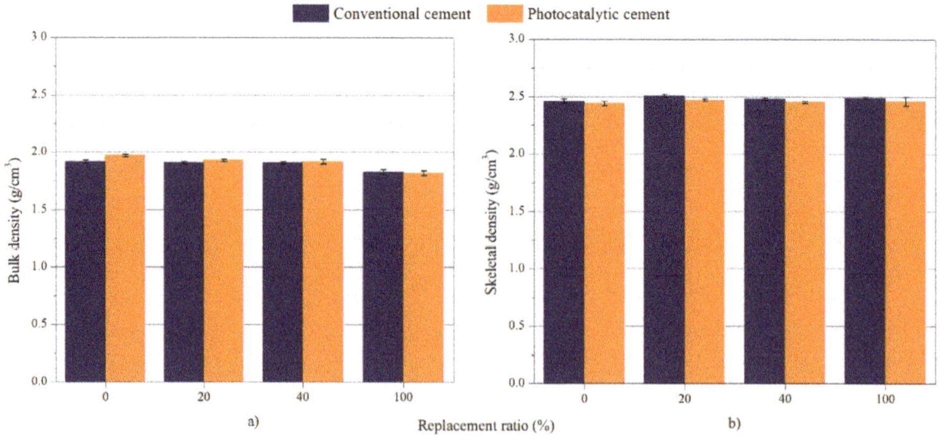

Figure 7. Comparison of results of (**a**) bulk density and (**b**) skeletal density.

4.6. Carbonation Depth

The results of the carbonation test are shown in Figure 8. The carbonation depth was greater when the RS increased, although this increase was slight, up to 40%. For 100% replacement, the increase in carbonation depth was much greater than in the reference mortar, possibly due to the greater porosity of this mixture, as observed in the previous section.

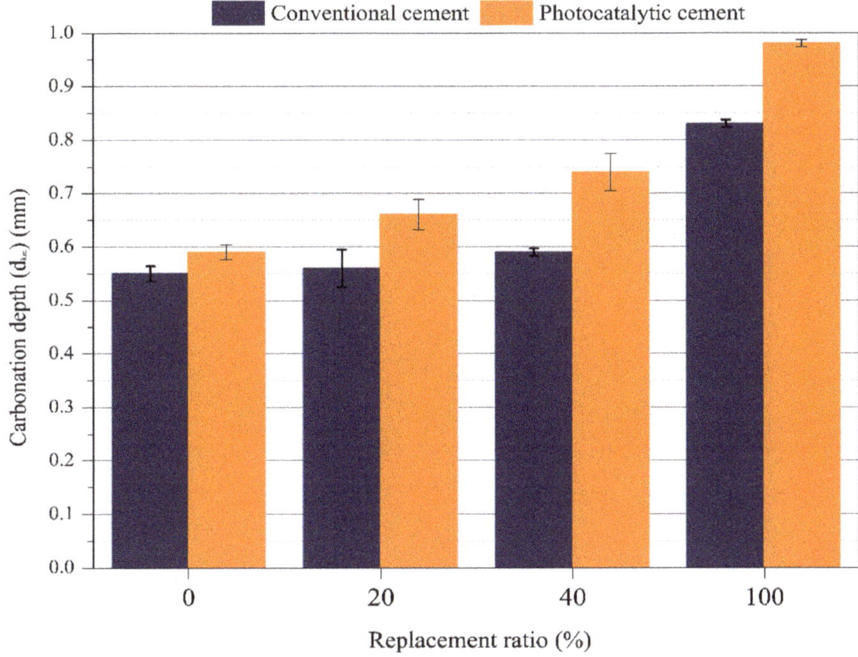

Figure 8. Carbonation depth comparison at 56 days.

This behavior was similar in both families, although it was more appreciable in the photocatalytic mortar family. Specimens of the photocatalytic mortar family showed an increase between 7% and 25% (17% average) higher than their counterparts in the conventional mortar family. This property showed the same trend as reported by Moro et al. [50].

Of all the mortars, PM100 is the one with the highest CO_2 absorption, with a 78% increase in carbonation depth as compared to the reference conventional mortar (M0), and a 66% increase as compared to the reference photocatalytic mortar (PM0). Thus, there is an added beneficial effect for the environment, since carbonation involves the absorption of CO_2 from the air. This reacts with the $Ca(OH)_2$ from cement hydrolysis, producing CO_3Ca and immobilizing CO_2. This process is detrimental to reinforced concrete, since it greatly affects durability due to the risk of corrosion of reinforcements by reducing the pH of the concrete. However, since mortars are not armored, that would not be a problem.

4.7. Photocatalytic Activity Test

According to the results of Figure 9, incorporating RS instead of NS slightly improved the decontaminating capacity. This can be explained by the greater porosity of the mortar as the percentage of recycled sand was increased, as observed in the previous sections, and agrees with the results reported by Poon and Cheung [59].

According to Spanish Standard UNE127197-1, most of the mixes were classified as category 1; that is, with decontaminating power, measured as the reduction of NO_x varying between 4$ and 6% (4.2%, 4.3%, and 4.8% for PM0, PM20, and PM40, respectively). However, the mix made with photocatalytic cement (PM100) and 100% RS showed a porous appearance and greater decontaminating capacity within category 2 (7.2%), which represents an increase of 71% of decontaminating power compared to mortar made with NS (PM0). This is consistent with Poon and Cheung [59], who indicated that the porosity of the surface layer is important, as it effectively increases the area available for reacting with pollutants.

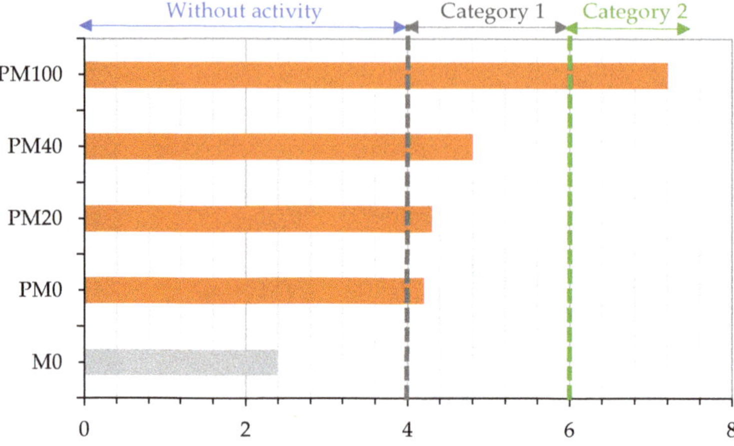

Figure 9. Elimination of NO_x (%).

5. Conclusions

This research produced mortars with decontaminating capacity by introducing waste into the productive cycle and contributed to the implementation of the circular economy model. The results show suitable mechanical behavior despite the incorporation of recycled aggregates, since total replacement of NS by RS meant a decrease in compressive and flexural strength of only approximately 18%, with average values of 46 MPa and 13 MPa, respectively. The strength obtained with photocatalytic cement was slightly higher compared to its counterpart made with traditional cement. The properties of water absorption by capillarity, water absorption capacity, and open porosity showed a slight increase for 40% and 100% RS, while in the results obtained for 20% RS, the values obtained were very similar to those of the reference mortar. The bulk and skeletal density showed very similar values in the two families and for all replacement ratios.

The penetration of CO_2 obtained in the carbonation test carried out showed a clear benefit with the incorporation of recycled sand. This could be a positive aspect, as it reduces the carbon footprint in the environment.

From a photocatalytic point of view, the incorporation of up to 40% RS slightly favored the elimination of NO_x, but the mortar with 100% RS had a significant increase relative to the conventional mortar, moving to a better classification (category 2). These mortars were more porous than conventional mortars, facilitating the entry of light into the interior and, consequently, the elimination of polluting gases. The results obtained add value to the use of recycled aggregates, clearing up the uncertainties that still exist in the use of this material.

Therefore, a photocatalytic mortar with 100% mixed recycled sand is proposed in order to produce the greatest environmental benefits, due to the greater absorption of CO_2 and NO_x and the greater use of recycled aggregates. Also, the strength is slightly lower but compatible with its use in low-requirement applications in contact with the atmosphere, such as pavement blocks or cladding mortar.

The findings of the present study prove that reducing natural sand mining, minimizing energy consumption and CO_2 emissions, reducing global warming, preventing illegal deposition and landfilling of the fine fraction of CDW, and complying with the limits of the European Waste Framework Directive are possible in order to achieve and promote cleaner production in the construction sector (eco-efficiency).

Future lines of research are intended to improve the photocatalytic capacity of mortars based on the good mechanical behavior of the mortars studied in this research by studying the influence of the content and origin of fine aggregates, different aggregates and their nature, reducing the amount of cement due to the good mechanical behavior obtained, or increasing the amount of water according to the percentage of increased RS, or pre-saturating it.

Author Contributions: Conceptualization, A.B., A.L.-L., and J.A.; Methodology, A.B. and A.L.-L.; Software and validation, A.B. and A.L.-L.; Formal analysis, A.P.G.; Investigation, A.L.-U. and A.P.G.; Data curation, A.B. and A.L.-L.; Writing—original draft preparation, A.B. and A.L.-L.; Writing—review and editing, A.B., A.L.-L., and J.A.; Visualization, A.L.-U. and A.P.G.; Supervision, A.B., A.L.-L., and J.A.; Funding acquisition, J.A.. All authors have read and agreed to the published version of the manuscript.

Funding: This research received no external funding.

Conflicts of Interest: The authors declare no conflict of interest.

References

1. Lozano-Lunar, A.; Barbudo, A.; Fernández, J.M.; Jiménez, J.R. Promotion of circular economy: steelwork dusts as secondary raw material in conventional mortars. *Environ. Sci. Pollut. Res.* **2020**, *27*, 89–100. [CrossRef]
2. Lozano-Lunar, A.; Da Silva, P.R.; De Brito, J.; Fernández, J.M.; Jiménez, J.R. Safe use of electric arc furnace dust as secondary raw material in self-compacting mortars production. *J. Clean. Prod.* **2019**, *211*, 1375–1388. [CrossRef]
3. European Commission 33. Report from the Commission to the European Parliament, the Council, the European Economic and Social Committee and the Committee of the Regions on the Implementation of the Circular Economy Action Plan. 2017. Available online: http://ec.europa.eu/environment/circular-economy/implementation_report.pdf (accessed on 17 October 2020).
4. Rattanashotinunt, C.; Tangchirapat, W.; Jaturapitakkul, C.; Cheewaket, T.; Chindaprasirt, P. Investigation on the strength, chloride migration, and water permeability of eco-friendly concretes from industrial by-product materials. *J. Clean. Prod.* **2018**, *172*, 1691–1698. [CrossRef]
5. Shen, L.Y.; Tam, V.W.Y.; Tam, C.M.; Drew, D. Mapping Approach for Examining Waste Management on Construction Sites. *J. Constr. Eng. Manag.* **2004**, *130*, 472–481. [CrossRef]
6. Tam, V.W.; Tam, C. Waste reduction through incentives: A case study. *Build. Res. Inf.* **2008**, *36*, 37–43. [CrossRef]
7. Rao, A.; Jha, K.N.; Misra, S. Use of aggregates from recycled construction and demolition waste in concrete. *Resour. Conserv. Recycl.* **2007**, *50*, 71–81. [CrossRef]
8. Europeo, P. DIRECTIVE 2008/98/EC del Parlamento Europeo y del Consejo de 19 de Noviembre de 2008 Sobre los Residuos y por la que se Derogan Determinadas Directivas. 2008. Available online: https://eur-lex.europa.eu/legal-content/EN/TXT/PDF/?uri=CELEX:32008L0098&from=ES (accessed on 17 October 2020).
9. Bravo, M.; De Brito, J.; Evangelista, L.; Pacheco, J. Durability and shrinkage of concrete with CDW as recycled aggregates: Benefits from superplasticizer's incorporation and influence of CDW composition. *Constr. Build. Mater.* **2018**, *168*, 818–830. [CrossRef]
10. Corinaldesi, V.; Moriconi, G. Behaviour of cementitious mortars containing different kinds of recycled aggregate. *Constr. Build. Mater.* **2009**, *23*, 289–294. [CrossRef]
11. Da Silva, S.R.; Andrade, J.J.D.O. Investigation of mechanical properties and carbonation of concretes with construction and demolition waste and fly ash. *Constr. Build. Mater.* **2017**, *153*, 704–715. [CrossRef]
12. Juan-Valdés, A.; Rodríguez-Robles, D.; García-González, J.; Guerra-Romero, M.I.; Morán-del Pozo, J.M. Mechanical and microstructural characterization of non-structural precast concrete made with recycled mixed ceramic aggregates from construction and demolition wastes. *J. Clean. Prod.* **2018**, *180*, 482–493. [CrossRef]
13. Kumar, R. Influence of recycled coarse aggregate derived from construction and demolition waste (CDW) on abrasion resistance of pavement concrete. *Constr. Build. Mater.* **2017**, *142*, 248–255. [CrossRef]

14. Ledesma, E.F.; Jiménez, J.R.; Ayuso, J.; Fernández, J.M.; De Brito, J. Maximum feasible use of recycled sand from construction and demolition waste for eco-mortar production—Part-I: ceramic masonry waste. *J. Clean. Prod.* **2015**, *87*, 692–706. [CrossRef]

15. Mehta, K.P. Reducing the environmental impact of concrete. *Concr. Int.* **2001**, *23*, 61–66. Available online: http://maquinamole.net/EcoSmartConcrete.com/docs/trmehta01.pdf (accessed on 17 October 2020).

16. EHE-08. *Instrucción de Hormigón Estructural*; Ministerio de Fomento, Secretaría General Técnica: Madrid, Spain, 2008; Available online: https://www.mitma.gob.es/organos-colegiados/mas-organos-colegiados/comision-permanente-del-hormigon/cph/instrucciones/ehe-08-version-en-ingles (accessed on 17 October 2020).

17. Agrela, F.; De Juan, M.S.; Ayuso, J.; Geraldes, V.; Jiménez, J. Limiting properties in the characterisation of mixed recycled aggregates for use in the manufacture of concrete. *Constr. Build. Mater.* **2011**, *25*, 3950–3955. [CrossRef]

18. Arbudo, A.; Agrela, F.; Ayuso, J.; Jiménez, J.; Poon, C. Statistical analysis of recycled aggregates derived from different sources for sub-base applications. *Constr. Build. Mater.* **2012**, *28*, 129–138. [CrossRef]

19. Rodríguez, C.; Parra, C.; Casado, G.; Miñano, I.; Albaladejo, F.; Benito, F.; Sánchez, I. The incorporation of construction and demolition wastes as recycled mixed aggregates in non-structural concrete precast pieces. *J. Clean. Prod.* **2016**, *127*, 152–161. [CrossRef]

20. Agrela, F.; Barbudo, A.; Ramírez, A.; Ayuso, J.; Carvajal, M.D.; Jiménez, J.R. Construction of road sections using mixed recycled aggregates treated with cement in Malaga, Spain. *Resour. Conserv. Recycl.* **2012**, *58*, 98–106. [CrossRef]

21. Tavira, J.; Jiménez, J.R.; Ayuso, J.; López-Uceda, A.; Ledesma, E.F. Recycling screening waste and recycled mixed aggregates from construction and demolition waste in paved bike lanes. *J. Clean. Prod.* **2018**, *190*, 211–220. [CrossRef]

22. Tavira, J.; Jiménez, J.R.; Ayuso, J.; Sierra, M.J.; Ledesma, E.F. Functional and structural parameters of a paved road section constructed with mixed recycled aggregates from non-selected construction and demolition waste with excavation soil. *Constr. Build. Mater.* **2018**, *164*, 57–69. [CrossRef]

23. Jiménez, J.R.; Ayuso, J.; Galvín, A.P.; López, M.; Agrela, F. Use of mixed recycled aggregates with a low embodied energy from non-selected CDW in unpaved rural roads. *Constr. Build. Mater.* **2012**, *34*, 34–43. [CrossRef]

24. Jiménez, J.R.; Ayuso, J.; Agrela, F.; López, M.; Galvín, A.P. Utilisation of unbound recycled aggregates from selected CDW in unpaved rural roads. *Resour. Conserv. Recycl.* **2012**, *58*, 88–97. [CrossRef]

25. Del Rey, I.; Ayuso, J.; Galvín, A.P.; Jiménez, J.R.; Barbudo, A. Feasibility of Using Unbound Mixed Recycled Aggregates from CDW over Expansive Clay Subgrade in Unpaved Rural Roads. *Materials* **2016**, *9*, 931. [CrossRef] [PubMed]

26. Evangelista, L.; De Brito, J. Mechanical behaviour of concrete made with fine recycled concrete aggregates. *Cem. Concr. Compos.* **2007**, *29*, 397–401. [CrossRef]

27. Pereira, P.; Evangelista, L.; De Brito, J. The effect of superplasticizers on the mechanical performance of concrete made with fine recycled concrete aggregates. *Cem. Concr. Compos.* **2012**, *34*, 1044–1052. [CrossRef]

28. Thomas, C.; Setién, J.; Polanco, J.; Lombillo, I.; Cimentada, A. Fatigue limit of recycled aggregate concrete. *Constr. Build. Mater.* **2014**, *52*, 146–154. [CrossRef]

29. Khatib, J.M. Properties of concrete incorporating fine recycled aggregate. *Cem. Concr. Res.* **2005**, *35*, 763–769. [CrossRef]

30. Evangelista, L.; de Brito, J. Durability performance of concrete made with fine recycled concrete aggregates. *Cem. Concr. Compos.* **2010**, *32*, 9–14. [CrossRef]

31. Vegas, I.; Azkarate, I.; Juarrero, A.; Frías, M. Design and performance of masonry mortars made with recycled concrete aggregates. *Materiales de Construcción* **2009**, *59*, 5–18. [CrossRef]

32. Fan, C.-C.; Huang, R.; Hwang, H.; Chao, S.-J. Properties of concrete incorporating fine recycled aggregates from crushed concrete wastes. *Constr. Build. Mater.* **2016**, *112*, 708–715. [CrossRef]

33. Kou, S.-C.; Poon, C.-S. Properties of concrete prepared with crushed fine stone, furnace bottom ash and fine recycled aggregate as fine aggregates. *Constr. Build. Mater.* **2009**, *23*, 2877–2886. [CrossRef]

34. Martínez, P.S.; Cortina, M.G.; Fernandez-Martinez, F.; Rodríguez, A. Comparative study of three types of fine recycled aggregates from construction and demolition waste (CDW), and their use in masonry mortar fabrication. *J. Clean. Prod.* **2016**, *118*, 162–169. [CrossRef]

35. Gayarre, F.L.; Boadella, Í.L.; Pérez, C.L.C.; López, M.S.; Cabo, A.D. Influence of the ceramic recycled aggregates in the masonry mortars properties. *Constr. Build. Mater.* **2017**, *132*, 457–461. [CrossRef]

36. Jiménez, J.; Ayuso, J.; López, M.; Fernández, J.; De Brito, J. Use of fine recycled aggregates from ceramic waste in masonry mortar manufacturing. *Constr. Build. Mater.* **2013**, *40*, 679–690. [CrossRef]

37. Rodríguez Rivas, F.A. Application of Layered Double Hydroxides as Photocatalysts in the Elimination of NOx Gases. Ph.D. Thesis, University of Cordoba, Córdoba, Spain, July 2019. (In Spanish).

38. Mendoza, C.; Valle, A.; Castellote, M.; Bahamonde, A.; Faraldos, M. TiO_2 and TiO_2–SiO_2 coated cement: Comparison of mechanic and photocatalytic properties. *Appl. Catal. B Environ.* **2015**, *178*, 155–164. [CrossRef]

39. Maury, A.; de Belie, N. State of the art of TiO_2 containing cementitious materials: self-cleaning properties. *Materiales de Construccion* **2010**, *60*, 33–50. [CrossRef]

40. Laplaza, A.; Jimenez-Relinque, E.; Campos, J.; Castellote, M. Photocatalytic behavior of colored mortars containing TiO_2 and iron oxide based pigments. *Constr. Build. Mater.* **2017**, *144*, 300–310. [CrossRef]

41. Jimenez-Relinque, E.; Llorente, I.; Castellote, M. TiO_2 cement-based materials: Understanding optical properties and electronic band structure of complex matrices. *Catal. Today* **2017**, *287*, 203–209. [CrossRef]

42. Jimenez-Relinque, E.; Castellote, M. Quantification of hydroxyl radicals on cementitious materials by fluorescence spectrophotometry as a method to assess the photocatalytic activity. *Cem. Concr. Res.* **2015**, *74*, 108–115. [CrossRef]

43. Bengtsson, N.; Castellote, M. Heterogeneous photocatalysis on construction materials: parametric study and multivariable correlations. *J. Adv. Oxid. Technol.* **2010**, *13*, 341–349.

44. Folli, A.; Jakobsen, U.H.; Guerrini, G.L.; Macphee, D.E. Rhodamine B Discolouration on TiO_2 in the Cement Environment: A Look at Fundamental Aspects of the Self-cleaning Effect in Concretes. *J. Adv. Oxid. Technol.* **2009**, *12*, 126–133. [CrossRef]

45. Goswami, D.Y.; Trivedi, D.M.; Block, S.S. Photocatalytic Disinfection of Indoor Air. *J. Sol. Energy Eng.* **1997**, *119*, 92–96. [CrossRef]

46. Sapiña, M.; Jimenez-Relinque, E.; Castellote, M. Controlling the Levels of Airborne Pollen: Can Heterogeneous Photocatalysis Help? *Environ. Sci. Technol.* **2013**, *47*, 11711–11716. [CrossRef]

47. Smits, M.; Chan, C.K.; Tytgat, T.; Craeye, B.; Costarramone, N.; Lacombe, S.; Lenaerts, S. Photocatalytic degradation of soot deposition: Self-cleaning effect on titanium dioxide coated cementitious materials. *Chem. Eng. J.* **2013**, *222*, 411–418. [CrossRef]

48. Rodrigues, F.; Carvalho, M.T.; Evangelista, L.; De Brito, J. Physical–chemical and mineralogical characterization of fine aggregates from construction and demolition waste recycling plants. *J. Clean. Prod.* **2013**, *52*, 438–445. [CrossRef]

49. Ledesma, E.F.; Jiménez, J.R.; Fernández, J.; Galvín, A.P.; Agrela, F.; Barbudo, A. Properties of masonry mortars manufactured with fine recycled concrete aggregates. *Constr. Build. Mater.* **2014**, *71*, 289–298. [CrossRef]

50. Moro, C.; Francioso, V.; Schrager, M.; Velay-Lizancos, M. TiO_2 nanoparticles influence on the environmental performance of natural and recycled mortars: A life cycle assessment. *Environ. Impact Assess. Rev.* **2020**, *84*, 106430. [CrossRef]

51. De Brito, J.; De Brito, J.; Dhir, R. Performance of cementitious renderings and masonry mortars containing recycled aggregates from construction and demolition wastes (Review). *Constr. Build. Mater.* **2016**, *105*, 400–415.

52. Joint Committee on Powder Diffraction Standards. *Powder Diffraction Database*; International Center for Diffraction Data: Swarthmore, PA, USA, 1975.

53. Silva, J.; De Brito, J.; Veiga, R. Recycled Red-Clay Ceramic Construction and Demolition Waste for Mortars Production. *J. Mater. Civ. Eng.* **2010**, *22*, 236–244. [CrossRef]

54. Corinaldesi, V. Mechanical behavior of masonry assemblages manufactured with recycled-aggregate mortars. *Cem. Concr. Compos.* **2009**, *31*, 505–510. [CrossRef]

55. SAC. *GB 28635-2012 Precast Concrete Paving Units*; Administration of Quality Supervision, Inspection and Quarantine of China: Beijing, China, 2012.

56. Poon, C.; Kou, S.; Lam, L. Use of recycled aggregates in molded concrete bricks and block. *Constr. Build. Mater.* **2002**, *6*, 281–289. [CrossRef]

57. Poon, C.S.; Chan, D. Paving blocks made with recycled concrete aggregate and crushed clay brick. *Constr. Build. Mater.* **2006**, *20*, 569–577. [CrossRef]

Appl. Sci. **2020**, *10*, 7305

58. Gonçalves, T.; Silva, R.V.; De Brito, J.; Fernández, J.; Esquinas, A. Mechanical and durability performance of mortars with fine recycled concrete aggregates and reactive magnesium oxide as partial cement replacement. *Cem. Concr. Compos.* **2020**, *105*, 103420. [CrossRef]

59. Poon, C.; Cheung, E. NO removal efficiency of photocatalytic paving blocks prepared with recycled materials. *Constr. Build. Mater.* **2007**, *21*, 1746–1753. [CrossRef]

60. Neno, C.; de Brito, J.; Veiga, R. Using fine recycled concrete aggregate for mortar production. *Mater. Res.* **2014**, *17*, 168–177. [CrossRef]

61. Dapena, E.; Alaejos, P.; Lobet, A.; Pérez, D. Effect of recycled sand content on characteristics of mortars and concretes. *J. Mater. Civ. Eng.* **2011**, *23*, 414–422. [CrossRef]

Publisher's Note: MDPI stays neutral with regard to jurisdictional claims in published maps and institutional affiliations.

 *applied sciences*

Article

Industrial Low-Clinker Precast Elements Using Recycled Aggregates

Carlos Thomas [1,*], Ana I. Cimentada [1], Blas Cantero [2], Isabel F. Sáez del Bosque [2] and Juan A. Polanco [1]

1   LADICIM (Laboratory of Materials Science and Engineering), University of Cantabria E.T.S. de Ingenieros de Caminos, Canales y Puertos, Av. Los Castros 44, 39005 Santander, Spain; anaisabel.cimentada@unican.es (A.I.C.); polancoa@unican.es (J.A.P.)
2   School of Civil Engineering, University of Extremadura, Institute for Sustainable Regional Development (INTERRA), Avda. de la Universidad, s/n, 10003 Cáceres, Spain; bcanteroch@unex.es (B.C.); cmedinam@unex.es (I.F.S.d.B.)
*   Correspondence: thomasc@unican.es

Received: 31 August 2020; Accepted: 21 September 2020; Published: 23 September 2020

Abstract: Increasing amounts of sustainable concretes are being used as society becomes more aware of the environment. This paper attempts to evaluate the properties of precast concrete elements formed with recycled coarse aggregate and low clinker content cement using recycled additions. To this end, six different mix proportions were characterized: a reference concrete; 2 concretes with 25%wt. and 50%wt. substitution of coarse aggregate made using mixed construction and demolition wastes; and others with recycled cement with low clinker content. The compressive strength, the elastic modulus, and the durability indicator decrease with the proportions of recycled aggregate replacing aggregate, and it is accentuated with the incorporation of recycled cement. However, all the precast elements tested show good performance with slight reduction in the mechanical properties. To confirm the appropriate behaviour of New Jersey precast barriers, a test that simulated the impact of a vehicle was carried out.

Keywords: recycled concrete; low clinker cement; precast; mechanical properties; physical properties; New Jersey barriers

---

1. Introduction

Construction and demolition waste (CDW) is non-hazardous, inert waste generated in any construction, rehabilitation or demolition work. The industrial and construction sectors generate practically the same amount of non-hazardous waste (industry 37,417 kt[i] and construction 35,869 kt[i]) in Spain [1]. The European Commission estimates that the volume of CDW comprises one third of all waste generated in the European Union, which constitutes the largest waste stream [2]. Recycling this CDW would lead to more sustainable growth, replacing a linear economy based on use of materials with a more circular economy. This is important, as aggregates are the second-most-used raw material by humans, behind only water [3]. There is European legislation to encourage recycling CDW [4] and many countries have specific norms for the use of recycled aggregates (RA) for concrete [5–8]. In addition, the use of RA could lead to cheaper concrete [9].

Several studies have corroborated that the inclusion of RA produces concrete with a lower density and increased heterogeneity [10–12]. RA normally has a higher porosity than natural aggregate (NA) [13]. In a fresh state, Silva et al. [11] concluded that recycled aggregate concrete (RAC) is less workable and, to achieve a workability equivalent to that of NA, RA could be pre-saturated, or water added during mixing to compensate [14]. However, the incorporation of completely saturated aggregates might cause an excessive water supply [15,16]. Once the RAC hardens, these aggregates make the concrete more susceptible to detrimental environmental effects, resulting in a lower durability [17,18], which should be taken into consideration. Consequently, Annex 15 of the Spanish Instruction for Structural Concrete EHE-08 [19] and other studies [14,20] propose solutions, such as increasing the cement content, reducing the water/cement ratio, or increasing the coating thickness in the case of reinforced concrete.

Generally, it is known that the incorporation of RA into concrete reduces its mechanical properties [21,22], due to the presence of contaminants such as plastics, glass, adhered mortar, etc., ref. [23] and the type of source material (crushed concrete, ceramic or mixed) of the RA [24–26]. The elastic modulus of RAC is lower than that of conventional concrete [15], reaching 45% less for 100% replacement [25]. The results obtained in the characterization of RAC with intermediate replacements present greater variation of results [20]. Other authors have demonstrated the viability of other types of recycled aggregates from waste, such as steel slag [27]. Moreover, the RA affects the fatigue behavior of the concrete [28–32], showing a greater loss of properties than with the static properties. Further research has evaluated the recycling of concrete which incorporates RA [33,34].

With regard to precast concrete elements, it should be noted that, according to the ANDECE (National Association of the Prefabricated Concrete Industry, based in Spain), although the initial cost of elements is higher, the final cost is lower [35]. Other studies such as López-Mesa et al. [36] indicate an almost 18% higher cost of precast slabs versus in situ slabs; although the former have a lower environmental impact and the quality may be higher. Normally, precast elements have a quality seal guaranteeing their properties. Due to a manufacturing process with complete exhaustive control, precast slabs can be: tailored with special properties more easily as they are not manufactured on site; designed with flexibility difficult to achieve in-situ; and incorporate RA in their fabrication. In the case of precast elements using RA, a lower density and strength is observed [37]. Poon et al. [37] investigated the factors that affect the properties of precast concrete blocks with RA, concluding that the compressive strength increases with the reduction in the aggregate/cement ratio (A/C), and that the water absorption of concrete blocks is significantly related to the absorption capacity of the aggregate. Katz [21] investigated the use of precast elements at different ages to produce RA for new precast elements, concluding that the mechanical properties (strength, modulus of elasticity, etc.) when using this type of aggregate in concrete, resemble those when using lightweight aggregates, such as those manufactured using fly ash.

This paper presents the effect on physical and mechanical properties of six types of mixes with different degrees of substitution. The physical properties and durability of these concretes will be analyzed first, then the mechanical properties will be assessed. Finally, the behavior of precast elements will be addressed.

2. Materials and Methodology

The natural siliceous aggregate used in this study is present in three different sizes: 6/0 mm (NS), 12/6 mm (NG-M), and 22/12 mm (NG-C). Mixed recycled aggregates (MRA) were used by substituting NG-M for MRA-M and NG-C for MRA-C. These MRA were obtained from CDW and were principally made up of concrete and mortars ($\approx$ 45%), unbound aggregate, and natural stone ($\approx$ 45%). Figure 1 shows the different size grading for each aggregate.

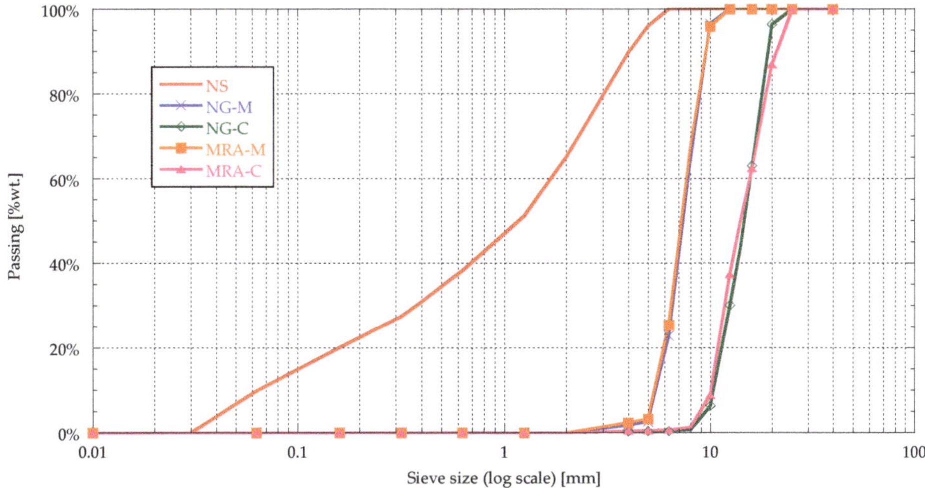

Figure 1. Grading of the aggregates.

Table 1 displays physical and mechanical properties: where *SSS* is the saturated dry surface density according to EN 1097-6 [38]; *A* is the water absorption by weight according to EN 1097-6 [38]; *LA* is the Los Angeles index according to EN 1097-2 [39]; and *FI* is the flakiness index according to EN 933-3 [40].

Table 1. Physical and mechanical properties of the aggregates.

Property	Aggregates				
	NS	**NG-M**	**NG-C**	**MRA-M**	**MRA-C**
SSS [g/cm^3]	2.76	2.74	2.74	2.42	2.45
A [%]	1.18	0.88	0.78	6.28	5.27
LA [%]	-	16	18	32	36
FI [%]	-	21	25	10	10

The conventional cement (OPC) was CEM I 42.5 R, and the low clinker content cement (RC) was constituted of 75% CEM I 42.5 R and 25% ceramic waste from CDW. The tests performed with the cement revealed a compressive strength 20% higher in the case of OPC.

Mixing the aggregates in different proportions with the two existing types of cement produced six concrete mixtures, as shown in Table 2. HP signifies a combination of natural aggregates and conventional cement. HPR is a mixture of natural aggregates and low clinker content cement. HR25 and HR50 were fabricated with conventional cement and substitutions of NA by 25%wt. and 50%wt. proportions of RA, respectively. Finally, HRR25 and HRR50 were obtained by amalgamating low clinker content cement with natural aggregates, substituted by 25%wt. and 50%wt. of recycled aggregates accordingly.

Table 2. Concrete mix proportions (by m^3).

Concrete:	HP	HPR	HR25	HR50	HRR25	HRR50
NS (6/0 mm) [kg]:	732	732	719	705	719	705
NG-M (12/6 mm) [kg]:	382	382	284	184	284	184
NG-C (22/12 mm) [kg]:	766	766	568	369	568	369
MRA-M (12/6 mm) [kg]:	-	-	89	178	89	178
MRA-C (22/12 mm) [kg]:	-	-	178	356	178	356
Cement [kg]:	400	-	400	400	-	-
Low clinker content cement [kg]:	-	400	-	-	400	400
Water [kg]:	193	193	202	211	202	211
Superplasticizer [kg]:	6.2	6.2	6.2	6.2	6.2	6.2
Water/cement ratio	0.48	0.48	0.50	0.53	0.50	0.53

2.1. Physical and Mechanical Properties

Densities were obtained according to EN-12390-7 [41]. Sub-specimens (10Ø × 10 cm) obtained by cutting 10Ø × 20 cm cylindrical specimens were used. The porosity coefficient is the result of comparing the absorbed water and specimen volume, while the absorption coefficient is the result of comparing the absorbed water and specimen weight. Compressive strength was determined using 10Ø × 20 cm cylindrical specimens according to EN-12390-3 [42], with an application strength rate of 0.5 MPa/s. Elastic modulus was determined with 10Ø × 20 cm cylindrical specimens according to EN-12390-13 [43], at a strength rate of 0.5 MPa/s.

2.2. Durability

A water penetration test was performed according to EN-12390-8 [44]. Sub-specimens (10Ø × 10 cm) obtained by cutting 10Ø × 20 cm cylindrical specimens were used. The samples were subjected to a pressure of 5 bar for 72 h. After 72 h water penetration under pressure, it was necessary to analyze how deep the water reached. To be able to observe the interior of the sample, it had to be opened. During this research, the Brazilian method (or indirect tensile strength method) was used to open the sample and analyze its interior. In general, when a cylindrical specimen is subjected to tension along its generatrix, it breaks into two halves, which allows the interior to be analyzed. Once the specimen had been opened, it was possible to measure the penetration depth of the water into the porous concrete. This technique also provided another interesting result: the indirect tensile strength of the concrete. For the determination of oxygen permeability, UNE-83981 [45] was taken as a reference. The 10Ø × 20 cm cylindrical specimens were cut to discard the upper and lower face obtaining a new sample of 10Ø × 10 cm. Silicone was impregnated perimetrically in the samples so that the oxygen could only pass longitudinally. A regulated oxygen pressure was applied on the upper face. Digital flow meters registered the oxygen escaping from the lower face.

2.3. Precast Element Preparation

Two different types of precast elements were manufactured: unreinforced concrete ditches and steel-reinforced New Jersey barriers. Both were manufactured with an industrial concrete mixer, poured in metallic molds and vibrated by hand (Figure 2). In the case of reinforced concrete, reinforcements were set into the mold before the pouring of concrete. In both cases, precast elements were unmolded and cured at ambient temperature.

Figure 2. Precast element manufacturing sequence.

2.4. Precast Element Mechanical Characterization

Concrete ditches have approximate measurements of 50 × 50 × 15 cm. In order to characterize concrete ditches, the tests were carried out by bending. The horizontality of the set was verified, and force was applied by a roller (10Ø × 22 cm) in the central section with a displacement rate of 0.1 mm/s (Figure 3).

Figure 3. Precast element characterization (concrete ditches left, New Jersey barriers right).

New Jersey barriers have a section with approximate measurements of 47 × 80 cm and a length of 100 cm. In order to characterize New Jersey barriers, a small crane was used to support the precast element on steel beams. These steel beams were placed at one end to correct the inclination of the face on which the test was to be performed, achieving horizontality on that face (Figure 3). The test consisted in applying a stress with a roller (3Ø × 40 cm). The time of the test was very short (0.1–0.2 s) to simulate an impact. The strength and displacement data of the actuator were recorded during the test.

3. Results and Discussion

3.1. Physical Properties

Figure 4 shows the relative and saturated densities of the concretes. As demonstrated, the density decreases as the percentage of NA replaced by RA increases. This is due to the lower density of RA. It also becomes clear that the use of this RC does not affect density significantly.

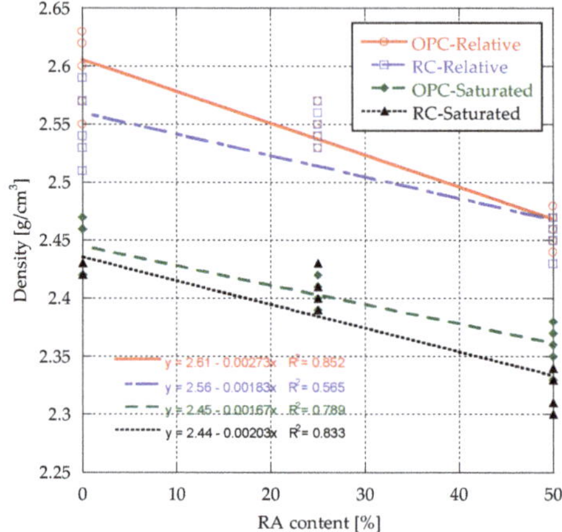

Figure 4. Density vs. RA content.

Figure 5a shows porosity, and Figure 5b shows the absorption coefficient vs. substitution of NA by RA. A decrease in both properties is found in the concretes containing OPC as the percentage of replacement of aggregate increases. However, in the case of concrete made with RC, both properties increase as the percentage of RA increases. This may be because this type of cement interacts more with RAs of different nature, making it difficult to fill all the gaps amongst aggregates. Alternatively, it may be because the RA is able to absorb more water during kneading, causing a small deficit in this type of cement, which is very susceptible to variations in the water dosage. It is possible that there may be another reason that has not been identified.

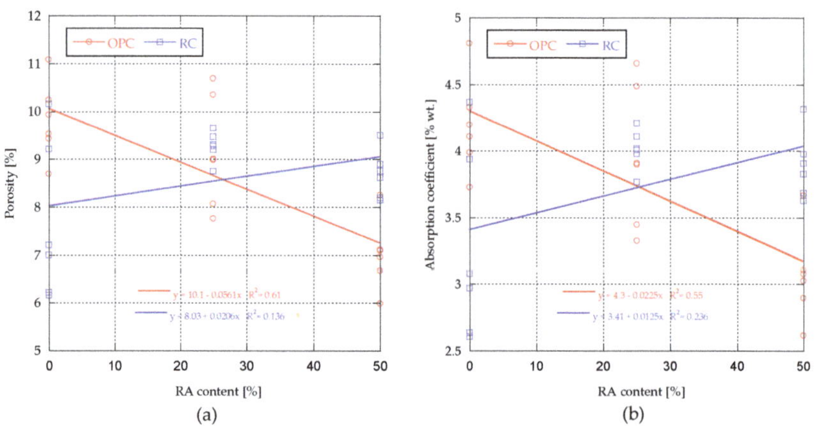

Figure 5. Porosity (**a**) and absorption coefficient (**b**) vs. RA content.

3.2. Compressive Strength and Modulus of Elasticity

Figure 6a shows the compressive strength-strain curves for each concrete at 160 days. Several studies [25,46,47] show that the concrete's compressive strength decreases with the degree of substitution of RA for NA, but in strain terms, concretes show similar values around 2500 μm/m for the failure. The exception is the HRR50 mix, which exceeds the values of the rest by almost 1000 μm/m. Figure 6b shows the same mixtures but at an age of 365 days. The decrease in strength may also be due to the randomness of the type of RA and its distribution into the mortar matrix, which causes greater uncertainty than conventional mixtures.

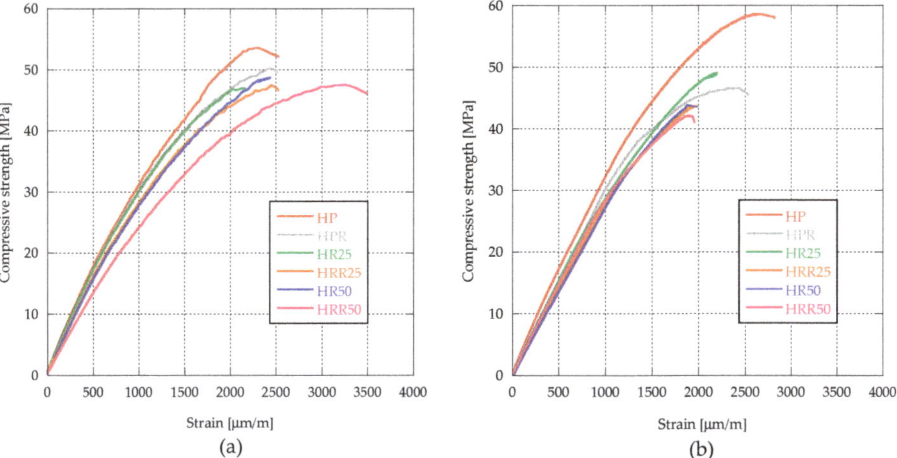

Figure 6. Compressive strength-strain at 160 (**a**) and 365 (**b**) days.

Table 3 shows the different values of compressive strength obtained at different ages.

Table 3. Compressive strength at different ages.

Concrete:	Compressive Strength [MPa]			
	28 days	**160 days**	**365 days**	Δ_{36-28} [%]
HP	51.2	53.5	56.8	+10.9
HPR	46.1	50.2	46.6	+1.1
HR25	51.7	47.0	45.1	−12.8
HR50	51.2	48.8	42.1	−17.7
HRR25	45.0	47.5	43.7	−2.9
HRR50	41.2	47.5	42.4	+2.9

Table 4 displays the modulus of elasticity, and shows that when using RC, the decrease in the elastic modulus is around 4%. The substitution of 25% by RA implies a decrease in elastic modulus of 5.6%, while the substitution of OPC in this case does not seem to have an influence. In the case of replacing 50% of aggregate by RA, the influence of the substitution of OPC by RC is meaningful, decreasing the elastic modulus by 15%. As for the loss of elastic modulus over time, a greater influence of the cement is observed than the type of aggregate, with a limit that tends to an asymptotic value of around 27 GPa.

Table 4. Modulus of elasticity.

Concrete:	Substitution [%]	Modulus of Elasticity [GPa]	Modulus of Elasticity at 365 days [GPa]	% of the Initial Elastic Modulus
HP	0	35.5	31.7	89.3
HPR	0	34.1	29.5	86.5
HR25	25	33.9	30.8	90.8
HR50	50	31.9	29.3	91.8
HRR25	25	34.2	27.9	81.6
HRR50	50	27.8	27.4	98.6

Some organizations such as EHE-08, ACI, and Eurocode present their expressions to predict elastic modulus at 28 days from the compressive strength. In Expressions (1)–(3): E is elastic modulus at 28 days [GPa] and f_{28} is the compressive strength at 28 days [MPa].

EHE-08 [48]

$$E = 8.5 \sqrt[3]{f_{28}} \tag{1}$$

ACI [49]

$$E = 4.7 \sqrt{f_{28}} \tag{2}$$

Eurocode 2 [50]

$$E = 22(f_{28}/10)^{0.3} \tag{3}$$

These expressions can be used to obtain the predictions and comparisons, with the experimental results shown in Table 5. The ACI method fits quite well in most cases but predicts higher values when the percentage of substitution is 50%. The EHE-08 method is safer, although when the substitution is 50% and the OPC is replaced by RC, higher values are produced due to the heterogeneity of the RA affecting the compressive strength. These types of expressions only satisfactorily fit ordinary concrete models.

Table 5. Elastic modulus obtained with different expressions.

Concrete:	Elastic Modulus [GPa]			
	Experimental	EHE-08	ACI	Eurocode 2	$\Delta_{\text{experimental–EHE-08}}$ [%]
HP	35.5	31.6	33.6	35.9	12.5
HPR	34.1	30.5	31.9	34.8	11.9
HR25	33.9	31.7	33.8	36.0	7.1
HR50	31.9	31.6	33.6	35.9	1.1
HRR25	34.2	30.2	31.5	34.5	13.1
HRR50	27.8	29.4	30.2	33.6	-5.3

Figure 7 shows that from approximately 48 MPa, concrete with RA achieved the same compressive strength as concrete with OPC. RA concrete increases its elastic modulus significantly. This might be due to the addition of a new variable, such as RA compared with OPC, which is much more standardized throughout its production process.

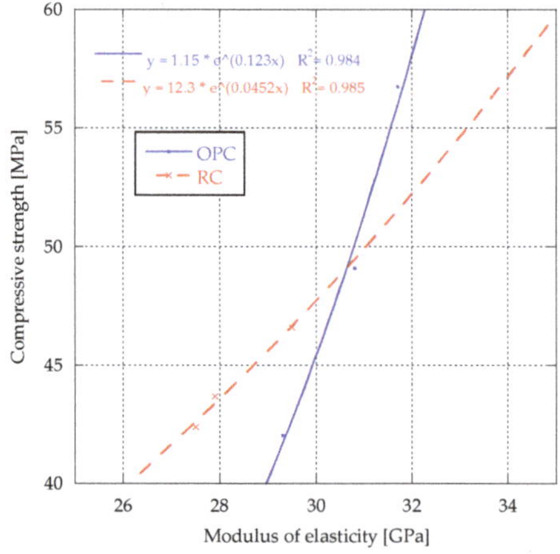

Figure 7. Compressive strength vs. modulus of elasticity.

3.3. Oxygen and Water Permeability

Figure 8a shows the oxygen permeability and Figure 8b shows the maximum penetration of water vs. percentage of substitution, respectively.

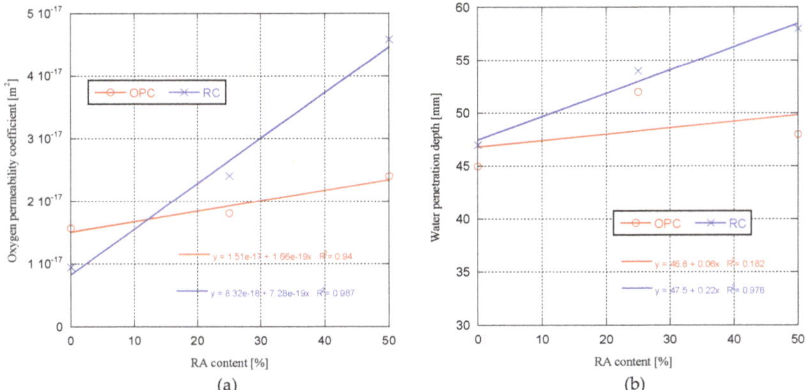

Figure 8. Oxygen permeability coefficient (**a**) and water penetration depth (**b**).

The oxygen permeability coefficient increases with the substitution of the NA by RA. This behavior has been reported in some studies, such as Ismail et al. [51], Medina et al. [52], and Thomas et al. [14]. This increase is higher in concrete with RC than OPC; the type of cement being used is an important factor.

The penetration of water increases with the increase in RA substitution. With these results, only HP and HPR comply with the standard EHE-08 [48] for structural concrete in the case of IIIa, IIIb, IV, etc. environment exposition, which requires an average penetration depth of 30mm, and maximum penetration depth of 50 mm. Penetration of water is related to typology and distribution of the RA, and its impurities with high absorption coefficients.

Figure 9 shows cross-sections of concrete where different colors can be seen. These are caused by the RC in HPR and HRR50 mixtures, and some kind of RA and impurities (such as wood or fired clay) in HRR50 mix.

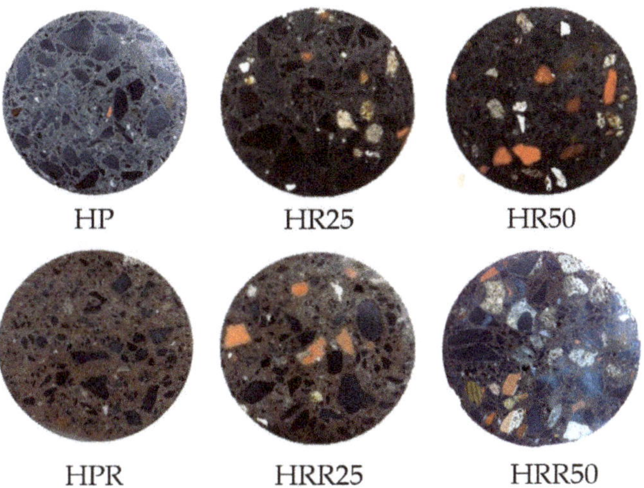

Figure 9. Concrete specimen sections.

3.4. Testing Precast Elements

Figure 10a shows the results of flexural tests on concrete ditches. It can be observed that the concrete composed of RC and RA (HRR50) behaves similarly to HP concrete, which is consistent with the results of splitting tensile strength shown in Table 6. Figure 10b shows the results of the impact test on reinforced precast New Jersey barriers, in which the force applied by the test machine and the position of the actuator are recorded. As expected, the concrete with OPC and NA displayed superior mechanical behavior than concrete with RC and RA. HRR50 could resist only 60% of the force, and 66% of the displacement that HP resisted.

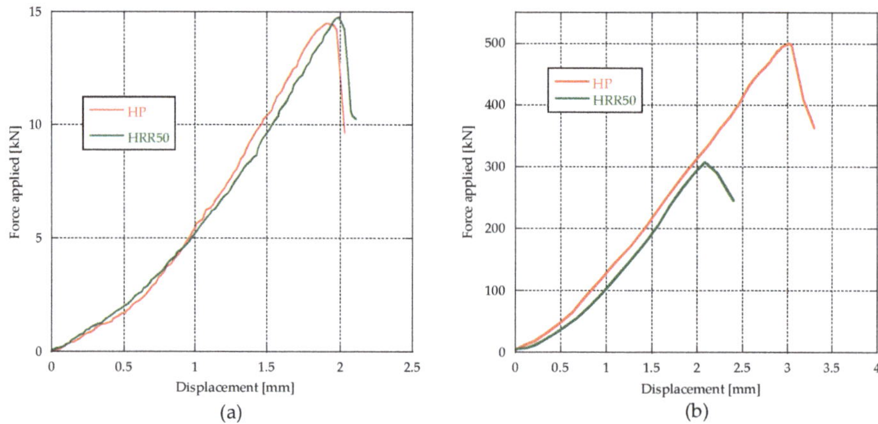

Figure 10. Mechanical characterization of precast elements: Bending test on ditches (**a**), impact test on barriers (**b**).

Table 6. Splitting tensile strength.

Splitting Tensile Strength [MPa]					
HP	**HPR**	**HR25**	**HR50**	**HRR25**	**HRR50**
3.36	3.51	3.48	-	3.30	3.58

Figure 11 shows the results of the test performed with both types of precast elements. Different sections of cracks in OPC and RC concrete ditches, and the fissure produced in a New Jersey barrier are visual results of the tests.

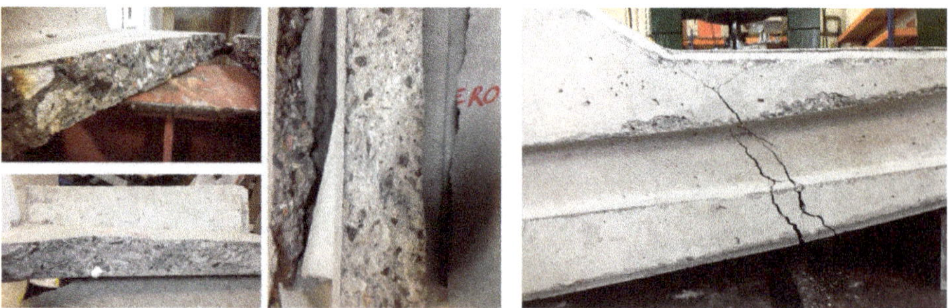

Figure 11. Precast test and cracking.

Equation (4) indicates whether a New Jersey barrier could withstand the perpendicular impact of a vehicle. Velocity and mass are variables, and it would be necessary to incorporate a restitution coefficient in order to avoid the elastic impact.

This coefficient relates the velocity before impact with the velocity after collision, considering the barrier is without velocity before and after impact.

$$C_R = -\frac{V_{1f} - V_{2f}}{V_{1i} - V_{2i}}; when\ V_{2f}, V_{2i} = 0 \rightarrow C_R = -\frac{V_f}{V_i} \tag{4}$$

García and Cabreiro [53] proposed a method for obtaining the coefficient of restitution based on experimental processes in "*Use of dynamic models in the investigation of road accidents*" (text in Spanish), for which they suggested two equations:

$$C_R = 0.45 \cdot e^{(-0.040278 \cdot v)},\ For\ v\ <\ 54\ km/h \tag{5}$$

$$C_R = 0.45 \cdot e^{(-0.015278 \cdot v)},\ For\ v\ \geq\ 54\ km/h \tag{6}$$

With Equations (4)–(6), considering the maximum force that a barrier resists, and the duration of the impact as 0.1 s, Equations (7) and (8) are obtained, shown in Figure 12.

$$m = \frac{0.1 \cdot F}{-\left(\frac{v_i}{3.6}\right) \cdot \left(0.45 \cdot e^{(-0.040278 \cdot v)} + 1\right)},\ For\ v\ <\ 54\ km/h \tag{7}$$

$$m = \frac{0.1 \cdot F}{-\left(\frac{v_i}{3.6}\right) \cdot \left(0.12 \cdot e^{(-0.015278 \cdot v)} + 1\right)},\ For\ v\ \geq\ 54\ km/h \tag{8}$$

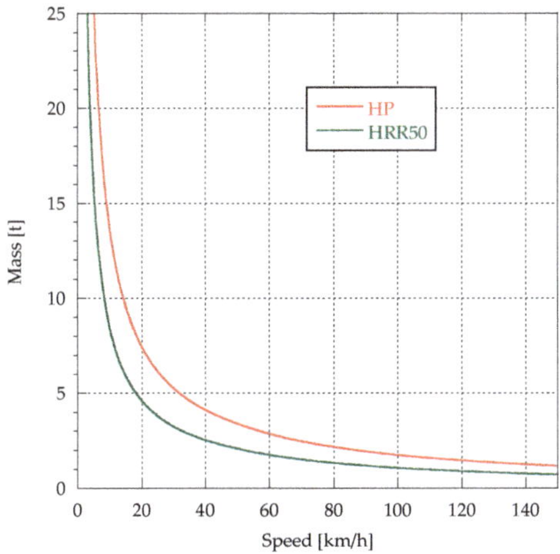

Figure 12. Simulated behavior of reinforced barriers.

These curves are conservative, as the barrier can withstand strains that absorb energy before cracking, and the parapet would not always be immobile (they are only anchored to the ground on viaducts).

4. Conclusions

Characterization tests on concrete specimens and precast elements have been carried out using low-clinker cements and recycled aggregates, obtaining the following conclusions. Firstly, the physical-mechanical properties of mixed recycled aggregates are suitable for the manufacture of concrete and precast elements when the medium and coarse fraction is used. Secondly, the use of mixed recycled aggregates causes a loss of density and compressive strength slightly higher than that which occurs when using recycled concrete aggregates. Recycled concretes made from low-clinker cement are slightly more porous than concretes made with ordinary Portland cement. Finally, regarding the mechanical properties of recycled concrete, a loss of around 10% of the compressive strength is observed when using low-clinker cement. In addition, recycled concrete made with ordinary Portland cement evolves slightly more when over 1 year of curing has elapsed.

Author Contributions: Conceptualization, C.T., A.I.C., J.A.P.; methodology, C.T., A.I.C., J.A.P., I.F.S.d.B., B.C.; validation, C.T., I.F.S.d.B., B.C.; formal analysis, C.T.; investigation, C.T., A.I.C., J.A.P.; resources, C.T., J.A.P.; writing—original draft preparation, C.T., I.F.S.d.B., B.C.; writing—review and editing C.T. All authors have read and agreed to the published version of the manuscript.

Funding: This research was funded by SODERCAN, S.A. (SODERCAN/FEDER) and BIA2013-48876-C3-2-R awarded by the Ministry of Science and Innovation.

Acknowledgments: The authors would like to express our gratitude to Jaime de la Fuente and César Medina for their support and participation in part of the project.

Conflicts of Interest: Authors declare no conflict of interest.

References

1. Instituto Nacional de Estadística (INE). Estadísticas Sobre Generación de Residuos. Available online: https://www.ine.es (accessed on 23 September 2020).
2. UE Comisión Europea. Protocolo de Gestión de Residuos de Construcción y Demolición en la UE. Available online: https://europa.eu (accessed on 23 September 2020).
3. ANEFA. Available online: http://www.aridos.org (accessed on 23 September 2020).
4. Directiva 2008/98/CE Sobre los Residuos. Available online: https://eur-lex.europa.eu/eli/dir/2008/98/oj (accessed on 23 September 2020).
5. Deutsches Institut für Normung. *DIN 4226-100 Aggregates for Concrete and Mortar—Part 100: Recycled Aggregates*; German Institute for Standardisation: Berlin, Germany, 2002.
6. Guobiao Standards. GB/T 25176-2010 Recycled Fine Aggregate from Concrete and Mortar. 2010. Available online: https://www.gbstandards.org (accessed on 23 September 2020).
7. Guobiao Standards. GB/T 25177-2010 Recycled Coarse Aggregate for Concrete. 2010. Available online: https://www.gbstandards.org. (accessed on 23 September 2020).
8. JIS. *JIS A 5021, Recycled Aggregate for Concrete*; Japanese Standards Association: Tokyo, Japan, 2005.
9. Dimitriou, G.; Savva, P.; Petrou, M.F. Enhancing mechanical and durability properties of recycled aggregate concrete. *Constr. Build. Mater.* **2018**, *158*, 228–235. [CrossRef]
10. Salesa, Á.; Pérez-Benedicto, J.A.; Colorado-Aranguren, D.; López-Julián, P.L.; Esteban, L.M.; Sanz-Baldúz, L.J.; Sáez-Hostaled, J.L.; Ramis, J.; Olivares, D. Physico-Mechanical properties of multi-Recycled concrete from precast concrete industry. *J. Clean. Prod.* **2017**, *141*, 248–255. [CrossRef]
11. Silva, R.V.; de Brito, J.; Dhir, R.K. Fresh-state performance of recycled aggregate concrete: A review. *Constr. Build. Mater.* **2018**, *178*, 19–31. [CrossRef]
12. Centro de Estudios y Experimentación de Obras Públicas (CEDEX). Ministerio de Fomento—Árido Reciclado Cerámico o Mixto. Available online: http://www.cedex.es/ (accessed on 23 September 2020).
13. Akhtar, A.; Sarmah, A.K. Construction and demolition waste generation and properties of recycled aggregate concrete: A global perspective. *J. Clean. Prod.* **2018**, *186*, 262–281. [CrossRef]
14. Thomas, C.; Setién, J.; Polanco, J.A.; Alaejos, P.; Sánchez De Juan, M. Durability of recycled aggregate concrete. *Constr. Build. Mater.* **2013**, *40*, 1054–1065. [CrossRef]
15. Etxeberria, M.; Vázquez, E.; Marí, A.; Barra, M. Influence of amount of recycled coarse aggregates and production process on properties of recycled aggregate concrete. *Cem. Concr. Res.* **2007**, *37*, 735–742. [CrossRef]
16. Cachim, P.B. Mechanical properties of brick aggregate concrete. *Constr. Build. Mater.* **2009**, *23*, 1292–1297. [CrossRef]
17. Guo, H.; Shi, C.; Guan, X.; Zhu, J.; Ding, Y.; Ling, T.-C.; Zhang, H.; Wang, Y. Durability of recycled aggregate concrete—A review. *Cem. Concr. Compos.* **2018**, *89*, 251–259. [CrossRef]

18. Gómez-Soberón, J.M.V. Porosity of recycled concrete with substitution of recycled concrete aggregate: An experimental study. *Cem. Concr. Res.* **2002**, *32*, 1301–1311. [CrossRef]
19. Ministerio de Fomento. *Gobierno de España EHE-08 ANEJO 15 Recomendaciones para la Utilización de Hormigones Reciclados*. Available online: https://www.mitma.gob.es/organos-colegiados/mas-organos-colegiados/comision-permanente-del-hormigon/cph/instrucciones/ehe-08-version-en-castellano (accessed on 23 September 2020).
20. Rahal, K. Mechanical properties of concrete with recycled coarse aggregate. *Build. Environ.* **2007**, *42*, 407–415. [CrossRef]
21. Katz, A. Properties of concrete made with recycled aggregate from partially hydrated old concrete. *Cem. Concr. Res.* **2003**, *33*, 703–711. [CrossRef]
22. Andreu, G.; Miren, E. Experimental analysis of properties of high performance recycled aggregate concrete. *Constr. Build. Mater.* **2014**, *52*, 227–235. [CrossRef]
23. Silva, R.V.; de Brito, J.; Dhir, R.K. Properties and composition of recycled aggregates from construction and demolition waste suitable for concrete production. *Constr. Build. Mater.* **2014**, *65*, 201–217. [CrossRef]
24. Hormigón reciclado. *Comisión 2 Grupo de Trabajo 2/5—Utilización de árido Reciclado para la Fabricación de Hormigón Estructural*; ACHE: Madrid, Spain, 2006; ISBN 84-89670-55-2.
25. Xiao, J.; Li, J.; Zhang, C. Mechanical properties of recycled aggregate concrete under uniaxial loading. *Cem. Concr. Res.* **2005**, *35*, 1187–1194. [CrossRef]
26. Poon, C.S.; Shui, Z.H.; Lam, L. Effect of microstructure of ITZ on compressive strength of concrete prepared with recycled aggregates. *Constr. Build. Mater.* **2004**, *18*, 461–468. [CrossRef]
27. Sosa, I.; Thomas, C.; Polanco, J.A.; Setién, J.; Tamayo, P. High performance self-compacting concrete with electric arc furnace slag aggregate and cupola slag powder. *Appl. Sci.* **2020**, *10*, 773. [CrossRef]
28. Sainz-Aja, J.; Carrascal, I.; Polanco, J.A.; Thomas, C. Fatigue failure micromechanisms in recycled aggregate mortar by μCT analysis. *J. Build. Eng.* **2020**, *28*. [CrossRef]
29. Sainz-Aja, J.; Thomas, C.; Carrascal, I.; Polanco, J.A.; de Brito, J. Fast fatigue method for self-compacting recycled aggregate concrete characterization. *J. Clean. Prod.* **2020**, *277*, 123263. [CrossRef]
30. Thomas, C.; Sosa, I.; Setién, J.; Polanco, J.; Cimentada, A.I. Evaluation of the fatigue behavior of recycled aggregate concrete. *J. Clean. Prod.* **2014**, *65*, 397–405. [CrossRef]
31. Thomas, C.; Setién, J.; Polanco, J.A.; Lombillo, I.; Cimentada, A. Fatigue limit of recycled aggregate concrete. *Constr. Build. Mater.* **2014**, *52*, 146–154. [CrossRef]
32. Sainz-Aja, J.; Thomas, C.; Polanco, J.A.; Carrascal, I. High-Frequency Fatigue Testing of Recycled Aggregate Concrete. *Appl. Sci.* **2019**, *10*, 10. [CrossRef]
33. Thomas, C.; de Brito, J.; Cimentada, A.I.A.I.; Sainz-Aja, J.A.J. Macro- and micro- properties of multi-recycled aggregate concrete. *J. Clean. Prod.* **2019**. [CrossRef]
34. Tamayo, P.; Pacheco, J.; Thomas, C.; de Brito, J.; Rico, J. Mechanical and Durability Properties of Concrete with Coarse Recycled Aggregate Produced with Electric Arc Furnace Slag Concrete. *Appl. Sci.* **2019**, *10*, 216. [CrossRef]
35. ANDECE. Asociación Nacional de la Industria del Prefabricado de Hormigón. Available online: https://www.andece.org (accessed on 1 September 2020).
36. López-Mesa, B.; Pitarch, Á.; Tomás, A.; Gallego, T. Comparison of environmental impacts of building structures with in situ cast floors and with precast concrete floors. *Build. Environ.* **2009**, *44*, 699–712. [CrossRef]
37. Poon, C.S.; Lam, C.S. The effect of aggregate-to-cement ratio and types of aggregates on the properties of pre-cast concrete blocks. *Cem. Concr. Compos.* **2008**, *30*, 283–289. [CrossRef]
38. European norm, EN 1097-6—Tests for Mechanical and Physical Properties of Aggregates. Part 6: Determination of Particle Density and Water Absorption. Available online: https://shop.bsigroup.com/ProductDetail?pid=000000000030218643 (accessed on 1 September 2020).
39. European norm, EN 1097-2—Tests for Mechanical and Physical Properties of Aggregates. Part 2: Methods for the Determination of Resistance to Fragmentation. Available online: https://shop.bsigroup.com/ProductDetail?pid=000000000030368676 (accessed on 1 September 2020).
40. European norm, EN 933-3—Tests for Geometrical Properties of Aggregates. Part 3: Determination of Particle Shape. Flakiness Index. Available online: https://shop.bsigroup.com/ProductDetail/?pid=000000000030241876 (accessed on 1 September 2020).

41. European norm, EN 12390-7:2009—Testing Hardened Concrete Part 7: Density of Hardened Concrete. Available online: https://shop.bsigroup.com/ProductDetail/?pid=000000000030164912 (accessed on 1 September 2020).
42. European norm, EN 12390-3—Testing Hardened Concrete—Part 3: Compressive Strength of Test Specimens. Available online: https://shop.bsigroup.com/ProductDetail?pid=000000000030360097 (accessed on 1 September 2020).
43. European norm, EN 12390-13:2013—Testing Hardened Concrete—Part 13: Determination of Secant Modulus of Elasticity in Compression. Available online: https://shop.bsigroup.com/ProductDetail/?pid=000000000030398745 (accessed on 1 September 2020).
44. European norm, EN 12390-8:2009/1M:2011—Testing Hardened Concrete—Part 8: Depth of Penetration of Water under Pressure. Available online: https://www.beuth.de/en/standard/une-en-12390-8/123990156 (accessed on 1 September 2020).
45. Spanish norm, UNE 83981—Concrete Durability. Test Methods. Determination to Gas Permeability of Hardened Concrete. Available online: https://standards.globalspec.com/std/1445618/une-83981 (accessed on 1 September 2020).
46. McGinnis, M.J.; Davis, M.; de la Rosa, A.; Weldon, B.D.; Kurama, Y.C. Strength and stiffness of concrete with recycled concrete aggregates. *Constr. Build. Mater.* **2017**, *154*, 258–269. [CrossRef]
47. López Gayarre, F.; Suárez González, J.; Blanco Viñuela, R.; López-Colina Pérez, C.; Serrano López, M.A. Use of recycled mixed aggregates in floor blocks manufacturing. *J. Clean. Prod.* **2017**, *167*, 713–722. [CrossRef]
48. Ministerio de Fomento—Gobierno de España. EHE-08: Code on Structural Concrete. 2008. Available online: http://www.fomento.gob.es/MFOM/LANG_CASTELLANO/ORGANOS_COLEGIADOS/CPH/instrucciones/EHE08INGLES/ (accessed on 23 September 2020).
49. ACI. The American Concrete Institute. Available online: https://www.concrete.org (accessed on 23 September 2020).
50. Eurocode 2: Design of concrete structures EN1992-1-1 1992. Available online: https://eurocodes.jrc.ec.europa.eu/doc/WS2008/EN1992_1_Walraven.pdf (accessed on 23 September 2020).
51. Ismail, S.; Kwan, W.H.; Ramli, M. Mechanical strength and durability properties of concrete containing treated recycled concrete aggregates under different curing conditions. *Constr. Build. Mater.* **2017**, *155*, 296–306. [CrossRef]
52. Medina, C.; Sánchez De Rojas, M.I.; Thomas, C.; Polanco, J.A.; Frías, M. Durability of recycled concrete made with recycled ceramic sanitary ware aggregate. Inter-indicator relationships. *Constr. Build. Mater.* **2016**, *105*, 480–486. [CrossRef]
53. García, A.; Cabreiro, J.P. Utilización de modelos dinámicos en la investigación de accidentes viales. In Proceedings of the Congreso Iberoamericano De Accidentología Vial, Avellaneda, Argentina, 9–11 October 2003.

 *applied sciences*

Article

Mechanical Properties and Flexural Behavior of Sustainable Bamboo Fiber-Reinforced Mortar

Marcus Maier [1],*, Alireza Javadian [2],*, Nazanin Saeidi [2], Cise Unluer [3], Hayden K. Taylor [4] and Claudia P. Ostertag [5]

[1] Berkeley Education Alliance for Research in Singapore (BEARS), Singapore 138602, Singapore
[2] Future Cities Laboratory, Singapore ETH-Centre, Singapore 138602, Singapore; saeidi@arch.ethz.ch
[3] School of Engineering, University of Glasgow, Glasgow G12 8LT, UK; Cise.Unluer@glasgow.ac.uk
[4] Department of Mechanical Engineering, University of California, Berkeley, CA 94720, USA; hkt@berkeley.edu
[5] Department of Civil and Environmental Engineering, University of California, Berkeley, CA 94720, USA; Ostertag@ce.berkeley.edu
* Correspondence: Mr.Marcus.Maier@gmail.com (M.M.); javadian@arch.ethz.ch (A.J.)

Received: 31 August 2020; Accepted: 17 September 2020; Published: 21 September 2020

Abstract: In this study, a sustainable mortar mixture is developed using renewable by-products for the enhancement of mechanical properties and fracture behavior. A high-volume of fly ash—a by-product of coal combustion—is used to replace Portland cement while waste by-products from the production of engineered bamboo composite materials are used to obtain bamboo fibers and to improve the fracture toughness of the mixture. The bamboo process waste was ground and size-fractioned by sieving. Several mixes containing different amounts of fibers were prepared for mechanical and fracture toughness assessment, evaluated via bending tests. The addition of bamboo fibers showed insignificant losses of strength, resulting in mixtures with compressive strengths of 55 MPa and above. The bamboo fibers were able to control crack propagation and showed improved crack-bridging effects with higher fiber volumes, resulting in a strain-softening behavior and mixture with higher toughness. The results of this study show that the developed bamboo fiber-reinforced mortar mixture is a promising sustainable and affordable construction material with enhanced mechanical properties and fracture toughness with the potential to be used in different structural applications, especially in developing countries.

Keywords: fiber-reinforced; natural fibers; bamboo; sustainable mortar; mechanical characterization; by-products; toughness

---

1. Introduction

The construction industry is a major consumer of energy and raw materials and contributes immensely to environmental pollution, especially to greenhouse gas (GHG) emissions [1]. Since the 1970s, annual GHG emissions have steadily increased and reached 53.5 GtCO$_2$e in 2017 [2]. Within the construction industry, concrete is the dominant building material with a global production of 20×10^{12} kg per annum which exceeds the amount of all other construction materials combined. With an increase in the demand for new infrastructure demonstrated by developing countries, the use of Portland cement (PC), the main component in concrete, has been rising rapidly [3]. Accordingly, the global use of PC has increased from 2.22 to 4.10 Gt/year within the past decade. The production of PC accounts for ~5% of the global anthropogenic CO$_2$ emissions [4]. Moreover, concrete is commonly reinforced with steel, whose production involves high energy emissions and consumption of fossil fuels that additionally contributes to CO$_2$ emissions. Furthermore, the fast pace of development in many developing countries has led to an increased demand for reinforced concrete for housing and infrastructure projects. Unfortunately, the majority of developing countries lack the resources to produce their own cement and steel for the production of reinforced concrete elements which forces them to import the majority of their needs from highly industrialized countries, and as a result of the import surge, trade deficits, economic slow-downs, and loss of jobs are prominent in those countries. Besides the economic challenges from the cement and steel import, environmental issues also need to be addressed. The construction industry is facing an urgent need for the use of sustainable materials incorporating locally available renewable resources as well as industrial by-products with lower environmental impacts.

One potential material is fly ash, a by-product of the combustion of coal, oil and biomass. Fly ash contains Silicon dioxide (SiO_2) and Aluminum oxide (Al_2O_3) as major components that can contribute to the hydration of cement. Furthermore, low-cost and renewable materials such as bamboo and wood can be found in abundant supply in many developing countries, where the bamboo and wood industry produce a large number of waste products. Replacement of cement with by-products such as fly ash or bamboo and wood waste can enable the reduction of the carbon footprint associated with the cement industry and improve the mechanical and thermal properties of the developed formulations. Other performance aspects such as the ductility of these mixes could be further enhanced via the use of other renewable materials such as natural fibers to avoid the brittle failure that is characteristic of plain concrete.

Previous studies [5–11] revealed improvements in the mechanical properties and durability of concrete mixtures, in which PC was partially replaced with wood ash or fly ash. Substituting aggregates with wood process waste such as wood chips, flax or hemp was also shown to enhance the mechanical or thermal properties of concrete mixtures [11–18]. Further research on fiber-reinforced concrete reported that the addition of synthetic fibers—such as polypropylene (PP), polyethylene (PE), polyvinyl alcohol (PVA)—or steel fibers could increase the fire resistance, ductility, tensile strength, impact resistance and toughness of concrete mixtures [19–22]. However, synthetic fibers, which are mainly derived from petroleum-based sources, and steel fibers require energy-intensive and expensive production processes. In contrast, natural fibers, such as those obtained from wood and bamboo industry by-products, can provide a low-cost and sustainable alternative for the construction industry. Challenges of resource scarcity and the negative environmental impacts of synthetic fiber production have led many researchers to search for alternative, green, sources of fibers for the production of fiber-reinforced concrete. Natural fibers represent a sustainable source of raw materials from renewable resources and can help to alleviate the need for synthetic fibers. While there is growing interest in the use of wood fibers to enhance the mechanical behavior and fracture toughness of concrete [23–25], there has so far been relatively little investigation of the use of bamboo fibers for this purpose. Only a few studies [26–30] investigated the performance of bamboo fiber-reinforced concrete and mortar mixtures through a series of mechanical tests. The bamboo fibers in those studies were obtained from bamboo forests and were subsequently processed as fibers for concrete mixtures. Furthermore, the studies showed that only concrete's tensile property had obvious improvement when bamboo fibers were added, while the enhancement to the compression property and flexural property was not obvious. The studies on the application of bamboo fiber-reinforced concrete and mortar mixtures are rather limited. Both bamboo fibers and fly ash present a great opportunity as sustainable and affordable replacements for cement and steel for developing countries. Bamboo belongs to the botanical family of grasses and shows high resistance to tensile stresses. The tensile strength of natural bamboo is superior to that of wood. This attribute marks bamboo as an attractive option to incorporate into fiber-reinforced concrete, especially in developing countries where demand for reinforced concrete is growing rapidly [31–33]. Bamboo is a gigantic grass, which belongs to the angiosperms (seed-bearing vascular plants) group and monocotyledon (flowering plants) subgroup. Bamboo attains maturity in 3 to 5 years, in favorable contrast to wood, which takes at least 20 years, depending on the species [34]. The growth behavior of bamboo culm and the extreme wind loads it has to sustain during its life cycle require a precise mechanical adaptation to the environment. Therefore, material optimization has to be achieved effectively from the bamboo fibers and their cell structures. This results in an optimized microstructure with superior material performance when compared to various wood species. Furthermore, bamboo can directly address global warming as it rapidly grows and sequesters carbon in biomass and soil faster than almost any wood species. The main components of bamboo culms are cellulose, hemicellulose and lignin. The minor components are resins, tannins, waxes and mineral salts. However, the percentage of each component differs from species to species and depends on the conditions of bamboo growth and the age of the bamboo, as well as the location of the section on the culm [31,34]. In general, cellulose in bamboo culms accounts for more than 50% of the bamboo chemical components. After cellulose, lignin is the next largest component, and normally accounts for more than 20% of the bamboo's mass. Bamboo displays a round-shaped cell cross-section, in contrast to the nearly rectangular and relatively large cells of wood species. Furthermore, bamboo culms have a particular multi-layered cell wall structure with alternating thick and thin layers of fibers, unlike the typical three-layered cell wall of wood species which have a structure with a dominating middle layer [34–36].

In recent years various methods have been developed to employ bamboo through new processing technologies for the fabrication of high-performance bamboo-composite materials in such a way that the inherent mechanical capacities of the fibers are retained, while the durability issues, specifically water absorption, swelling, shrinking and chemical resistance, of the composite could be enhanced for application as structural elements in buildings [31–33,37]. The bamboo-composite materials display high mechanical properties and have been used as either reinforcement in concrete, replacing steel or as structural elements in the form of a beam or column. However, the process through which natural bamboo culms transform into bamboo-composite materials employs only certain sections of the culms and therefore the remaining parts usually become part of the waste of the production process which can be safely utilized for applications as sustainable and affordable fibers in fiber-reinforced mortar.

Therefore, the objective of this study is to develop a sustainable and affordable mortar mixture incorporating by-products and renewable materials (i.e., fly ash and bamboo fibers) that show improved mechanical properties and fracture behavior and could be employed for the construction of low-cost and low-rise housing solutions in developing countries. The developed mixture was characterized via compression, splitting and bending tests, whereby the fracture properties including toughness and absorption energy were also assessed. The findings generated through this work set the foundation for further research on bamboo fibers and bamboo-reinforced concrete and mortar mixtures for structural applications.

2. Materials and Methodology

2.1. Bamboo Plant

There are about 1200 species of bamboo under some 90 genera. The physical and mechanical properties of bamboo culms are correlated with the specific gravity and the fiber content. Therefore, the physical and mechanical properties of bamboo differ from species to species and even within the same species or same culm, due to changes in chemical composition as well as specific gravity [38]. For the purpose of this study *Dendrocalamus asper*, known as Petung Putih bamboo, was selected from a bamboo forest on the Java island of Indonesia. *Dendrocalamus asper* is widely used for low-rise and low-cost housing across Indonesia.

2.1.1. Bamboo Fibers

A new processing technology was developed to process bamboo culms into fibers that were suitable for use in a novel engineered bamboo composite materials [33]. Accordingly, the fibers were obtained by processing entire bamboo culms. They were then added to epoxy resin and fabricated into high-tensile-strength bamboo composite materials by using a hot-press fabrication method. The process yielded an engineered bamboo composite material, which was then cut into different sizes to be used in concrete as reinforcement in place of steel bars in previous studies [39,40].

Throughout this process, some portions of the fibers are not used and remain as the waste by-products of the fabrication process. This study adopts the use of these waste portions of these fibers to reinforce the mortar matrix. In an earlier study [37], the mechanical properties of *Dendrocalamus asper* bamboo from Indonesia were investigated with respect to culm physical properties including culm diameter, wall thickness, height, moisture content and specific density. Correlations were drawn between the culm's physical properties and the resulting mechanical properties, including tensile strength, modulus of rupture and modulus of elasticity in flexure and tension.

Bamboo fibers were ground and sieved with a sieve tower using sieve sizes of 1 mm and 500, 300 and 125 μm. The final bamboo fibers used in this study were those remained in the 500 μm and 300 μm sieves (Figure 1) and are referred to as "500 μm" and "300 μm" fibers throughout this study, respectively. Bigger and smaller sieves were used for a better separation of the fiber sizes. This ensured more constant fiber sizes and avoided clogging of the 500-micron sieves with bigger fibers. Trail mixes with 212 μm fibers showed no promising results and were therefore not considered in this article.

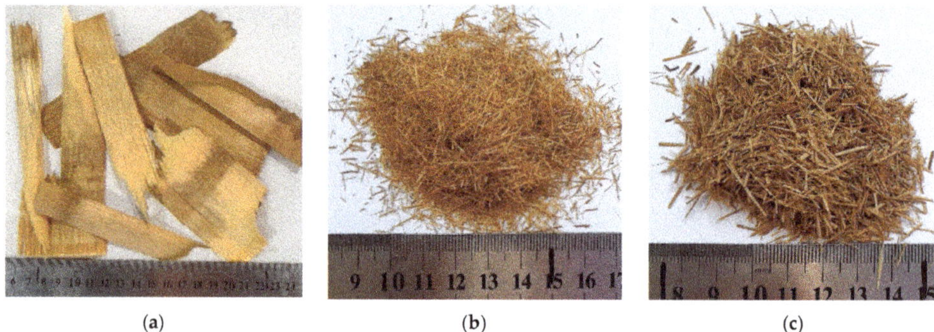

(a)   (b)   (c)

Figure 1. Bamboo fibers. (**a**) Raw waste material; (**b**) 300 μm-diameter fibers after grinding and sieving; and (**c**) 500 μm-diameter fibers after grinding and sieving.

2.1.2. Bamboo Fiber Treatment

A variety of treatments to improve the durability and the bonding of natural fibers embedded in concrete and mortar matrix are available, with different levels of complexity including but not limited to cement surface coating, lime surface coating, cement–lime coating and oil impregnation [41–44]. For the present study, two treatments were selected for the bamboo fibers. To enhance the durability, lignin was partially removed by simply heating the bamboo fibers in water at 85 °C for 72 h and drying them at 80 °C for 24 h [45]. In addition, the bond between the fibers and the mortar matrix was improved by an alkaline treatment, which involved the stirring of the bamboo fibers in a lime $(Ca(OH)_2)$ solution for 2 h. The solution contained 40 g of lime per liter of water. This suspension was stirred repeatedly during the entire treatment duration to avoid the sedimentation of the fibers and any undissolved lime particles. After the lime treatment, the bamboo fibers were dried at 80 °C for 24 h. This treatment is known to modify the surface of the fibers and improve their mechanical strength [45].

2.1.3. Characterization of the Bamboo Fiber Geometry

The fiber geometry was assessed by microscopy, during which a total of 200 fibers within each size category were measured with an optical microscope and analyzed with the Zeiss software "ZEN 2 (Blue edition). Table 1 presents the mean value of the length and diameter as well as the standard deviation obtained from these microscopy measurements. The mean value $\bar{x}$ is calculates by $\bar{x} = \frac{1}{n} \sum_{i=1}^{n} x_i$ whereby the standard deviation is calculated by $\sigma = \sqrt{\frac{1}{n} \sum_{i=1}^{n} (x_i - \bar{x})^2}$.

The intentionally chosen simplified manufacturing and sieving process leads to the relatively high standard deviation of the fibers, but allows replication with minimal equipment and labor costs, especially in developing countries with limited access to such facilities. The average density of the bamboo fibers used in this study was 840 kg/m^3 and was obtained from [31].

Table 1. Geometry of the bamboo fibers.

Fiber Category	300 μm	500 μm
Mean value of fiber length [mm]	7.2	7.8
Standard deviation [mm]	2.4	2.5
Standard deviation [%]	32.7	32.3
Mean value of fiber diameter [μm]	415.7	680.1
Standard deviation [μm]	79.4	167.7
Standard deviation [%]	19.1	24.7
Mean value of Aspect ratio [-]	17.3	11.5
Standard deviation [mm]	6.4	6.2
Standard deviation [%]	37.10	54.0

2.2. Mortar Mixtures

A mortar mixture containing 547 kg/m^3 of Ordinary Portland cement, CEM I, according to Singapore Standard SS EN 197-1 [46] and 656 kg/m^3 of fly ash Class C, according to ASTM C 618 [47], was used as a control mixture. Different contents of bamboo fibers (4, 6, and 8% by volume of the concrete) were added to investigate the resulting mechanical and fracture behavior. Fine sand with a maximum grain size of 2 mm was used as aggregates. The sieve curve of the sand is shown in Figure 2. To adjust the workability of the mixtures, a superplasticizer (ACE 8538, BASF), was added to the mixture. To ensure a constant mix procedure and fiber distribution trail mixes were prepared to evaluate mixing time and the amount of superplasticizer needed to achieve good workability. At first Cement, Fly Ash and aggregates were mixed for 90 s before water containing superplasticizer was added and mixed for 5 min. Finally, the bamboo fibers were gradually added and the constituents were mixed for another 3 min before filled into the molds. To evaluate the workability of the mixtures a flow table test was carried out according to ASTM C1437 [48]. All mixtures showed a spread of more than 255 mm diameter (maximum diameter of the flow table test apparatus according to [49]) and therefore could be filled well in the molds and compacted.The compositions of all mixtures used in this study are provided in Table 2.

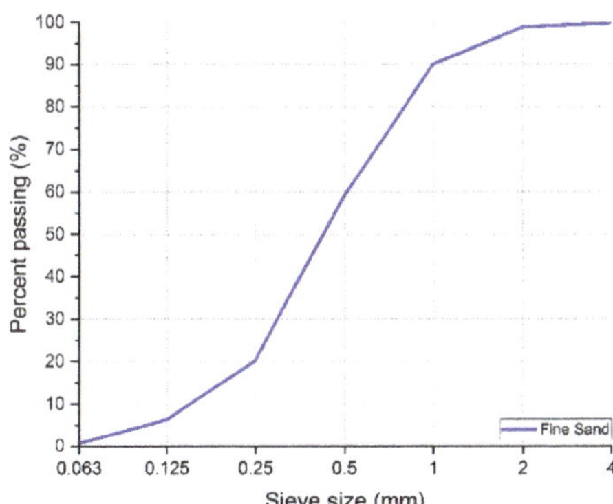

Figure 2. Particle size distribution of the fine aggregates used in this study.

Table 2. Compositions of mortar mixtures.

Mixture	Unit	Control	Bamboo [1] 300 μm	Bamboo [1] 500 μm
Fiber content	[Vol %]	0	4/6/8	4/6/8
Cement: Cem I	[kg/m^3]	547	547	547
Fly ash	[kg/m^3]	656	656	656
Fine aggregates	[kg/m^3]	541	435/382/329	435/382/329
Bamboo fibers	[kg/m^3]	-	32.6/48.9/65.2	32.6/48.9/65.2
Water/binder—ratio	[-]	0.30	0.30	0.30
Water	[kg/m^3]	361	361	361
Superplasticizer	[w% of binder]	0.42	0.53/0.61/0.70	0.53/0.61/0.70

[1] Density of bamboo fibers: 840 kg/m^3 obtained from [31].

2.3. Test Procedure

The experimental program included compressive and splitting tensile tests for the mechanical characterization of the prepared mixtures. A four-point bending test was also used to evaluate the post-cracking behavior of the bamboo fiber mixtures. These were assessed on 50 mm cubes and beams with dimensions of 50 × 50 × 300 mm. Compression tests were performed at a loading rate of 55 kN/min, following the ASTM C109 standard [50]. Splitting tests were performed in accordance to ASTM C496 [51], at a loading rate of 1.0 mm/min. Four-point bending tests were performed according to ASTM C1609 [52] on prisms with a span, *L*, of 150 mm. The load was applied in the 1/3 points at a rate of 0.2 mm/min. The deflection of the beam was measured with two Linear Variable Differential Transducers (LVDTs) at the mid-span (Figure 3).

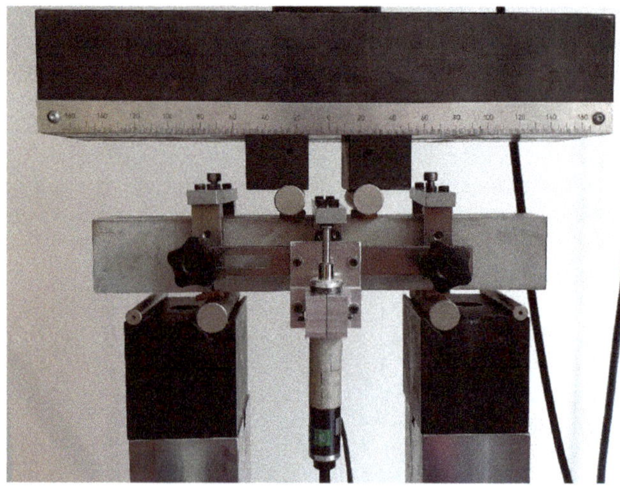

Figure 3. Four-point bending test performed according to ASTM C1609 [52].

With the recorded load–deflection data at hand, the load at the limit of proportionality, F_{max}, and the corresponding modulus of rupture (MOR) could be assessed according to the RILEM TC 162-TDF [53] recommendation. According to the RILEM recommendation, the limit of proportionality is equal to the maximum load recorded up to 0.05 mm. The modulus of rupture, corresponding to the maximum force in a four-point bending test, can be calculated by the following expression:

$$\text{MOR} = F_{max}\, l/(b\, h^2)\ (\text{MPa}) \tag{1}$$

where b, h and l are the width, height and span of the tested specimens and equal to 50 mm, 50 mm and 150 mm respectively.

The density as well as the compressive and tensile splitting tests were evaluated on three specimens after 28 days of curing. The notation throughout this study was chosen as XXX-YP, where XXX referred to the bamboo fiber diameter in microns and Y referred to the volumetric fiber content. Therefore, notation 300-6P referred to the mixture containing 6 V% (volume %) of bamboo fibers with a nominal diameter of 300 μm.

3. Results and Discussion

3.1. Mechanical Properties

3.1.1. Density and Compressive Strength

The control mix exhibited a density of 2132 kg/m^3, which was slightly higher than the mixes containing fibers. The addition of bamboo fibers had little effect on the density, revealing a reduction ranging between 0.7% and 2.5%, as shown in Table 3. In terms of performance, the control mixture achieved the highest 28-day compressive strength of 75.1 MPa, as shown in Figure 4. The inclusion of fibers led to a reduction in strength, which was directly correlated with the number of bamboo fibers added to the mix design as shown in Table 3. Accordingly, higher fiber contents resulted in lower compressive strengths. This behavior can be attributed to the difference in the compressive strength and E-modulus of the bamboo fibers and the cement matrix. A previous study of the authors on different grades of the bamboo culms revealed a compressive strength between 43.2 and 68.4 MPa and an E-modulus between 18.1 and 28.2 GPa [31]. The incorporated bamboo fibers reduce the overall strength of the matrix resulting in a higher loss of strength with increasing fiber content which is in compliance with the findings of other researchers [20,54,55]. In addition, a study of Li et al. on natural fibers found that air pockets were formed at some of the fibers resulting in a reduced compressive strength compared to the control mixture [56]. Furthermore, a lower aspect ratio, thicker and longer fibers (i.e., "500 μm" batch), generally indicated a greater reduction in the compressive strength than thinner and shorter fibers (i.e., "300 μm" batch), which was also found in [54]. In this respect, the reduction of the compressive strength for the 4 V% to 8 V% of fibers was within a range of 7.8–19.9% for the 300 μm fibers and 9.1–27% for the 500 μm fibers (see Figure 4 and Table 3). An average compressive strength of 60.2 MPa was recorded for the 8 V% mixture with 300 μm fibers, while the corresponding figure was 54.8 MPa for the 8 V% mixture with 500 μm fibers. Although a slight reduction in strength was observed, the overall findings reveal the comparable performance of the bamboo fiber reinforced mortar mixtures with those of commonly used mix designs for structural members in similar studies [20,31,54].

Table 3. Physical and mechanical properties of the Bamboo Fiber-reinforced mortar mixtures.

Mix	Density	SD [1]	SD [2]	Compressive Strength	SD [1]	SD [2]	Splitting Tensile Strength	SD [1]	SD [2]
	kg/m^3	kg/m^3	%	MPa	kg/m^3	%	MPa	kg/m^3	%
Control	2132	11.2	0.5	75.1	1.1	1.5	7.2	0.50	6.9
300-8P	2105	28.3	1.3	60.2	0.8	1.3	5.3	0.76	14.4
300-6P	2117	12.3	0.6	63.9	0.5	0.7	5.4	0.47	8.8
300-4P	2111	27.4	1.3	69.3	4.0	5.9	6.7	0.42	6.4
500-8P	2101	12.1	0.6	54.8	6.3	6.3	4.9	0.57	11.6
500-6P	2089	6.7	0.3	63.4	3.2	3.2	5.0	0.19	3.8
500-4P	2129	13.5	0.6	68.2	2.1	2.1	5.5	0.31	5.7

[1] SD: Standard deviation calculated from mean value in kg/m^3 or MPa; [2] SD: Standard deviation calculated from mean value in percentage.

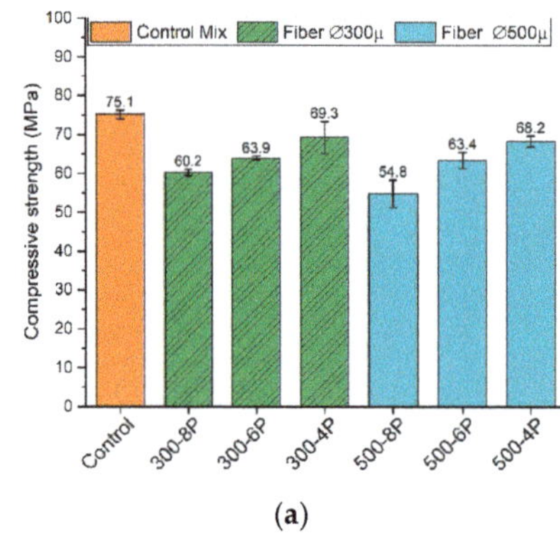
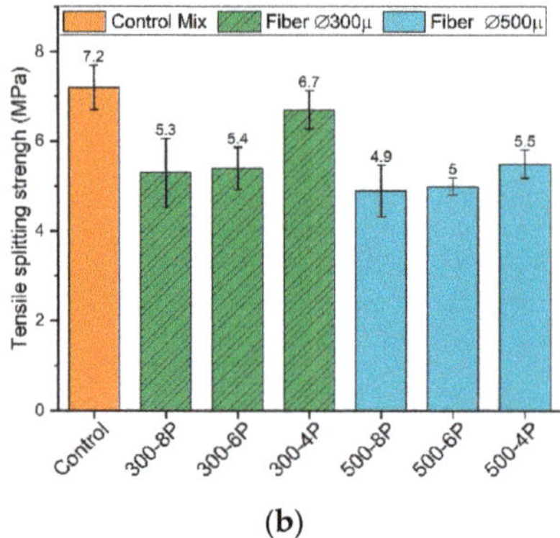

Figure 4. Compressive strength (**a**) and splitting tensile strength (**b**) of the bamboo fiber-reinforced mortar mixtures.

3.1.2. Splitting Tensile Strength

The control mixture showed a splitting tensile strength of 7.2 MPa. Similar to the trends observed in the compressive strength results, the splitting tensile strength of the samples with 8/6/4 V% of fibers was reduced by 6.9/25.0/26.4% for the 300 μm fibers and 23.6/30.6/31.9% for the 500 μm fibers (see Figure 4 and Table 3). It is worth noting that the splitting test setup compression is applied to a small area along the specimen resulting in tensile stress (lateral force) within the matrix which causes splitting. The splitting tensile strength is the maximum load before cracks appear and therefore the crack bridging behavior of the fibers is not considered in this test. This is investigated with bending tests and discussed later on. According to [57], the tensile strength of a mortar matrix can be assumed to be in the range of 10% of the compressive strength which matches the results obtained in this study.

3.1.3. Flexural Tensile Behavior

The control mix showed linear elastic behavior up to the peak load, which was followed by an abrupt failure, resulting in complete separation of the specimens into two parts. In contrast, the bamboo fiber mixtures exhibited improved post-crack behavior, resulting in a strain-softening behavior, where a single crack occurred within the central third of the prismatic samples. This effect can be seen in Figures 5 and 6 where the pictures of the tested specimens taken after the bending tests are shown for both sets of samples containing 300 μm and 500 μm fibers. Within these figures, it is possible to visualize the fibers bridging the developed crack, which was much more evident as the fiber content increased from 4 V% to 8 V%.

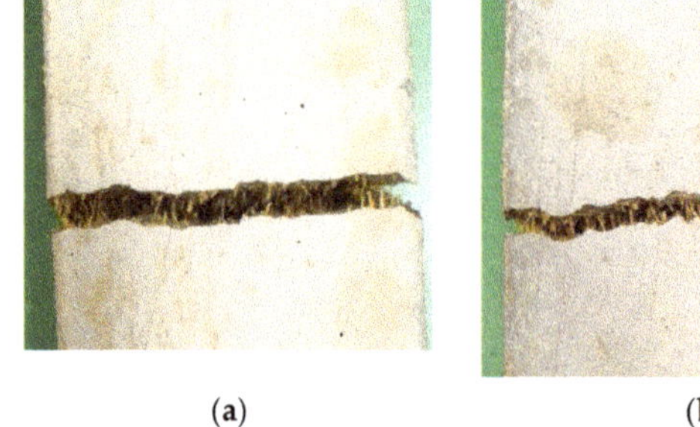

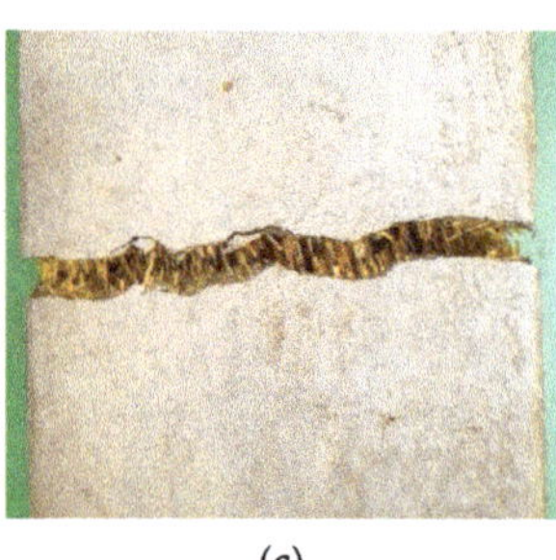

Figure 5. Photographs of test samples after bending tests involving 300 μm fibers. Fiber content: (**a**) 4 V%, (**b**) 6 V% and (**c**) 8 V%.

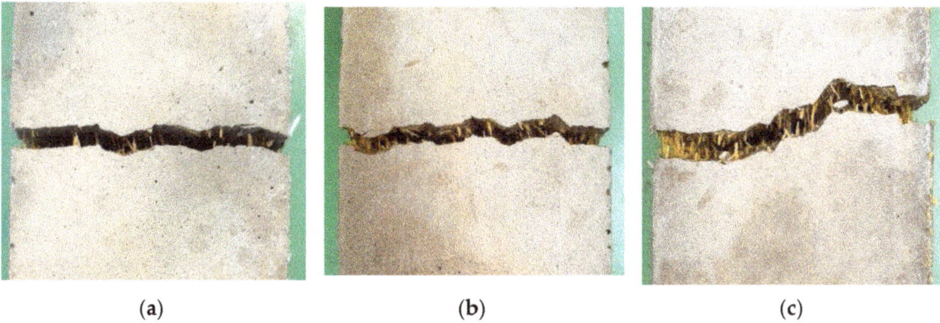

(a) (b) (c)

Figure 6. Photographs of test samples after bending tests involving 500 μm fibers. Fiber content: (a) 4 V%, (b) 6 V% and (c) 8 V%.

The load–deflection behaviors of all test specimens are shown in Figure 7. Test results with different fiber contents are plotted with an offset of 0.1 mm from each other for better visual comparison. Both fiber groups involving the use of 300 μm and 500 μm fibers, revealed similar trends, during which 8 V% mixtures exhibited the highest peak load and the mixtures with 4 V% the lowest. Furthermore, the bamboo fibers were able to bridge the developing crack, resulting in a strain-softening behavior in all bamboo mixtures. The post-crack behavior was more pronounced for mixtures with higher fiber contents, which was in line with the findings with Aydin Serdar and Hameed et.al. where it was shown that a higher fiber content results in a higher load-carrying capacity due to the crack bridging behavior of the fibers [58,59].

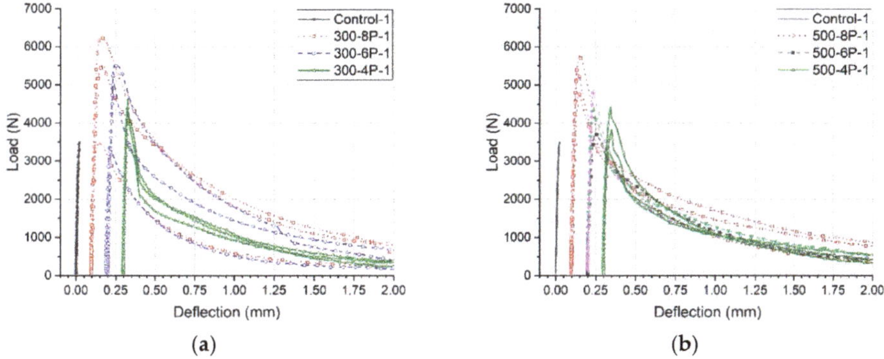

(a) (b)

Figure 7. Bending test results of mixtures with 4 V%, 6 V% and 8 V% of (a) 300 μm fibers and (b) 500 μm fibers.

3.1.4. Fracture Properties and Toughness

The limit of proportionality and the modulus of rupture are shown in Figure 8. It should be noted that one out of the 3 samples of the 300 μm with 8 V% and 6 V% exhibited a significantly lower bending performance as can be seen in Figure 7. This resulted in a high standard variation shown in Figure 8. Samples with the 500 μm fiber batch revealed a lower modulus of rupture compared to those containing 300 μm fibers. These values were around 0.0%, 7.7% and 13.0% lower for the 8/6/4 V% fibers specimens, respectively. These findings were in agreement with ACI 544.1R-96, where a lower MOR was reported with decreasing aspect ratio of the fibers [60]. Longer and thinner fibers exhibit a higher bond within a concrete or mortar matrix resulting in a higher pull-out resistance compared to shorter and thicker fibers as shown which was also found by other researchers in [55,61].

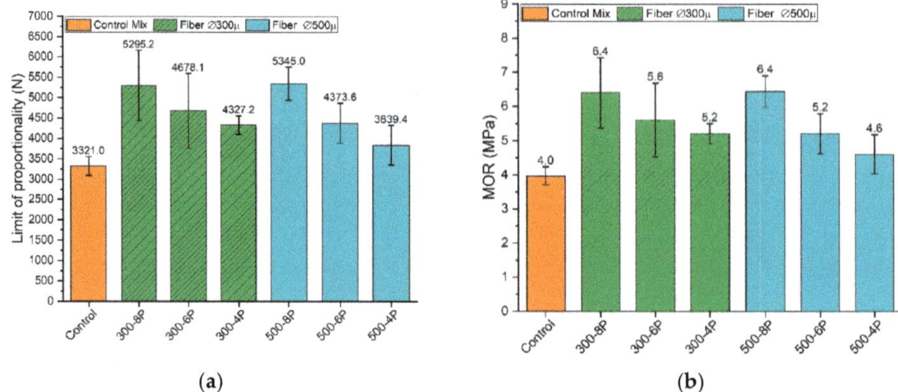

Figure 8. Bending test assessment: (**a**) limit of proportionality F_{max}; (**b**) modulus of rupture.

The toughness was calculated as the area under the load–deflection curve and expressed as energy in Joules. Accordingly, the resulting cumulative energy up to 2 mm deflection is shown in Figures 9 and 10. As expected, a higher fiber content resulted in higher toughness. Thicker and longer fibers, as found in the 500 μm batch, showed an overall reduction in toughness when compared to the smaller and shorter fibers in the 300 μm batch. This is explained due to the higher bridge bearing capacity of the mixtures with a higher amount of fibers and a higher pull-out resistance for fibers with a higher aspect ratio as discussed earlier and shown in [59,61].

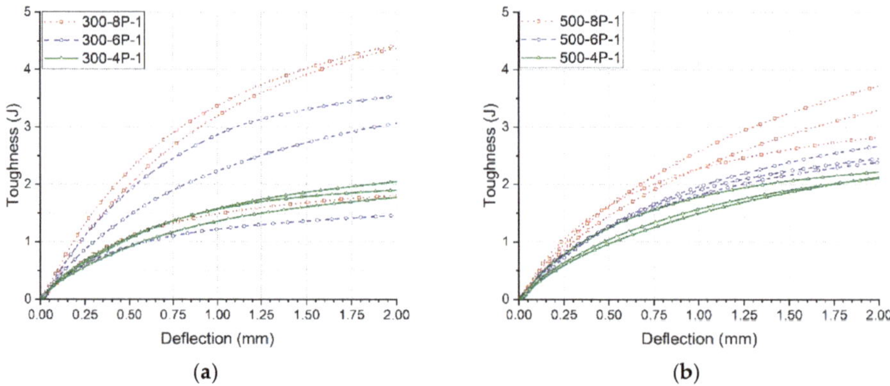

Figure 9. Cumulative toughness of the bamboo fiber reinforced specimens up to 2 mm deflection: (**a**) 300 μm batch; (**b**) 500 μm batch evaluated from the four-point bending test according to ASTM C1609 [52].

To characterize the post-cracking behavior of the different mixes, two toughness parameters were evaluated in accordance with RILEM RC 162-TDF [53]. The toughness values at a mid-span deflection of span/150, which corresponded to 1 mm deflection, and the toughness at a mid-span deflection of span/75 (2 mm deflection) were evaluated. The corresponding results, presented in Figure 10 and Table 4, revealed that a higher fiber content enhanced the ductility and toughness of the prepared mixes. The toughness values of the 300 μm fiber specimens at 1 mm deflection were 40.0% and 73.3% higher for the 6 V% and 8 V% mixes when compared to 4 V% mixes, respectively. Assessing the 500 μm fiber batch, an increase of 16% and 47% increase in toughness was recorded for the 8 V% and 6 V% mixes when compared to 4 V% mixes, respectively (see Figure 10). Furthermore, the results of the mixes containing 500 μm fibers showed lower energy up to both 1.0 mm and 2.0 mm deflection when compared to the 300 μm fiber results and showed the influence of the aspect ratio of the fibers on the bond strength of the specimen which is in compliance with findings in [55,61]. The high standard deviation of the 300 μm with 8 V% and 6 V% is the result of the lower bending performance of one out of three tested specimens shown in Figure 7.

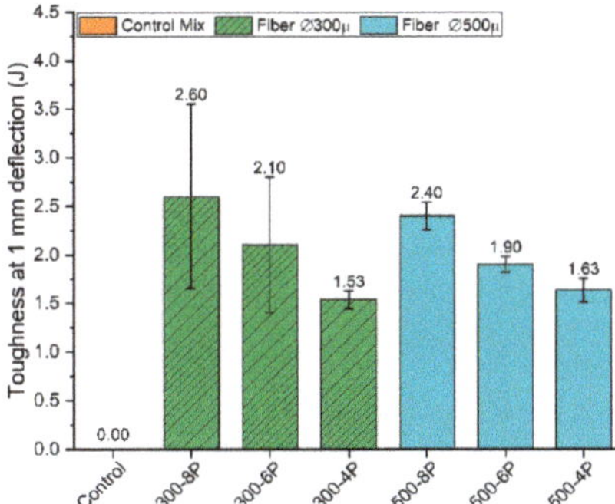

Figure 10. Toughness of the mixtures at 1 mm deflection measured according to RILEM RC 162-TDF [53].

Table 4. Flexural strength and toughness parameters of the four-point bending test.

Mix	Limit of Propor-Tionality	SD	SD	MOR	SD	SD	Tough-Ness at 1 mm	SD	SD	Tough-Ness at 2 mm	SD	SD
	[N]	[N]	[%]	MPa	[MPa]	[%]	[J]	[J]	[%]	[J]	[J]	[%]
Contr.	3321	231.7	7.0	4.0	0.26	6.6	0.0	0.00	0.0	0.0	0.00	0.0
300-8P	5295	860.9	16.3 *	6.4	1.03	16.2 *	2.6	0.95	35.9 *	3.4	1.35	39.8 *
300-6P	4678	914.2	19.5 *	5.6	1.08	19.3 *	2.1	0.70	33.2 *	2.7 *	0.86	32.0 *
300-4P	4327	226.5	5.2	5.2	0.29	5.7	1.5	0.09	6.1	1.9	0.12	6.5
500-8P	5345	406.5	7.6	6.4	0.46	7.2	2.4	0.14	5.9	3.3	0.37	11.3
500-6P	4374	491.9	11.2	5.2	0.59	11.3	1.9	0.08	4.3	2.5	0.12	4.9
500-4P	3839	484.5	12.6	4.6	0.57	12.4	1.6	0.12	7.6	2.1	0.04	2.2

* The high standard deviation is a result of a significantly lower bending performance of one specimen.

4. Conclusions

This study focused on the development of a sustainable and affordable mortar mixture consisting of a high amount of fly ash (656 kg/m^3) and varying contents of bamboo fibers (4/6/8 V%) from waste by-products of engineered bamboo-composite fabrication. The bamboo fibers, obtained from bamboo composite production waste, were categorized into two groups of 300 μm and 500 μm and were incorporated at three different volumes of 4/6/8 V% within each mix. The mechanical performance of the developed formulations was assessed via the measurement of the mechanical and fracture properties. From these results, the following conclusions can be drawn:

The addition of 4/6/8 V% of bamboo fibers has a reasonably modest effect on the compressive strength of the mixtures. Compared to the control mixture without fibers, the reduction of the compressive strength of the mixtures with 300 μm fibers is between 7.8% and 19.9%, with larger fiber volume fractions corresponding to a greater reduction of strength.

- The mixtures with 500 μm fibers were reduced in compressive strength between 9.1% and 27.0% by the same range of 4/6/8 V% of fiber content. The reduction of the compressive strength for both the 300 μm and 500 μm fibers could be related to the lower mechanical properties of the bamboo fibers as well as the influence of their aspect ratio.

- The compressive strengths of the mixtures with 8 V% of fibers were shown to be 60.2 MPa and 54.6 MPa for the 300 μm and 500 μm mixtures, respectively, and is within a reasonable range for use in structural members.

- The splitting tensile strength of the bamboo fiber-reinforced mixtures displays a reduction of between 6.9% and 31.9% compared to the control mixture depending on the fiber volume and aspect ratio. The mixtures containing the 500 μm fibers (lower aspect ratio) show lower strength in comparison with the 300 μm mixtures.

- All bamboo fiber-reinforced mortar mixtures display a strain-softening behavior. The mixtures with 8 V% of fibers show better crack-bridging effects than those with lower volume fractions, resulting in higher residual strength.

- The toughness of the mixtures was evaluated at a mid-span deflection of L/150. The 300 μm fiber mixtures show values of 2.6, 2.1 and 1.5 Joule for 8, 6 and 4 V% of fibers respectively, whereas the 500 μm mixtures show, in general, lower toughness of 2.40, 1.90 and 1.63 Joule.

- Mixtures containing 300 μm fibers demonstrate an overall enhanced mechanical performance and post-crack behavior compared to the mixtures with 500 μm fibers as a result of the higher bond strength due to their higher aspect ratio.

The findings emerging from this study demonstrate the suitability of using natural bamboo fibers, obtained from process waste, to improve the ductility of high-volume fly ash mortar. The resulting formulations can enable the development of a sustainable and low-cost mixture for structural members. These results can be utilized for the construction of low-cost and low-rise housing units in developing countries, especially in Southeast Asia, Latin and Central America, where there is access to bamboo and low-cost cementitious materials with low demand for ductility. Further studies on the durability of bamboo fibers and the replacement of steel reinforcement with engineered bamboo composites and natural bamboo fiber members are being performed to broaden the application range of these materials on a larger scale.

Author Contributions: Conceptualization, M.M.; Methodology, M.M.; Validation, M.M.; Formal analysis, M.M.; Investigation, M.M., A.J., and N.S.; Resources: M.M., A.J., and N.S.; Data curation, M.M., A.J., and N.S.; Writing—original draft preparation, M.M.; Writing—review and editing, A.J.; N.S.; H.K.T.; C.U. Visualization, M.M.; Supervision, H.K.T. and C.P.O.; Project administration, M.M. and A.J.; Funding acquisition, M.M., A.J., and N.S. All authors have read and agreed to the published version of the manuscript.

Funding: This research is funded by the Republic of Singapore's National Research Foundation through a grant to the Berkeley Education Alliance for Research in Singapore (BEARS) for the Singapore-Berkeley Building Efficiency and Sustainability in the Tropics (SinBerBEST) Program. BEARS has been established by the University of California, Berkeley, as a center for intellectual excellence in research and education in Singapore. The research was conducted at the Future Cities Laboratory at the Singapore-ETH Centre, which was established collaboratively between ETH Zurich and Singapore's National Research Foundation (FI 370074016) 766 under its Campus for Research Excellence and Technological Enterprise program.

Conflicts of Interest: The authors declare no conflict of interest.

References

1. Onuaguluchi, O.; Banthia, N. Plant-based natural fibre reinforced cement composites: A review. *Cem. Concr. Compos.* **2016**, *68*, 96–108. [CrossRef]

2. Nations, U. *Emission Gap Report 2018*; UNEP DTU Partnership; 2018. Available online: https://core.ac.uk/reader/189891060 (accessed on 30 August 2020).

3. Krausmann, F.; Gingrich, S.; Eisenmenger, N.; Erb, K.-H.; Haberl, H.; Fischer-Kowalski, M. Growth in global materials use, GDP and population during the 20th century. *Ecol. Econ.* **2009**, *68*, 2696–2705. [CrossRef]

4. Boden, T.; Marland, G.; Andres, R. *Global, Regional, and National Fossil-Fuel CO$_2$ Emissions*; Carbon Dioxide Information Analysis Center, Oak Ridge National Laboratory: Oak Ridge, TN, USA, 2013. [CrossRef]

5. Siddique, R. Utilization of wood ash in concrete manufacturing. *Resour. Conserv. Recycl.* **2012**, *67*, 27–33. [CrossRef]

6. Cheah, C.B.; Ramli, M. The implementation of wood waste ash as a partial cement replacement material in the production of structural grade concrete and mortar: An overview. *Resour. Conserv. Recycl.* **2011**, *55*, 669–685. [CrossRef]

7. Elinwa, A.U.; Mahmood, Y.A. Ash from timber waste as cement replacement material. *Cem. Concr. Compos.* **2002**, *24*, 219–222. [CrossRef]

8. Guo, Z.; Jiang, T.; Zhang, J.; Kong, X.; Chen, C.; Lehman, D.E. Mechanical and durability properties of sustainable self-compacting concrete with recycled concrete aggregate and fly ash, slag and silica fume. *Constr. Build. Mater.* **2020**, *231*, 117115. [CrossRef]

9. Uthaman, S.; Vishwakarma, V.; George, R.P.; Ramachandran, D.; Kumari, K.; Preetha, R.; Premila, M.; Rajaramand, R.; KamachiMudali, U.; Amarendra, G. Enhancement of strength and durability of fly ash concrete in seawater environments: Synergistic effect of nanoparticles. *Constr. Build. Mater.* **2018**, *187*, 448–459. [CrossRef]

10. Velandia, D.F.; Lynsdale, C.J.; Provis, J.L.; Ramirez, F. Effect of mix design inputs, curing and compressive strength on the durability of Na2SO4-activated high volume fly ash concretes. *Cem. Concr. Compos.* **2018**, *91*, 11–20. [CrossRef]

11. Singh, N.; Kumar, P.; Goyal, P. Reviewing the behaviour of high volume fly ash based self compacting concrete. *J. Build. Eng.* **2019**, *26*, 100882. [CrossRef]

12. Bederina, M.; Marmoret, L.; Mezreb, K.; Khenfer, M.M.; Bali, A.; Quéneudec, M. Effect of the addition of wood shavings on thermal conductivity of sand concretes: Experimental study and modelling. *Constr. Build. Mater.* **2007**, *21*, 662–668. [CrossRef]

13. Coatanlem, P.; Jauberthie, R.; Rendell, F. Lightweight wood chipping concrete durability. *Constr. Build. Mater.* **2006**, *20*, 776–781. [CrossRef]

14. Guo, A.; Aamiri, O.B.; Satyavolu, J.; Sun, Z. Impact of thermally modified wood on mechanical properties of mortar. *Constr. Build. Mater.* **2019**, *208*, 413–420. [CrossRef]

15. Akkaoui, A.; Caré, S.; Vandamme, M. Experimental and micromechanical analysis of the elastic properties of wood-aggregate concrete. *Constr. Build. Mater.* **2017**, *134*, 346–357. [CrossRef]

16. Corinaldesi, V.; Mazzoli, A.; Siddique, R. Characterization of lightweight mortars containing wood processing by-products waste. *Constr. Build. Mater.* **2016**, *123*, 281–289. [CrossRef]

17. Taoukil, D.; Sick, F.; Mimet, A.; Ezbakhe, H.; Ajzoul, T. Moisture content influence on the thermal conductivity and diffusivity of wood–concrete composite. *Constr. Build. Mater.* **2013**, *48*, 104–115. [CrossRef]

18. Viel, M.; Collet, F.; Lanos, C. Development and characterization of thermal insulation materials from renewable resources. *Constr. Build. Mater.* **2019**, *214*, 685–697. [CrossRef]

19. Hager, I.; Zdeb, T.; Krzemień, K. The impact of the amount of polypropylene fibres on spalling behaviour and residual mechanical properties of Reactive Powder Concretes. *MATEC Web Conf.* **2013**, *6*, 02003. [CrossRef]

20. Bencardino, F.; Rizzuti, L.; Spadea, G.; Swamy, R.N. Experimental evaluation of fiber reinforced concrete fracture properties. *Compos. Part B Eng.* **2010**, *41*, 17–24. [CrossRef]

21. Ding, Y.; Yu, K.-Q.; Yu, J.-T.; Xu, S.-L. Structural behaviors of ultra-high performance engineered cementitious composites (UHP-ECC) beams subjected to bending-experimental study. *Constr. Build. Mater.* **2018**, *177*, 102–115. [CrossRef]

22. Pan, Z.; Wu, C.; Liu, J.; Wang, W.; Liu, J. Study on mechanical properties of cost-effective polyvinyl alcohol engineered cementitious composites (PVA-ECC). *Constr. Build. Mater.* **2015**, *78*, 397–404. [CrossRef]

23. Li, M.; Khelifa, M.; Khennane, A.; El Ganaoui, M. Structural response of cement-bonded wood composite panels as permanent formwork. *Compos. Struct.* **2019**, *209*, 13–22. [CrossRef]

24. Fonseca, C.S.; Silva, M.F.; Mendes, R.F.; Hein, P.R.G.; Zangiacomo, A.L.; Savastano, H., Jr.; Tonoli, G.H.D. Jute fibers and micro/nanofibrils as reinforcement in extruded fiber-cement composites. *Constr. Build. Mater.* **2019**, *211*, 517–527. [CrossRef]

25. Ali, M.; Liu, A.; Sou, H.; Chouw, N. Mechanical and dynamic properties of coconut fibre reinforced concrete. *Constr. Build. Mater.* **2012**, *30*, 814–825. [CrossRef]

26. Li, M.; Zhou, S.; Guo, X. Effects of alkali-treated bamboo fibers on the morphology and mechanical properties of oil well cement. *Constr. Build. Mater.* **2017**, *150*, 619–625. [CrossRef]

27. Dewi, S.M.; Wijaya, M.N. The use of bamboo fiber in reinforced concrete beam to reduce crack. *AIP Conf. Proc.* **2017**, *1887*, 020003.

28. Zhang, X.; Pan, J.Y.; Yang, B. Experimental study on mechanical performance of bamboo fiber reinforced concrete. *Appl. Mech. Mater.* **2012**, *174*, 1219–1222. [CrossRef]

29. Zhang, C.; Huang, Z.; Chen, G.W. Experimental research on bamboo fiber reinforced concrete. *Appl. Mech. Mater.* **2013**, *357*, 1045–1048. [CrossRef]

30. Terai, M.; Minami, K. Basic Study on Mechanical Properties of Bamboo Fiber Reinforced Concrete. *Glob. Think. Struct. Eng. Recent Achiev.* **2012**, *8*, 17–24.

31. Javadian, A. Composite Bamboo and its Application as Reinforcement in Structural Concrete. Ph.D. Thesis, ETH Zurich, Zurich, Switzerland, 2017.

32. Rahman, N.; Shing, L.W.; Simon, L.; Philipp, M.; Alireza, J.; Ling, C.S.; Wuan, L.H.; Valavan, S.; Nee, S.S. Enhanced bamboo composite with protective coating for structural concrete application. *Energy Procedia* **2017**, *143*, 167–172. [CrossRef]

33. Hebel, D.E.; Javadian, A.; Heisel, F.; Schlesier, K.; Griebel, D.; Wielopolski, M. Process-controlled optimization of the tensile strength of bamboo fiber composites for structural applications. *Compos. Part B Eng.* **2014**, *67*, 125–131. [CrossRef]

34. Liese, W. *The Anatomy of Bamboo Culms*; BRILL: Boston, MA, USA, 1998.

35. Liese, W. *Bamboos Biology, Silvics, Properties, Utilization (Schriftenreihe der GTZ, no. No 180)*; Deutsche Gesellschaft für Technische Zusammenarbeit (GTZ): Eschborn, Germany, 1985; p. 132.

36. Wang, X.; Keplinger, T.; Gierlinger, N.; Burgert, I. Plant material features responsible for bamboo's excellent mechanical performance: A comparison of tensile properties of bamboo and spruce at the tissue, fibre and cell wall levels. *Ann. Bot.* **2014**, *114*, 1627–1635. [CrossRef]

37. Javadian, A.; Smith, I.F.C.; Saeidi, N.; Hebel, D.E. Mechanical Properties of Bamboo Through Measurement of Culm Physical Properties for Composite Fabrication of Structural Concrete Reinforcement. *Front. Mater.* **2019**, *6*, 15. [CrossRef]

38. Ray, K.A.; Mondal, S.; Das, S.K.; Ramachandrarao, P. Bamboo—A functionally graded composite-correlation between microstructure and mechanical strength. *J. Mater. Sci.* **2005**, *40*, 5249–5253. [CrossRef]

39. Javadian, A.; Wielopolski, M.; Smith, I.F.; Hebel, D.E. Bond-behavior study of newly developed bamboo-composite reinforcement in concrete. *Constr. Build. Mater.* **2016**, *122*, 110–117. [CrossRef]

40. Grüner Stahl—Hochfeste Naturfaserverbundwerkstoffe für strukturelle Anwendungen in der Industrie. In *Projekt abschlussbericht KTI-Commission for Technology and Innovation*; ETH Zurich, Empa Dubendorf: Bern, Switzerland, 2016; Volume 1.

41. Xu, R.; He, T.; Da, Y.; Liu, Y.; Li, J.; Chen, C. Utilizing wood fiber produced with wood waste to reinforce autoclaved aerated concrete. *Constr. Build. Mater.* **2019**, *208*, 242–249. [CrossRef]

42. Bederina, M.; Gotteicha, M.; Belhadj, B.; Dheily, R.M.; Khenfer, M.M.; Queneudec, M. Drying shrinkage studies of wood sand concrete—Effect of different wood treatments. *Constr. Build. Mater.* **2012**, *36*, 1066–1075. [CrossRef]

43. Claramunt, J.; Ardanuy, M.; García-Hortal, J.A.; Filho, R.D.T. The hornification of vegetable fibers to improve the durability of cement mortar composites. *Cem. Concr. Compos.* **2011**, *33*, 586–595. [CrossRef]

44. Ardanuy, M.; Claramunt, J.; Toledo Filho, R.D. Cellulosic fiber reinforced cement-based composites: A review of recent research. *Constr. Build. Mater.* **2015**, *79*, 115–128. [CrossRef]

45. Traore, Y.B.; Messan, A.; Hannawi, K.; Gerard, J.; Prince, W.; Tsobnang, F. Effect of oil palm shell treatment on the physical and mechanical properties of lightweight concrete. *Constr. Build. Mater.* **2018**, *161*, 452–460. [CrossRef]

46. *SS-EN197-1—Cement: Compositoin, Specification and Conformity Criteria for Common Cements*; Enterprise Singapore: Singapore, 2011.

47. *ASTM C618-19 Standard Specification for Coal Fly Ash and Raw or Calcined Natural Pozzolan for Use in Concrete*; ASTM International: West Conshohocken, PA, USA, 2020.

48. *ASTM C1437-07—Standard Test Method for Flow of Hydraulic Cement Mortar*; ASTM International: West Conshohocken, PA, USA, 2009.

49. *ASTM C230/C230M—Standard Specification for Flow Table for Use in Tests of Hydraulic Cement*; ASTM International: West Conshohocken, PA, USA, 2009.

50. *ASTM C109, Compressive Strength of Hydraulic Cement Mortars*; ASTM International: West Conshohocken, PA, USA, 2002.

51. *ASTM C496, Standard Test Method for Splitting Tensile Strength of Cylindrical Concrete Specimens*; ASTM International: West Conshohocken, PA, USA, 2017.

52. *ASTM C1609, Standard Test Method for Flexural Performance of Fiber-Reinforced Concrete*; ASTM International: West Conshohocken, PA, USA, 2011.

53. Vandewalle, L. RILEM TC 162-TDF: Test and design methods for steel fibre reinforced concrete. *Mater. Struct.* **2000**, *33*, 3–5.
54. Yazıcı, Ş.; İnan, G.; Tabak, V. Effect of aspect ratio and volume fraction of steel fiber on the mechanical properties of SFRC. *Constr. Build. Mater.* **2007**, *21*, 1250–1253. [CrossRef]
55. Chen, H.; Cheng, H.; Wang, G.; Yu, Z.; Shi, S.Q. Tensile properties of bamboo in different sizes. *J. Wood Sci.* **2015**, *61*, 552–561. [CrossRef]
56. Li, Z.; Wang, L.; Wang, X. Compressive and flexural properties of hemp fiber reinforced concrete. *Fibers Polym.* **2004**, *5*, 187–197. [CrossRef]
57. Mohamed, O.A.; Syed, Z.I.; Najm, O.F. Splitting Tensile Strength of Sustainable self-consolidating Concrete. *Procedia Eng.* **2016**, *145*, 1218–1225. [CrossRef]
58. Aydın, S. Effects of fiber strength on fracture characteristics of normal and high strength concrete. *Period. Polytech. Civ. Eng.* **2013**, *57*, 191–200. [CrossRef]
59. Hameed, R.; Turatsinze, A.; Duprat, F.; Sellier, A. Metallic fiber reinforced concrete: Effect of fiber aspect ration on the flexural properties. *ARPN J. Eng. Appl. Sci.* **2009**, *4*, 67–72.
60. *ACI 544.1R-96: Report on Fiber Reinforced Concrete*; ACI World Headquarters: Farmington Hills, MI, USA, 2002.
61. Pacheco-Torgal, F.; Jalali, S. Cementitious building materials reinforced with vegetable fibres: A review. *Constr. Build. Mater.* **2011**, *25*, 575–581. [CrossRef]

applied
sciences

Article

Durability Assessment of Recycled Aggregate HVFA Concrete

Valeria Corinaldesi [1,*], **Jacopo Donnini** [2], **Chiara Giosué** [2], **Alessandra Mobili** [2] and **Francesca Tittarelli** [2]

1   Dipartimento di Scienze e Ingegneria della Materia, dell'Ambiente ed Urbanistica, Università Politecnica delle Marche, 60131 Ancona, Italy
2   Engineering Faculty, Università Politecnica delle Marche, 60131 Ancona, Italy; jacopo.donnini@staff.univpm.it (J.D.); c.giosue@staff.univpm.it (C.G.); a.mobili@staff.univpm.it (A.M.); f.tittarelli@univpm.it (F.T.)
*   Correspondence: v.corinaldesi@staff.univpm.it; Tel.: +39-071-220-4428

Received: 22 August 2020; Accepted: 14 September 2020; Published: 16 September 2020

Abstract: The possibility of producing high-volume fly ash (HVFA) recycled aggregate concrete represents an important step towards the development of sustainable building materials. In fact, there is a growing need to reduce the use of non-renewable natural resources and, at the same time, to valorize industrial by-products, such as fly ash, that would otherwise be sent to the landfill. The present experimental work investigates the physical and mechanical properties of concrete by replacing natural aggregates and cement with recycled aggregates and fly ash, respectively. First, the mechanical properties of four different mixtures have been analyzed and compared. Then, the effectiveness of recycled aggregate and fly ash on reducing carbonation and chloride penetration depth has been also evaluated. Finally, the corrosion behavior of the different concrete mixtures, reinforced with either bare or galvanized steel plates, has been evaluated. The results obtained show that high-volume fly ash (HVFA) recycled aggregate concrete can be produced without significative reduction in mechanical properties. Furthermore, the addition of high-volume fly ash and the total replacement of natural aggregates with recycled ones did not modify the corrosion behavior of embedded bare and galvanized steel reinforcement.

Keywords: fly ash; HVFA; recycled aggregate; RAC; sustainable building; reinforced concrete; corrosion of concrete

---

1. Introduction

In order to contribute to sustainable construction processes, some building materials, no longer able to fulfill their original task, can be reused as aggregate for concrete after being adequately processed [1]. Replacing natural aggregate (Nat) with recycled aggregate (Rec) in concrete allows the protection of the environment, since it reduces both the impact of quarries from which virgin aggregates are extracted and the volume of rubble disposed to landfills.

Similarly, the employment of fly ash (FA) in concrete enables the recycling of an industrial waste product. In particular, due to its pozzolanic activity, FA can partially replace cement, thus reducing the energy consumption and carbon dioxide emissions related to cement production [2].

Unlikely, replacing Nat with Rec can significantly reduce the performances of concrete in terms of workability. Moreover, Rec, due to its higher porosity with respect to Nat [3,4], also penalizes the concrete's compressive and tensile strength, the stiffness, the permeability, and the adherence between steel reinforcing bars and cement paste.

However, the literature has reported that the addition of mineral admixtures as fly ash (FA), metakaolin, silica fume, and ground granulated blast furnace slag in the mix is able to mitigate these worsening effects both in traditional [5] and in self compacting concretes (SCC) [6]. In fact, generally, these additions seem able to improve more the properties of Rec concrete than those of Nat concrete [7].

Previous experiments [1,8] have already shown the feasibility of manufacturing structural concretes with Rec and high-volume fly ash (HVFA) since FA, by refining the pore structure, reduces the macro-pores volume. In this way, performance similar to Nat concrete can be achieved except for somewhat lower stiffness of the Rec mixture.

Water absorption [9], chloride ion penetration [10], sulphate attack [11], and shrinkage [12] increase with the increasing incorporation level of Rec. However, the addition of HVFA counteracts this effect [5] thanks to the chemical reaction between some particles of FA, that act as a pozzolanic addition instead of a filler, with Rec [12].

Wei et al. [13] have indicated that an adequate amount of Rec can even increase the frost resistance of concrete, especially when a low amount of FA is added, thanks to optimization of the concrete pore distribution [14].

Moreover, since the thermal expansion coefficient of the new cement paste is similar to that of the cement paste adhered to Rec, Rec concrete deteriorates less in terms of mechanical and durability properties than Nat after high temperature exposures [15], especially when FA is used as mineral admixture [16,17]. FA as bacteria immobilizer also improves the crack healing capacity of Rec concrete [18].

Concerning carbonation resistance, Rec and HVFA concrete suffer a deeper carbonation depth with respect to Nat concrete [19], also in SCC [20]. However, again, the incorporation of FA in Rec concrete allows the counteracting of this problem thanks to a synergistic effect between Rec and FA [9,21,22].

According to Limbachiya et al. [11], the best amount of coarse Rec in concrete is 30%, whereas up to 30% Rec does not significantly affect the concrete's properties. Regarding FA, European standards EN 197-1 and EN 206 limit the incorporation level of FA to 35% by cement mass, since at higher amounts, FA behave as a filler rather than as a binder. However, these two limits for Rec and FA can be exceeded in concrete mixes incorporating both FA and Rec [5]. After 90 days of curing, concretes manufactured with about 50% Rec and 50% FA can be generally classified at the same strength class of the control mix.

Rawaz Kurda et al. [22], thanks to a multicriteria decision method for concrete optimization (CONCRETOP), have shown that the best concrete mixes in terms of both concrete properties and cost and environmental impact, are those manufactured with both FA and Rec additions, rather than with only FA or Rec. In particular, the Global Warming Potential (GWP) of concrete mixes depends on the FA and Rec dosage ratio rather than the dosage of the single materials [23]. Moreover, the GWP of Rec strongly depends on the transportation scenario, but this effect significantly decreases with FA addition [24].

Therefore, it has been already widely proved that replacement of Nat with Rec and the replacement of cement with HVFA, given a little bit of compromise towards strength and durability aspect, can give great benefits to both economic and ecological aspects.

As reported above, many researchers have already studied the different properties of Rec concrete with HVFA. However, durability, which is a key property to ensure sustainable application of these materials in the construction sector, still needs more research to be fully investigated.

In this field, in particular, the literature still reports very few works on the protection offered by HVFA and Rec concrete to the corrosion of reinforcing bars. Stambaugh et al. [25], thanks to the theoretical development, validation, and implementation of a 1D numerical service-life prediction model for RCA, have affirmed that the use of either FA or slag allows the achievement of a 50-year service life for Rec concrete in chloride-laden environments. By the salt ponding test, Rehvati et al. [26] have stated that impermeable and high-quality Rec concrete, able to give high corrosion resistance to reinforcements, can be produced by replacing 20–30% cement with FA. In Gurdián et al. [27], no significant differences in the corrosion resistance of reinforced Rec concretes, manufactured with

15% of spent fluid catalytic cracking catalyst and 35% of FA, and Nat concretes under a natural chloride attack have been observed.

Moreover, in our knowledge, the corrosion behavior of galvanized steel reinforcements in reinforced RCA in HVFA has never been investigated.

Therefore, the purpose of this work is to determine whether the sustainability issue introduced in concrete design by Rec and HVFA would have any adverse effect on the durability of reinforced concrete in terms of penetration speed of chloride and carbon dioxide, and in terms of corrosion of bare or galvanized steel reinforcement embedded in concrete, if cracked.

To investigate the single and combined effect of Rec and HVFA addition on concrete properties, four different concrete mixes were prepared and compared:

1. Nat, as reference;
2. Nat + FA with HVFA added at equal amounts as cement;
3. Rec, by replacing the 100% of Nat with Rec;
4. Rec + FA with HVFA.

The different mixtures were compared in terms of mechanical performances, carbonation and chlorides penetration, and corrosion behavior of embedded bare and galvanized steel reinforcements.

2. Materials and Methods

2.1. Materials

As cementitious binder, Portland limestone blended cement type CEM II/A-L 42.5 R was used. The cement's Blaine fineness was 0.418 m^2/g and its density was 3.04 kg/m^3. The cement's chemical composition is reported in Table 1.

Table 1. Chemical composition of cement and fly ash.

Oxide (%)	CEM II/A-L 42.5 R	Fly Ash
SiO_2	29.7	59.9
Al_2O_3	3.7	22.9
Fe_2O_3	1.8	4.7
TiO_2	0.1	0.9
CaO	59.3	3.1
MgO	1.1	1.6
SO_3	3.2	0.3
K_2O	0.8	2.2
Na_2O	0.3	0.6
Loss on Ignition (L.o.I.)	11.6	3.3

Two virgin aggregate fractions were used: limestone aggregate from a quarry (up to 15 mm particle size) and quartz sand (up to 6 mm particle size). Their grain size distribution is shown in Figure 1 and their physical properties are reported in Table 2.

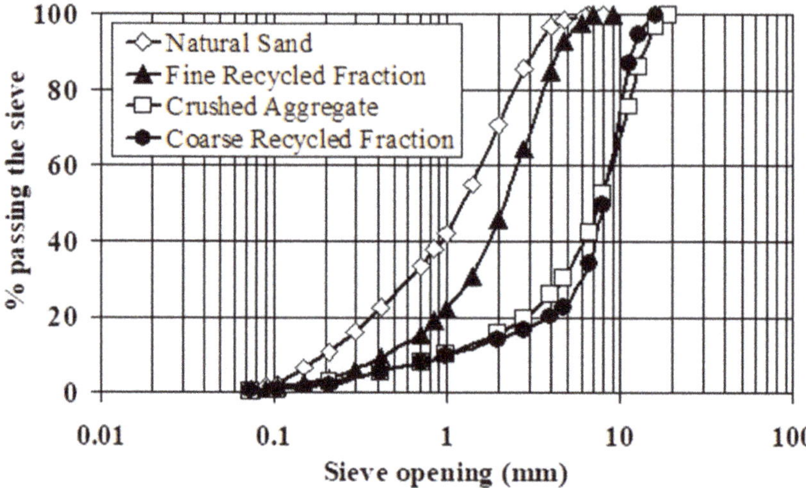

Figure 1. Grain size distribution curves of the aggregate fractions.

Table 2. Physical properties of the aggregate fractions.

Aggregate Fractions	Bulk Density (SSD), kg/m^3	Water Absorption, %	Passing 75 µm, %
Natural Sand	2620	3	0.9
Crushed Aggregate	2680	2	0.2
Fine Recycled Fraction	2150	10	0.5
Coarse Recycled Fraction	2320	8	0.3

Two recycled aggregate fractions were used: coarse aggregate (up to 15 mm) and fine aggregate (up to 6 mm). The origin of these recycled aggregates was a recycling plant in Villa Musone (Italy), where debris coming from building demolition has been selected, crushed, cleaned, and finally, sieved. The grain size distribution of recycled aggregates is shown in Figure 1 and their physical properties are reported in Table 2.

As water-reducing admixture, an acrylic-based superplasticizer (in the form of 30% aqueous solution) was added, when required, to reach the optimal workability degree.

Low calcium fly ash (according to the definition of ASTM C 618 Class F) coming from a thermal power plant located in La Spezia (Italy), with Blaine fineness of 0.458 m^2/g and density of 2.23 kg/m^3, was used. Fly ash chemical composition is detailed in Table 1.

2.2. Mixture Proportions

Four different concrete mixtures were designed as reported in Table 3. The two different kinds of aggregate particles, either recycled or natural, were used by maintaining the same grain size distribution (up to 15 mm). The optimization of grain size particle distribution was achieved by suitable combining of fine and coarse aggregate fractions, according to Bolomey [12]. All mixtures were designed so to have the same workability, with a slump value in the range 150–180 mm. According to that, when recycled aggregates and fly ash have been used, an acrylic-based superplasticizer was added at dosages up to 2.0% by weight of cement.

Table 3. Concrete mixture proportions (kg/m^3).

	Nat-0.6	**Nat + FA-0.6**	**Rec-0.3**	**Rec + FA-0.6**
Water/Cement	0.6	0.6	0.3	0.6
Water/Binder	0.6	0.3	0.3	0.3
Water	250	250	180	180
Cement	420	420	600	300
Fly ash	-	420	-	300
Superplasticizer	-	8.4	6.0	5.4
Natural sand	290	-	-	-
Fine recycled aggregate	-	-	125	125
Crushed aggregate	1280	1280	-	-
Coarse recycled aggregate	-	-	1030	1030

The first reference mixture (Nat-0.6) was designed with only the addition of virgin aggregates and a water-to-cement ratio (*w/c*) of 0.60.

In the second mixture (Nat + FA-0.6), fly ash (FA) at the same dosage of cement was added, in replacement of natural sand. However, to reach the same workability of the reference mixture, acrylic-based superplasticizer as water-reducing admixture was added at a dosage of 2.0% by weight of cement.

In the third concrete mixture (Rec-0.3), virgin aggregates were fully substituted with recycled ones, while a lower *w/c* equal to 0.3 was adopted, to recover the strength loss due to the addition of weaker aggregate.

Finally, in the fourth mixture (Rec + FA-0.6), fine and coarse recycled aggregates were used in complete substitution of natural aggregates, while cement was partially replaced with fly ash. The use of superplasticizer at dosage of 1.8% by weight of cement allowed us to keep unchanged the water-to-cement ratio (equal to 0.6).

3. Preparation and Testing of Specimens

3.1. Compression Tests

Compression tests were performed on cubic specimens (100 mm edge) after 3, 7, 28, and 56 days of wet curing at 20 °C (according to the procedure of UNI EN 12390-1 [28]). Compression tests were carried out according to the procedure described in the Italian Standards UNI EN 12390-3 [29]. Three specimens for each curing time and each type of mixture were tested.

3.2. Carbonation Depth

Carbonation depth was evaluated through a phenolphthalein test (following the indication reported in RILEM CPC-18 [30]) on three cubic concrete specimens (100 mm edge) for each mixture, exposed to the open air at an average temperature of 20 °C (only for the first day of curing, wet curing was adopted).

3.3. Chloride Penetration

Chloride penetration speed into concrete was evaluated by means of silver nitrate and fluorescein test [13]. Both solutions were sprayed on the two cracked surfaces obtained by splitting the cubic (100 mm edge) concrete specimens. These specimens were previously wet-cured for 7 days, air-cured for 21 days at a temperature of 20 °C, and finally, exposed to 10% sodium chloride aqueous solution.

3.4. Corrosion Tests

Concrete specimens with dimensions of $280 \times 70 \times 70$ mm^3 were manufactured for electrochemical tests. Each specimen was reinforced with a bare or galvanized steel plate ($210 \times 40 \times 1$ mm^3), embedded within a 3 cm concrete cover. Steel plates were used instead of common bars because they can allow specimen cracking without splitting and they offer a higher anodic area at the crack apex. The galvanized steel plates, obtained by molten zinc immersion, were covered by a 100 μm thick zinc layer, with an outer pure zinc layer (η phase) about 20 μm thick. The galvanized reinforcements, just before being embedded in the fresh concrete, were submerged for 5 s in a 15% sodium hydroxide solution to dissolve the $ZnCO_3$ layer formed during air exposure. The electric contacts between the metallic plates and the measurement equipment were arranged as reported in [14].

After 1 month of air curing at T = 20 ± 3 °C and RH = 50 ± 5%, a crack width of 1 mm was produced in a pre-formed notch area of the specimens by applying a flexural stress with the apex crack reaching the metallic plates. Then, the specimens were exposed to weekly wet–dry cycles (2 days dry and 5 days wet) in a 10% NaCl solution.

The corrosion risk of the reinforcement in the concrete specimens exposed to the chloride environment was evaluated by corrosion potential measurements by using a saturated calomel electrode (SCE) as a reference. The kinetics of the corrosion process was followed by polarization resistance measurements through the galvanodynamic method, where an external graphite bar was used as a counter-electrode. The polarization resistance was calculated as the average value between the anode and the cathode branch.

In the following graphs, the reported electrochemical values are averages of the measurements carried out on 3 specimens of each type during the full immersion period.

In order to validate the electrochemical tests, after 7 wet–dry cycles in the chloride solution, the concrete specimens were saw-cut and all the metallic plates were removed after splitting the concrete specimens, to evaluate the reinforcement corrosion by visual observation. The surface of the corroded area on bare steel plates was evaluated after pickling, whereas metallographic analysis was carried out on the cross-section of the galvanized steel plates to evaluate the coating thickness decrease due to the corrosive attack.

4. Results and Discussion

4.1. Compression Tests

The experimental results of compressive tests on the four mixtures at different curing times are reported in Figure 2. All mixtures showed a compressive strength greater than 27 MPa at 28 days. At early ages (3 and 7 days), the different w/c influenced the compressive strength more than the kind of aggregate. The total substitution of natural aggregates with recycled ones, simultaneously with the reduction in w/c and the use of superplasticizer (Rec-0.3), did not substantially modify the compressive strength with respect to the concrete mixture with natural aggregates (Nat-0.6). When cement was partially replaced with FA (Rec + FA-0.6), the mixture showed a lower compressive strength at early ages (3 and 7 days) and a slightly higher compressive strength after 56 days of curing, thanks to the fly ash pozzolanic activity which develops at long ages. The use of lighter aggregate (i.e., recycled aggregate) increasingly influenced the compressive strength value as the cement matrix became stronger, since it represents the "weak link" in the chain [15]. The best results have been obtained for the mixture realized with natural aggregates and fly ash (Nat + FA-0.6), which showed a compressive strength of about 44 MPa after 56 days of curing.

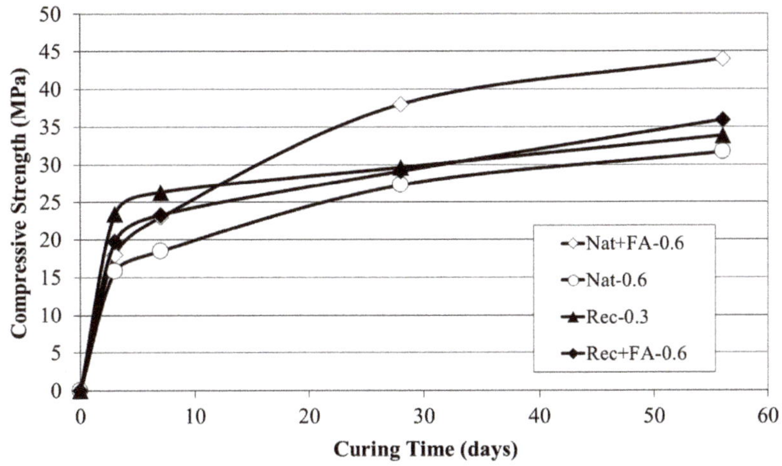

Figure 2. Compressive strength vs. curing time.

4.2. Carbonation Depth

In Figure 3, the measured values of carbonation depth (*x*, in mm) were reported vs. days of air exposure (*t*). Experimental data can be quite accurately described by a linear relationship between the carbonation depth and square root of time (following the law $x = k \cdot \sqrt{t}$). The higher carbonation depth, equal to about 7 mm after 1 year of exposure, was found for the mixture with natural aggregates (Nat-0.6), while the lower carbonation depth, equal to 2.8 mm, was found in the concrete mixture realized with natural aggregates and fly ash. These results confirm that the addition of high volumes of fly ash is able to significantly reduce the carbonation process, even when a porous aggregate (such as the recycled aggregate) is used, thanks to the refinement of the pores and a consequent improved microstructure.

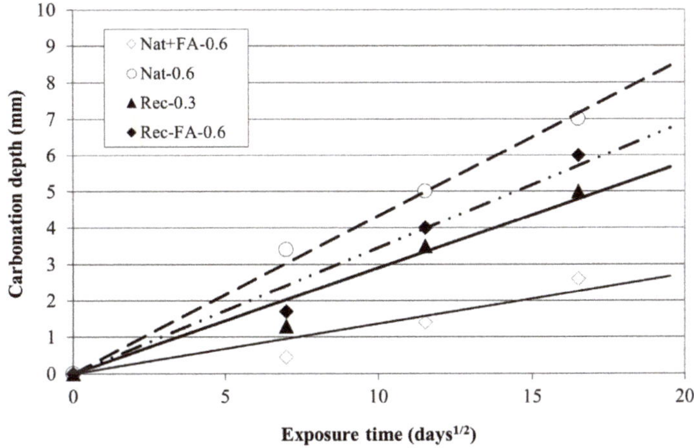

Figure 3. Carbonation depth vs. time of air exposure.

4.3. Chloride Penetration

In Figure 4, chloride penetration depth values are reported vs. time of exposure to 10% NaCl aqueous solution after saturation with water of concrete specimens. Figure 4 shows only data obtained after about 28-day water immersion due to initial instability phenomena (in practice, the chloride binding with cement matrix can interfere with the chloride migration [16]).

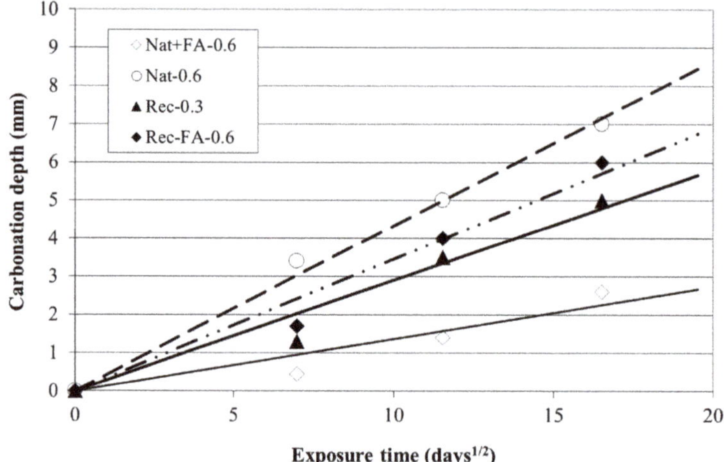

Figure 4. Chloride penetration depth vs. immersion time.

Collepardi et al. [13] demonstrated that chloride penetration depth (x) varies with immersion time (t) according to Equation (1), which can be obtained from the solution of Fick's second law under non-steady-state conditions for diffusion in a semi-finite solid:

$$x = 4 \cdot \sqrt{D \cdot t} \tag{1}$$

where D is the diffusion coefficient of Cl ions into wetted concrete pores, expressed as $cm^2 \cdot s^{-1} \cdot 10^{-8}$. The D values, coming from (1) by interpolation of the results reported in Figure 4, have been calculated (see Table 4). It can be noticed a significant beneficial effect due to fly ash addition on the chloride penetration depth, which has been measured for Rec + FA-0.6 and Nat + FA-0.6 concretes, respectively. As a matter of fact, Cl diffusion coefficients into these mixtures are 10 times less than those measured for the other concrete mixtures.

Table 4. Chloride diffusion coefficients at 20 °C.

	Nat-0.6	Nat + FA-0.6	Rec-0.3	Rec + FA-0.6
D ($cm^2 \cdot s^{-1} \cdot 10^{-6}$)	1.90	0.87	0.72	0.46

Collepardi et al. [13] found experimental results, which support the hypothesis that the different structure of pore surfaces created with the addition of pozzolanic materials played an important role in influencing concrete porosity while chloride ions penetrate it. In fact, if concrete is prepared by adding fly ash, chloride binding operated by the cement matrix significantly increases.

A lower water/cement resulted beneficial in terms of hindering of chloride penetration. In fact, the Cl diffusion coefficient into the Nat-0.6 mixture is double than Rec-0.3, even if recycled instead of virgin aggregate was employed. Nevertheless, aggregate pore structure could have a significant effect on concrete permeability to chloride penetration (as you can see by comparing Nat + FA-0.6 and Rec + FA-0.6 mixtures in Figure 4). Zhang et al. [17] experimented several lightweight aggregate concretes and the conclusion was that concrete permeability depends more on cement matrix porosity than lightweight aggregate porosity.

4.4. Corrosion Tests

4.4.1. Bare Steel Plates

Figures 5 and 6 show, respectively, the free corrosion potential values and the corrosion rates of bare steel plates embedded in cracked concrete as a function of wet–dry cycles. Just after the exposure to the chloride environment, all the steel plates assumed activation values lower than −500 mV/SCE, reflecting a general high corrosion risk, regardless of the type of concrete mixture. At the same time, the related polarization resistance values did not change significantly with the addition of FA or when natural aggregates were replaced with recycled ones, thus indicating similar corrosion rates for the different concretes. Therefore, the total replacement of natural aggregates, with or without high FA volume, does not seem to negatively affect the corrosion behavior of embedded steel reinforcements when an adequate strength class is guaranteed. Moreover, despite the reduction in concrete pore solution alkalinity due to the pozzolanic reaction of fly ash, the corrosion behavior of steel reinforcements in high-volume fly ash concrete seems not to be negatively affected, at least for cracked concrete.

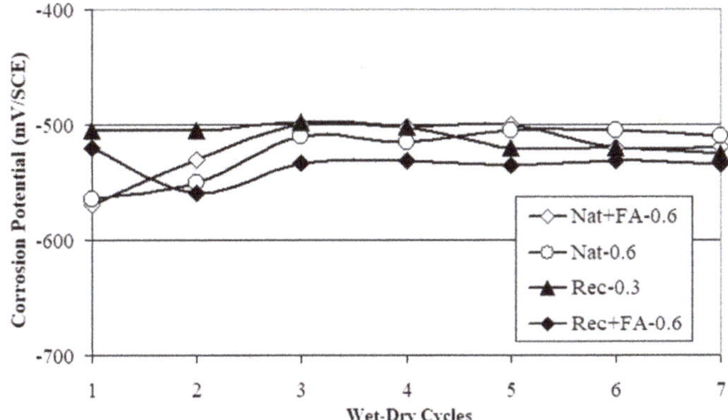

Figure 5. Corrosion potential of bare steel plates in cracked concrete as a function of wet–dry cycles.

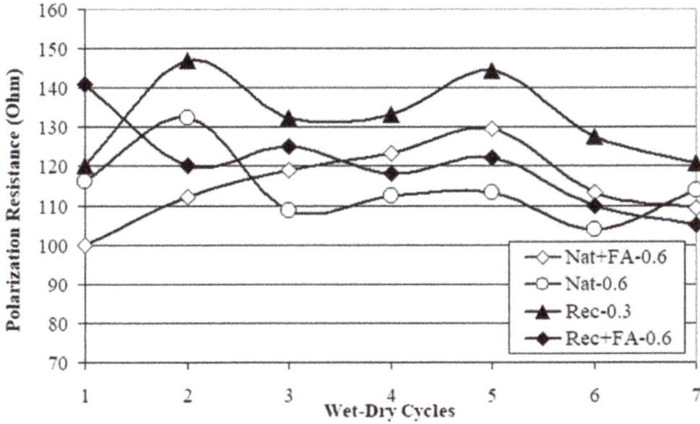

Figure 6. Polarization resistance of bare steel plates in cracked concrete as a function of wet–dry cycles.

To better analyze the electrochemical behavior, all specimens were autopsied after 7 wet–dry cycles in order to visually evaluate the corroded area and to assess the weight loss of the bare steel plates after pickling. A visual observation of the corrosive attack on the steel plates embedded in the different concrete mixtures is reported in Figure 7. All the bare steel plates showed significant corrosive attack in the crack area, due to the high Cl concentration detected on the steel reinforcement surface (2–5% by weight of cement), thus overcoming the concentration threshold (0.4% by cement weight), which is considered the critical value able to induce the corrosion of bare steel. However, from a morphological point of view, in the presence of recycled aggregate (Rec-0.3) or when a high amount of fly ash is used (Nat + FA-0.6), the corrosive attack appears more diffuse and less penetrating (see Figure 7).

Figure 7. Visual observation of the corrosive attack on the bare steel plates in reference natural aggregate concrete (Nat-0.6, on the left), in recycled aggregate concrete (Rec-0.3, in the middle), and high-volume fly ash concrete (Nat + FA-0.6, on the right).

4.4.2. Galvanized Steel Plates

The free corrosion potential and the polarization resistance of galvanized steel plates embedded in cracked concrete, as a function of wet–dry cycles, are reported in Figures 8 and 9, respectively. The electrochemical tests showed no significant increase in the corrosion rate of galvanized steel embedded in concrete mixtures containing recycled aggregate and/or FA. Indeed, the corrosion risk seems to be even lower in recycled aggregate concrete (see Rec-0.3 in Figure 8) than in the other mixtures.

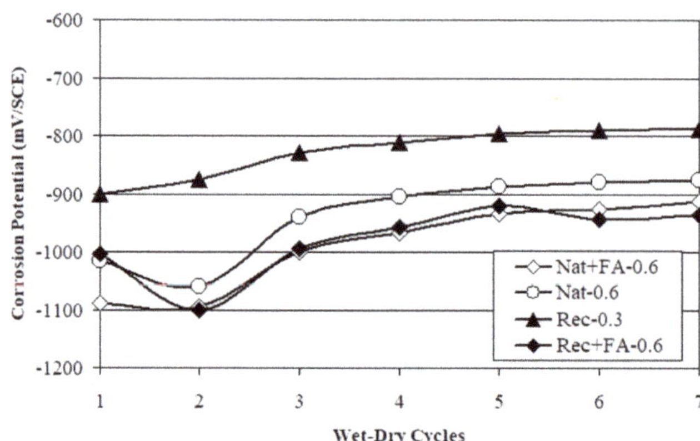

Figure 8. Corrosion potential of galvanized steel plates in cracked concrete vs. wet–dry cycles.

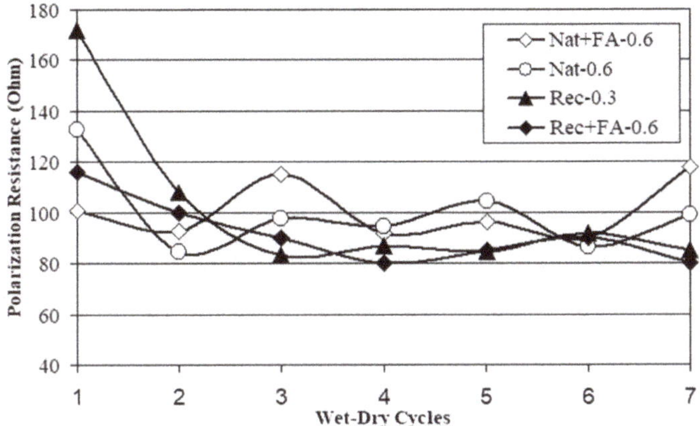

Figure 9. Polarization resistance of galvanized steel plates in cracked concrete vs. wet–dry cycles.

The visual observation of the interface between galvanized steel plates and concrete added unexpected information to that obtained by the electrochemical tests. Indeed, in the concrete without additions (Nat-0.6, Figure 10a), far from the crack apex, a Fe-Zn alloy appeared on the surface of the reinforcement, meaning total consumption of the η zinc layer due to the corrosive attack, as later confirmed by metallographic observation (Figure 10b). On the other hand, the galvanized steel plates extracted from recycled aggregate concrete (Rec-0.3) showed a less deep corrosive attack. Zinc grains were still visible on the galvanized surface after exposure to wet–dry cycles (Figure 11a). Moreover, the metallographic observation revealed that a continuous thick η zinc layer was still present on the metallic plate (Figure 11b). Similar results were observed when high-volume fly ash was added in the mix.

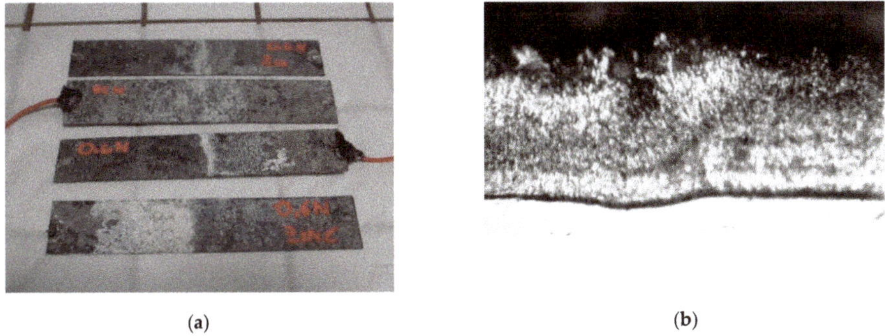

(a)   (b)

Figure 10. Visual obs. (**a**) and metallographic cross section (**b**) of galvanized steel plate in Nat-0.6

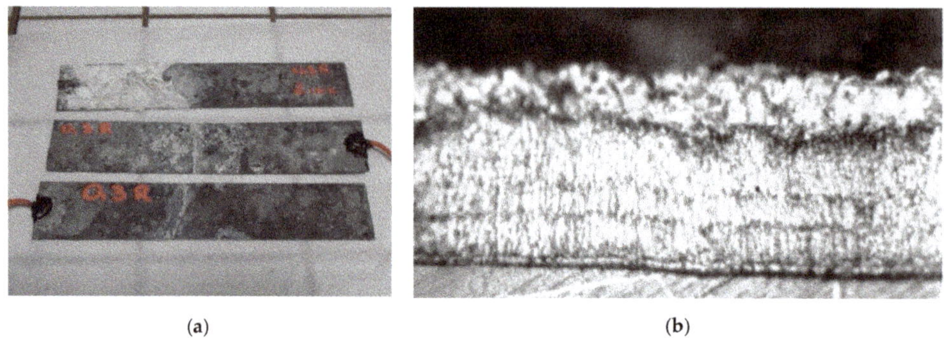

(a)   (b)

Figure 11. Visual obs. (**a**) and metallographic cross section (**b**) of a galvanized steel plate in Rec-0.3

5. Conclusions

Based on the experimental investigation carried out on recycled aggregate HVFA concrete to study its mechanical properties and durability characteristics, the following conclusions can be drawn:

- The use of recycled aggregate fractions (up to 15 mm) proved to be suitable for manufacturing structural concrete, even when employed in total replacement of fine and coarse natural aggregates.
- If natural aggregates are substituted by recycled ones, fly ash is added to the mixture in partial substitution of cement, and the w/c is decreased with the aid of a superplasticizer, the compressive strength is equal or even greater than that of natural aggregate concrete.
- The use of fly ash proved to be very effective in reducing carbonation and Cl penetration depths in concrete, even if recycled instead of virgin aggregate was used.
- When the concrete cover was cracked, the addition of fly ash and the use of recycled aggregates (Rec + FA-0.6) did not reduced the corrosion resistance of steel reinforcement, while it appeared very effective to protect the galvanized steel reinforcement.
- An improved attention to the sustainability issue in concrete manufacturing, promoted by using recycled aggregate and high-volume fly ash, did not cause bad side effects on durability performance of reinforced concrete specimens under testing.

Author Contributions: Investigation, V.C., J.D., C.G., A.M. and F.T.; Methodology, J.D. and F.T.; Supervision, V.C.; Writing—review and editing, J.D. and F.T. All authors have read and agreed to the published version of the manuscript.

Funding: This research received no external funding.

Conflicts of Interest: The authors declare no conflict of interest.

References

1. Corinaldesi, V.; Moriconi, G. Influence of mineral additions on the performance of 100% recycled aggregate concrete. *Constr. Build. Mater.* **2009**, *23*, 2869–2876. [CrossRef]
2. Siddique, R. Performance characteristics of high-volume class F fly ash concrete. *Cem. Concr. Res.* **2004**, *34*, 487–493. [CrossRef]
3. Lima, C.; Caggiano, A.; Faella, C.; Martinelli, E.; Pepe, M.; Realfonzo, R. Physical properties and mechanical behaviour of concrete made with recycled aggregates and fly ash. *Constr. Build. Mater.* **2013**, *47*, 547–559. [CrossRef]
4. Mobili, A.; Giosue', C.; Corinaldesi, V.; Tittarelli, F. Bricks and Concrete Wastes as Coarse and Fine Aggregates in Sustainable Mortars. *Adv. Mater. Sci. Eng.* **2018**, *2018*, 1–11. [CrossRef]
5. Faella, C.; Lima, C.; Martinelli, E.; Pepe, M.; Realfonzo, R. Mechanical and durability performance of sustainable structural concretes: An experimental study. *Cem. Concr. Compos.* **2016**, *71*, 85–96. [CrossRef]

6. Singh, N.; Kumar, P.; Goyal, P. Reviewing the behaviour of high volume fly ash based self compacting concrete. *J. Build. Eng.* **2019**, *26*, 100882. [CrossRef]

7. Kou, S.C.; Poon, C.S.; Agrela, F. Comparisons of natural and recycled aggregate concretes prepared with the addition of different mineral admixtures. *Cem. Concr. Compos.* **2011**, *33*, 788–795. [CrossRef]

8. Palaniraj Saravanakumar, P.; Dhinakaran, G. Durability aspects of HVFA-based recycled aggregate concrete. *Mag. Concr. Res.* **2014**, *66*, 186–195. [CrossRef]

9. Kurda, R.; de Brito, J.; Silvestre, J.D. Water absorption and electrical resistivity of concrete with recycled concrete aggregates and fly ash. *Cem. Concr. Compos.* **2019**, *95*, 169–182. [CrossRef]

10. Kurda, R.; Silvestre, J.D.; de Brito, J.; Ahmed, H. Optimizing recycled concrete containing high volume of fly ash in terms of the embodied energy and chloride ion resistance. *J. Clean. Prod.* **2018**, *194*, 735–750. [CrossRef]

11. Limbachiya, M.; Meddah, M.S.; Ouchagour, Y. Use of recycled concrete aggregate in fly-ash concrete. *Constr. Build. Mater.* **2012**, *27*, 439–449. [CrossRef]

12. Kurda, R.; de Brito, J.; Silvestre, J.D. Combined influence of recycled concrete aggregates and high contents of fly ash on concrete properties. *Constr. Build. Mater.* **2017**, *157*, 554–572. [CrossRef]

13. Wei, Y.; Chai, J.; Qin, Y.; Xu, Z.; Zhang, X. Performance evaluation of green-concrete pavement material containing selected C&D waste and FA in cold regions. *J. Mater. Cycles Waste Manag.* **2019**, *21*, 1550–1562. [CrossRef]

14. Li, Y.; Wang, R.; Li, S.; Zhao, Y.; Qin, Y. Resistance of recycled aggregate concrete containing low- and high-volume fly ash against the combined action of freeze–thaw cycles and sulfate attack. *Constr. Build. Mater.* **2018**, *166*, 23–34. [CrossRef]

15. Kou, S.C.; Poon, C.S.; Etxeberria, M. Residue strength, water absorption and pore size distributions of recycled aggregate concrete after exposure to elevated temperatures. *Cem. Concr. Compos.* **2014**, *53*, 73–82. [CrossRef]

16. Bui, N.K.; Satomi, T.; Takahashi, H. Effect of mineral admixtures on properties of recycled aggregate concrete at high temperature. *Constr. Build. Mater.* **2018**, *184*, 361–373. [CrossRef]

17. Ren, Q.; Wu, Y.; Zhang, X.; Wang, Y. Effects of fly ash on the mechanical and impact properties of recycled aggregate concrete after exposure to high temperature. *Eur. J. Environ. Civ. Eng.* **2019**, 1–17. [CrossRef]

18. Khushnood, R.A.; Qureshi, Z.A.; Shaheen, N.; Ali, S. Bio-mineralized self-healing recycled aggregate concrete for sustainable infrastructure. *Sci. Total Environ.* **2020**, *703*, 135007. [CrossRef]

19. Carević, V.; Ignjatović, I.; Dragaš, J. Model for practical carbonation depth prediction for high volume fly ash concrete and recycled aggregate concrete. *Constr. Build. Mater.* **2019**, *213*, 194–208. [CrossRef]

20. Singh, N.; Singh, S.P. Carbonation resistance and microstructural analysis of Low and High Volume Fly Ash Self Compacting Concrete containing Recycled Concrete Aggregates. *Constr. Build. Mater.* **2016**, *127*, 828–842. [CrossRef]

21. Kurda, R.; De Brito, J.; Silvestre, J.D. Carbonation of concrete made with high amount of fly ash and recycled concrete aggregates for utilization of CO2. *J. Co2 Util.* **2019**, *29*, 12–19. [CrossRef]

22. Kurda, R.; de Brito, J.; Silvestre, J.D. CONCRETop method: Optimization of concrete with various incorporation ratios of fly ash and recycled aggregates in terms of quality performance and life-cycle cost and environmental impacts. *J. Clean. Prod.* **2019**, *226*, 642–657. [CrossRef]

23. Kurad, R.; Silvestre, J.D.; de Brito, J.; Ahmed, H. Effect of incorporation of high volume of recycled concrete aggregates and fly ash on the strength and global warming potential of concrete. *J. Clean. Prod.* **2017**, *166*, 485–502. [CrossRef]

24. Kurda, R.; de Brito, J.; Silvestre, J.D. A comparative study of the mechanical and life cycle assessment of high-content fly ash and recycled aggregates concrete. *J. Build. Eng.* **2020**, *29*, 101173. [CrossRef]

25. Stambaugh, N.D.; Bergman, T.L.; Srubar, W.V. Numerical service-life modeling of chloride-induced corrosion in recycled-aggregate concrete. *Constr. Build. Mater.* **2018**, *161*, 236–245. [CrossRef]

26. Revathi, P.; Nikesh, P. Effect of Fly-Ash on Corrosion Resistance Characteristics of Rebar Embedded in Recycled Aggregate Concrete. *J. Inst. Eng. India Ser. A* **2018**, *99*, 473–483. [CrossRef]

27. Gurdián, H.; García-Alcocel, E.; Baeza-Brotons, F.; Garcés, P.; Zornoza, E. Corrosion Behavior of Steel Reinforcement in Concrete with Recycled Aggregates, Fly Ash and Spent Cracking Catalyst. *Materials* **2014**, *7*, 3176–3197. [CrossRef] [PubMed]

28. UNI EN 12390-1:2012. *Prova Sul Calcestruzzo Indurito—Parte 1: Forma, Dimensioni ed Altri Requisiti per Provini e per Casseforme*; Ente Nazionale Italiano di Unificazione: Milan, Italy, 2012.
29. UNI EN 12390-3:2019. *Prove sul Calcestruzzo Indurito—Parte 3: Resistenza alla Compressione dei Provini*; Ente Nazionale Italiano di Unificazione: Milan, Italy, 2019.
30. CPC-18. Measurement of hardened concrete carbonation depth. *Mater. Struct.* **1988**, *21*, 453–455. [CrossRef]

Article

Effect of Fly Ash as Cement Replacement on Chloride Diffusion, Chloride Binding Capacity, and Micro-Properties of Concrete in a Water Soaking Environment

Jun Liu, Jiaying Liu, Zhenyu Huang *, Jihua Zhu, Wei Liu and Wei Zhang

Guangdong Provincial Key Laboratory of Durability for Marine Civil Engineering, Shenzhen University, Shenzhen 518060, China; liujun@szu.edu.cn (J.L.); 1910472060@email.szu.edu.cn (J.L.); zhujh@szu.edu.cn (J.Z.); liuwei@szu.edu.cn (W.L.); zhangwdnv@gmail.com (W.Z.)

* Correspondence: huangzhenyu@szu.edu.cn; Tel.: +86-755-8697-5402; Fax: +86-755-2673-2850

Received: 16 August 2020; Accepted: 7 September 2020; Published: 9 September 2020

Featured Application: The study on the diffusion, bonding and micro-properties of chloride penetration in concrete in a water soaking environment can provide fundamental information to evaluate the durability of the marine concrete.

Abstract: This paper experimentally studies the effects of fly ash on the diffusion, bonding, and micro-properties of chloride penetration in concrete in a water soaking environment based on the natural diffusion law. Different fly ash replacement ratio of cement in normal concrete was investigated. The effect of fly ash on chloride transportation, diffusion, coefficient, free chloride content, and binding chloride content were quantified, and the concrete porosity and microstructure were also reported through mercury intrusion perimetry and scanning electron microscopy, respectively. It was concluded from the test results that fly ash particles and hydration products (filling and pozzolanic effects) led to the densification of microstructures in concrete. The addition of fly ash greatly reduced the deposition of chloride ions. The chloride ion diffusion coefficient considerably decreased with increasing fly ash replacement, and fly ash benefits the binding of chloride in concrete. Additionally, a new equation is proposed to predict chloride binding capacity based on the test results.

Keywords: fly ash; carbon dioxide emission; chloride diffusion; binding capacity of chlorine

---

1. Introduction

Since Portland cement was invented in 1824, cement has become the most important and irreplaceable building material in infrastructure construction. However, problems associated with cement are becoming increasingly prominent. For example, cement production consumes a great amount of energy and released a great deal of toxic pollutants, such as dust, soot, sulfur dioxide, and carbon dioxide, into the environment [1,2]. The production of Portland cement causes about 7% of the world's carbon dioxide emissions, which is essential to be reduced. Using supplement materials in concrete (e.g., fly ash, and slag) is effective in reducing the amount of cement consumption without sacrificing any properties [3]. Fly ash is the main coal combustion by-product from power plants [4–6]. It has become the fifth largest raw material resource in the world and has been used for more than 50 years in concrete [7,8]. Replacing part of the cement admixture in concrete can save cement clinker, thus reducing the environmental pollution from cement production [9,10]. It has been reported that, per ton, replacing Portland cement by fly ash can reduce one ton of carbon dioxide emissions [11]. In addition, fly ash can obviously enhance the workability of fresh concrete, reduce hydration heat in the early stages, improve corrosion resistance [9,10,12–14], and improve concrete's durability performance [15–19].

It is generally believed that chloride ion penetration in concrete is one of the most critical causes of steel reinforcement corrosion that may further cause structural failure [20–22]. Chloride ions are present in many environments, including marine, road de-icing salt, salinized soil, and industrial wastewater environments. Chloride ions penetrate through pores into concrete and accumulate around steel reinforcements, which destroys the passive film on the steel surface and activates steel reinforcement corrosion [23–25]. For existing reinforced concrete structures, the penetration process of chloride ions is an important parameter in durability life assessment [26]. Liu et al. [24] examined the microstructures of fly ash concrete and normal concrete in a chloride atmosphere environment; the fly ash content was discovered to be sensitive to the chloride binding capacity and diffusion coefficient. Wu et al. [27] found that chloride transportation was affected by drying–wetting cycles and compressive stress ratios—the former significantly increased the chloride ion content, and the latter decreased the chloride ion content. Mehta and Monteriropj [28] pointed out that a pore size of <100 nm has little or no effect on concrete strength and permeability, while a pore size of >100 nm negatively affects concrete strength and permeability. The permeability of fly ash concrete seems physically different from that of normal concrete. Unlike that in salt-spray environments, concrete in seawater environments has different chloride ion concentrations, in which migration speed of chloride ion varies considerably. The chloride ion distributions in concrete are quite different, which is related to the way that chloride ions penetrate concrete. In fact, chloride ion penetration is a complex process involving extensive transportation mechanisms, such as diffusion under a concentration difference [27], penetration under water pressure [29], and capillary action under a humidity gradient [30]. In a soaking environment, when the concrete is saturated, chloride ions are mainly transferred through diffusion to the concrete. Meanwhile, in a salt-spray environment, the migration of chloride ions is achieved mainly by capillary suction and the diffusion effect. Therefore, the chloride profile of concrete can be quite different in different environments, which may cause different corrosion mechanisms of reinforcement in concrete. Thus, the knowledge of such durability behavior of fly ash concrete remains unclear, even though it is essential for promoting durable material design in construction under water. Due to filling and pozzolanic effects [31–33], adopting fly ash to partly replace cement can increase chloride ion resistance and lead to the better durability of concrete [34–36]. Moreover, there are very few references on the distribution of chloride ions in concrete soaked with seawater, thus indicating the demand for a comprehensive study that covers not only the micro-properties and transportation mechanism but also the chloride binding capacity of fly ash concrete.

The current investigation aimed to systematically examine the effect of fly ash on chloride ion diffusion in concrete in a water soaking environment. Specifically, the effect was quantified by the chloride ion diffusion coefficient and chloride ion deposit amounts at different layers. The study also compared the amount of free and binding chloride ions, as well as binding resistances, under different dosages of fly ash replacement. Through a mercury intrusion method and SEM, the study explored the effects of fly ash on the pore structure and cement hydration products.

2. Experimental Investigation

2.1. Materials

ASTM (American Society for Testing and Materials) type I Portland cement, Class F fly ash, river sand (fineness: 2.3–3.0), and a coarse aggregate (size: 5–20 mm in diameter) were used for concrete preparation. Industrial chloride salt (purity: 99%) was adopted to simulate the NaCl solution. Table 1 lists the proportions of three mixtures of normal concrete with fly ash replacements of 0%, 15%, and 30% (by weight). Table 2 lists the chemical compositions of the cement and fly ash.

Table 1. Mix proportions of concrete. kg/m³.

NO	OPC	Sand	CA	FA	W	W/C
PCFA0	409.0	720.0	1079.0	0.0	192	0.47
PCFA15	347.7	697.3	1054.0	61.3	192	0.47
PCFA30	286.3	689.0	1041.0	122.7	192	0.47

OPC = ordinary Portland cement; CA = coarse aggregate; FA = fly ash; and W/C = water to cement ratio; PCFA0 represents plain concrete with 0% fly ash.

Table 2. Chemical compositions (mass percentages of oxides, %).

Material	CaO	SiO$_2$	Al$_2$O$_3$	Fe$_2$O$_3$	MgO	SO$_3$	K$_2$O
Cement	64.67	18.59	4.62	4.17	2.35	3.32	0.92
Fly ash	4.74	62.32	23.95	1.33	2.04	1.25	0.76

2.2. Immersion Test Methodology

To simulate the concrete deterioration process in a seawater soaking environment, immersion tests of concrete samples were conducted based on NT (NORDTEST) Build 443 [37]. Three standard concrete cylinders were cast for each concrete mix and demolded 48 hours later. After standard curing under the environment with a constant temperature of 20 ± 3 °C and a relative humidity of >95% for 90 days, these specimens were first saturated with a Ca(OH)$_2$ solution, ensuring that the weight change of all samples was less than 0.1%. After that, the samples were exposed to a saturated-surface-dry state environment under ambient temperature before the top surface was selected as the exposed surface, while the remaining surfaces were sealed with an epoxy material to guarantee the unidirectional diffusion of the chloride ions during the test. Then, the samples were soaked in an NaCl solution with a 16.5% concentration at a temperature of 23 ± 2 °C for 35 days. Figure 1a shows the concrete samples immersed in the NaCl solution.

2.3. Measurement of Chloride Content

After the immersion process, the crystal salt on the concrete surface was removed. Then, the samples were grinded by a specially-designed pulveriser into powder from the exposed surface, layer by layer. From the height of 0–16 mm, the grinding thickness was 1 mm, while from the height of 16–40 mm, the grinding thickness was 2 mm, as shown in Figure 1b. The powder samples were then placed in an oven at up to 105 °C to dry to a constant weight. The determination of the free and total chloride ion contents was based on AASHTO T260 (2009) [38], while the concentration of chloride ions was determined by the automatic potentiometric titrator as shown in Figure 1c.

2.4. Microstructural Characterization

The study adopted mercury intrusion porosimetry to evaluate the porosity and pore size distribution characteristics of concrete [39,40]. For each mix, three sample tests were conducted, and the average results were used for analysis. The concrete samples were crushed into small sizes and dried by a solvent replacement method, changing the solvent every 6 hours for the first few days and then every day for one week. After that, the samples were vacuum-dried and subjected to mercury intrusion up to 210 MPa. The contact angle between mercury and concrete was set to 130°. Assuming that the shape of the porous material was cylinder, the applied pressure p (in mN/m²) was converted to the pore diameter D (in m) by the Washburn equation [41]:

$$D = (-4\sigma \cos \theta)/p, \tag{1}$$

where σ is the mercury surface tension force in mN/m and θ is the contact angle between mercury and the capillary surface.

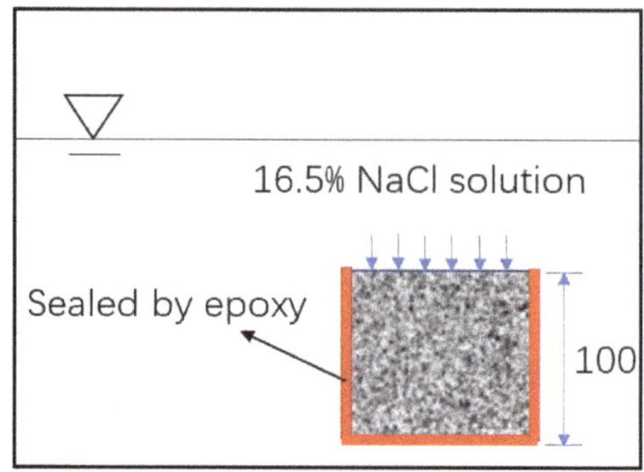

(**a**) Concrete sample immersion in NaCl solution.

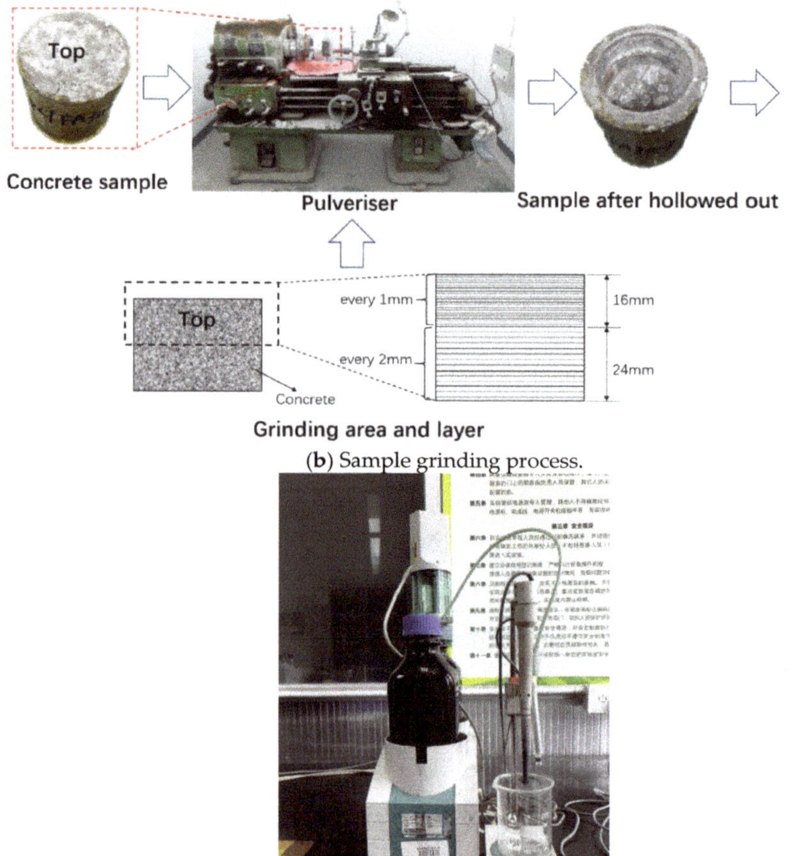

(**b**) Sample grinding process.

(**c**) Concentration titration of chloride ions.

Figure 1. Measurement of chloride content in concrete.

Microstructural studies of the concrete sample after exposure to seawater immersion were performed using SEM. The concrete samples were taken from the cylinder surface with a diameter of <25 mm and a thinness of <20 mm. The specimens were washed and dried to a constant weight in an oven under 105 °C. The specimens were then coated with a gold-plated film due to the poor electric conductive performance of concrete. Ultra-high resolution SEM was adopted to identify the hydration products, observe the fly ash, and evaluate the pore structures.

3. Results and Discussion

3.1. Effect of Fly Ash on Chloride Transportation

The free and total chloride contents were determined in accordance with the AASHTO T260 standard [38]. The former was obtained based on the water-soluble extraction method, while the latter was obtained by acid-soluble extraction in a nitric acid solution. After extraction, the chloride content was measured with an automatic potentiometer titrator. Figure 2 shows that the concentration of free chloride content from each concrete layer reduced significantly as the fly ash replacement ratio increased. Two individual zones of chloride deposition, namely the descending (0–15 mm) and flat zones (15–40 mm), were observed. The curve descended significantly at the 0.0–7.5 mm zone, in twhich the chloride concentration of plain concrete with 0% fly ash (PCFA0) reduced by 0.426% with a reduction percentage of 54.5%. For PCFA15 and PCFA30, the related reductions were 0.4722% and 0.4876% with reduction percentages of 65.9% and 69.4%, respectively. It was found that concrete with fly ash had a lower chloride content, which indicated a larger reduction percentage compared to the normal concrete without fly ash. For the zone of 17–27 mm, the reduction of chloride concentration was negligible for all samples since the chloride ion concentration became lower as concrete depth increased.

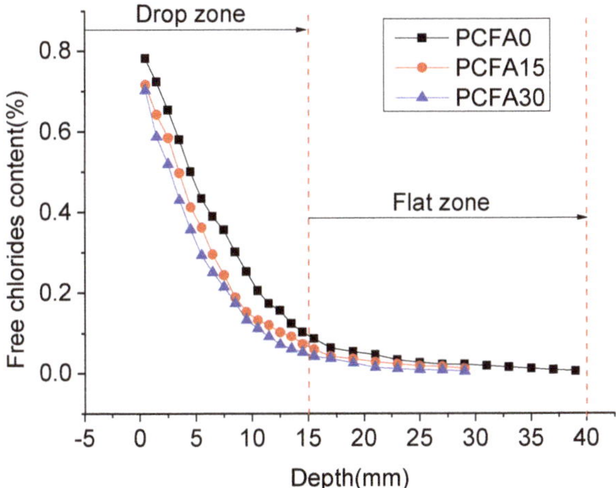

Figure 2. Free chloride deposition of concrete with different fly ash replacement ratios.

Figure 2 shows that the chloride ion distribution curves in the soaking environment satisfied Fick's second law, while Figure 3 shows that the curves in the salt-spray environment did not follow Fick's second law. In the salt-spray environment, there was a rising branch called the convection zone for the chloride ion on the concrete surface. Beyond this convection zone, the chloride ion concentration decreased. Chloride ion distribution in concrete depends on the way chloride ions penetrate into the concrete. In fact, chloride ion penetration is a complex process including extensive transportation mechanisms, such as diffusion under a concentration difference [27], penetration under water pressure [29], and capillary action under a humidity gradient [30]. In a soaking environment, when concrete is saturated, chloride ions are transferred mainly through diffusion to the concrete. Meanwhile, in a salt-spray environment, the migration of chloride ions is mainly attributed to capillary suction and the diffusion effect. Chloride aerosols deposited on the surface of concrete are absorbed through capillary suction. Then, the free chloride ions are diffused into the concrete due to a concentration difference. When the water evaporates from the concrete pore, the chloride ions left in pores gather to a peak amount through a series of drying and wetting cycles. This results in the convection of the chloride ion migration under the salt-spray environment, which can be seem in Figure 2 [24]. According to the results of chloride penetration in the soaking and salt-spray environments in our experiments, the distribution of chloride ions on the concrete surface varied, but the trend approached consistency as the concrete depth increased.

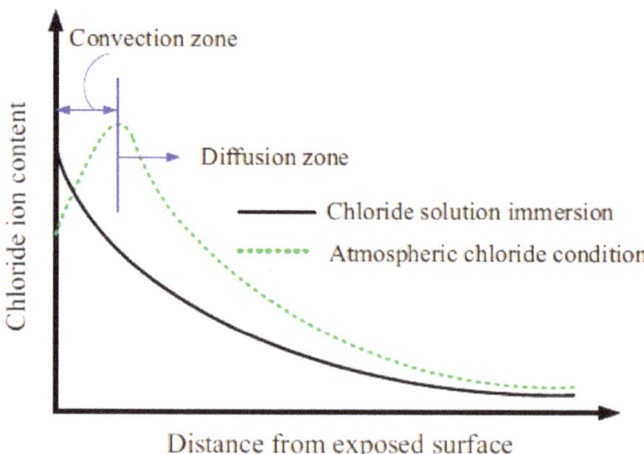

Figure 3. Qualitative expression of the chloride penetration curves [24].

The deposition of chloride ions gradually approached zero at concrete depths over 15 mm. This was mainly due to the maximum exposure day (35 days), and this transition depth could be increased if the number of exposure days increases. In addition, the ingress depth also depended on the specific concrete mix used (water to cement ratio: W/C). The curve in the descending zone in Figure 2 indicates that the free chloride deposition of the samples with 15% and 30% fly ash replacement considerably decreased compared to that of the normal PCFA0 concrete. This is because fly ash has three main effects on concrete, namely the pozzolanic, morphological, and microaggregate filling effects. The pozzolanic effect is triggered by $Ca(OH)_2$ formed from Portland cement hydration. However, the pozzolanic reaction at the beginning of cement hydration is very slow because of the presence of less $Ca(OH)_2$. With the increase of curing age, the secondary hydration reaction between the activated ingredients, such as SiO_2 and Al_2O_3, in fly ash, and $Ca(OH)_2$ produces C–S–H gel, calcium aluminate hydrate products, and so on. These products may absorb more free chloride ions, reducing the deposition of chloride ions in concrete. This also indicates that chloride binding capacity is improved with the addition of fly ash. These hydrates can fill large size pores in cement matrixes and thus reduce porosity, narrow pore diameter, and block the connectivity of the pores, all of which subsequently slow the diffusion and migration of chloride ions. In addition, 70% of fly ash particles are inert, compact hollow microsphere (cenospheres) with smooth surfaces [31,32], which are like activated nanomaterials that improve the microstructures of concrete. Consequently, a concrete with fly ash replacement has a much lower chloride deposition than normal concrete.

3.2. Influence of Diffusion Coefficient

It was found that the chloride deposition curve in Figure 2 generally followed Fick's second law (Equation (2)) [42,43]. Based on the data in Figure 2, Figure 4 plots the regression curves of the chloride coefficient using the Origin software. The chloride ion diffusion coefficient D and surface concentration of chloride C for each group can be determined by Equation (2), as shown in Table 3.

$$C(x,t) = C_0 + (C_s - C_0)[1 - erf(\frac{x}{2\sqrt{Dt}})],\tag{2}$$

where $C(x,t)$ is the chloride percentage in concrete depth x at exposure time t (%), C_s is the chloride percentage at the concrete surface (%), C_0 is the initial percentage of chloride in concrete, and $erf(z)$ is the error function, $erf(z) = \frac{2}{\sqrt{\Pi}}\int_0^z \exp(-z^2)dz$.

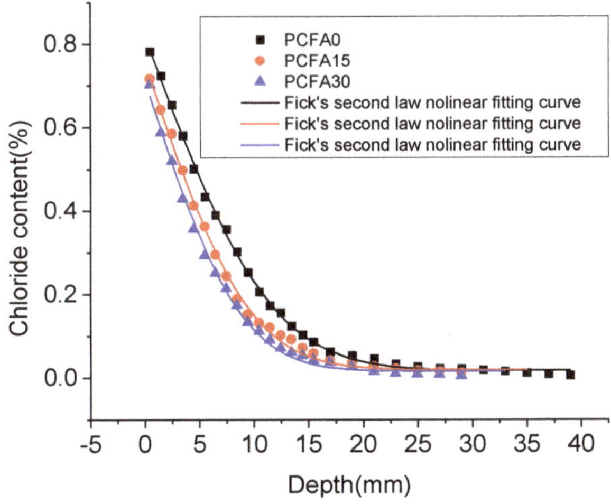

Figure 4. Regression curves of chloride coefficient using Fick's second law (by mass%).

Table 3. Regression results of chloride diffusion coefficient. D: chloride ion diffusion coefficient; C_S: chloride percentage at the concrete surface; and σ^2: the mercury surface tension force.

ID	D/($\times 10^{-12}$ m^2/s)	C_s/%	R^2	σ^2
PCFA0	12.2617	0.8210	0.9988	0.0006
PCFA15	9.0633	0.7614	0.9970	0.0014
PCFA30	7.6176	0.7187	0.9966	0.0014

Figure 4 shows the regression curves of the chloride diffusion coefficient for PCFA0, PCFA15, and PCFA30 using Fick's second law. The chloride diffusion coefficient significantly decreased with the 15% and 30% replacements of cement by mass, with reduction ratios of about 26.1% and 37.9%, respectively, compared to normal concrete. The related coefficient reductions were 3.1984×10^{-12} and 4.6441×10^{-12} m^2/s. This indicated that the chloride diffusion coefficient reduced more as the fly ash dosage rose. However, the reduction rate was slower as more fly ash was added. This was clear from the fact that the chloride diffusion coefficient of the PCFA30 sample was not proportionally reduced, which was less than half of that (1.4457×10^{-12} m^2/s) for the PCFA15 sample. Two main reasons for the reduction of the chloride diffusion coefficient were the (1) pozzolanic effect formed tobermorite and calcium aluminate hydrate product [44], which could have increased the chloride binding resistance, and (2) Al_2O_3 in the fly ash was the essential ingredient to generate Friedel's salt. The chloride ions in concrete are mainly in the form of Friedel's salt. Al_2O_3 reacts with $Ca(OH)_2$ and may lower the C/A (calcium/aluminum) product and accelerate AFM (monosulphate) to form Friedel's salt, as explained in Equation (3). This process physically and chemically benefits the chloride binding capacity. Nevertheless, differentiating between the chemical and physical absorption process with the chosen test setup was difficult in this study since chloride penetration have extensive transport mechanisms in concrete. Consequently, the decrease of free chloride content leads to reduce the chloride diffusion coefficient [45]. However, differentiating between the chemical and physical adsorption process needs further study.

3.3. Influence of Free and Binding Chlorides

Binding chloride is the solidifying process of chloride ions by cement-based materials. The binding chloride in the cement composite could not transport in water in concrete. Binding chloride can be divided into two types: chemical bound and physical absorption. Chemical bound results from the reaction of chloride and cement, while physical absorption depends on the electrostatic interactions (or van der Waals forces) [46,47] between ions. Equation (3) shows that chemical bound chloride ions follow the formation of Friedel's salt [48]:

$$C_3A + CaCl_2 + 10H_2O \rightarrow C_3A \cdot CaCl_2 \cdot 10H_2O, \tag{3}$$

$C_3A \cdot CaCl_2 \cdot 10H_2O$ reduces the porosity of concrete, thus increasing the resistance of chloride penetration and reducing the chloride ion binding capacity. According to Table 2, the contents of Al_2O_3 in the fly ash and cement were 23.95% and 4.62%, respectively. Since Al_2O_3 is a raw material for salt formation, adding fly ash into the concrete increased Friedel's salt formation, which further improved the chloride binding capacity.

Figure 5 shows the distribution of total chloride, free chloride, and bound chloride of the concrete samples. The bound chloride amount was much lower than the total and free chloride amounts. Figure 6 shows the relationship between bound chloride and concrete depth, in which the bound chloride amount decreased as concrete depth increased, but then the chloride amount went slightly up. PCFA15 and PCFA30 had much higher bound chloride amounts compared to that without any fly ash addition (PCFA0). The bound chloride content decreased significantly within the 0.0–12.5 mm depth, while it slightly increased to a constant value beyond 12.5 mm. In this case, free chloride could be reduced in the pores of concrete, which deceased the potential risk of rebar corrosion. The free chloride content kept a linear relationship with the total chloride content [49], as shown in Figure 7. Equation (4) introduces the index R to represent the chloride binding capacity:

$$R = \frac{\partial C_b}{\partial C_f}, \tag{4}$$

where C_b is the concentration of the bound chloride and C_f is the concentration of the free chloride.

Equation (5) gives a linear formulation to represent the linear relationship between the free and total chloride contents:

$$C_t = KC_f + b, \tag{5}$$

where C_t is the concentration of total chloride and K and b are constants.

Substituting Equation (1) into Equation (2) results in Equation (6):

$$C_b = (K-1)C_f + b, \tag{6}$$

Equation (7) introduces chloride binding capacity index R shown below:

$$R = K - 1, \tag{7}$$

Based on the calculations of Equation (7), Figure 8 shows the relationship between the chloride bound capacity amount and concrete depth. The chloride binding capacity increased as concrete depth rose. Based on Equation (4), the bound capacity indexes of chloride were 0.1426, 0.1882, and 0.2134, respectively, for PCFA0, PCFA15, and PCFA30. This calculation indicates that the chloride binding capacity improves as fly ash addition increases.

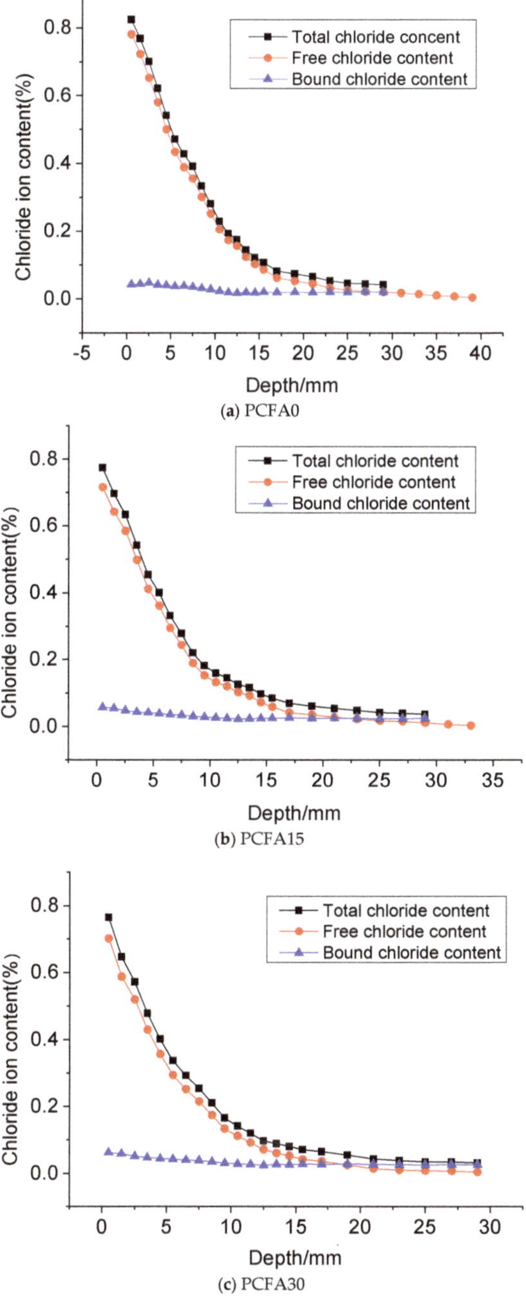

(**a**) PCFA0

(**b**) PCFA15

(**c**) PCFA30

Figure 5. Distribution of total chloride, free chloride, and bound chloride in concrete samples (by mass%).

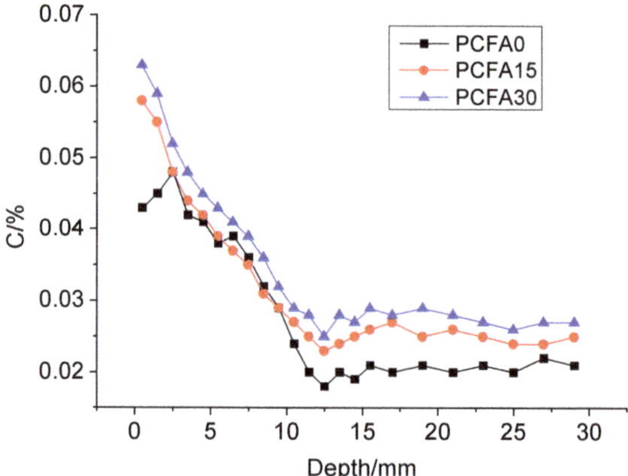

Figure 6. Relationship between bound chloride amount and concrete depth (by mass%).

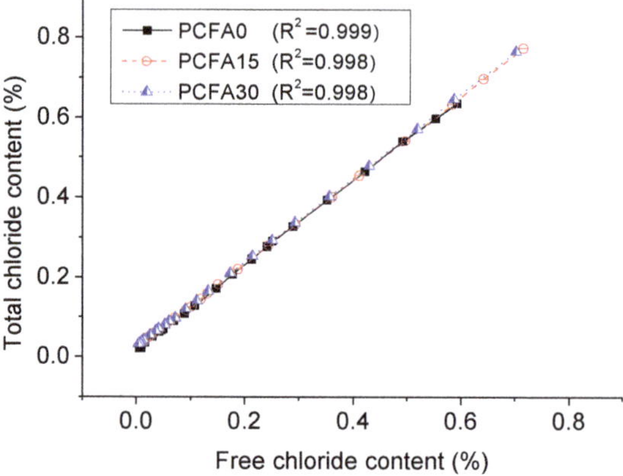

Figure 7. Relation between free chloride and total chloride depth (by mass%).

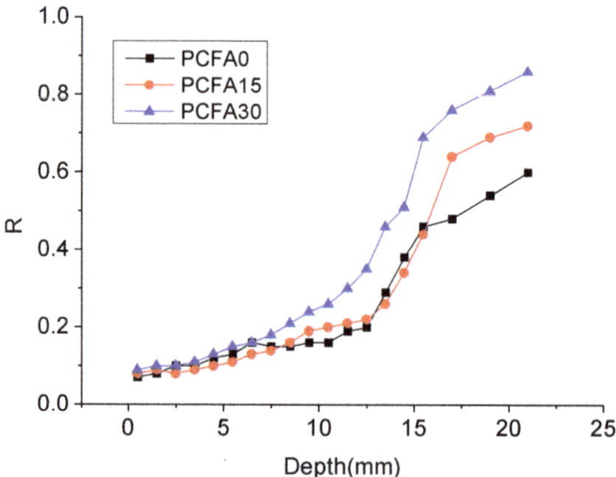

Figure 8. Relation between chloride bound capacity amount and concrete depth.

3.4. Effect of Fly Ash on Porosity and Microstructure

Physical and chemical reactions may significantly affect the porosity and microstructure of concrete due to fly ash addition. Figure 9 shows the cumulated porosity of the concrete samples. It was found that the cumulative pore volume reduced as the fly ash addition increased. Figure 10 shows the pore size distribution of each sample after 90 days of curing. The cumulative volume of pores reduced as fly ash content increased. The mean pore size of the PCFA0, PCFA15, and PCFA30 samples were about 40, 26, and 13 nm, respectively, which indicated that the fly ash addition led to a denser concrete microstructure.

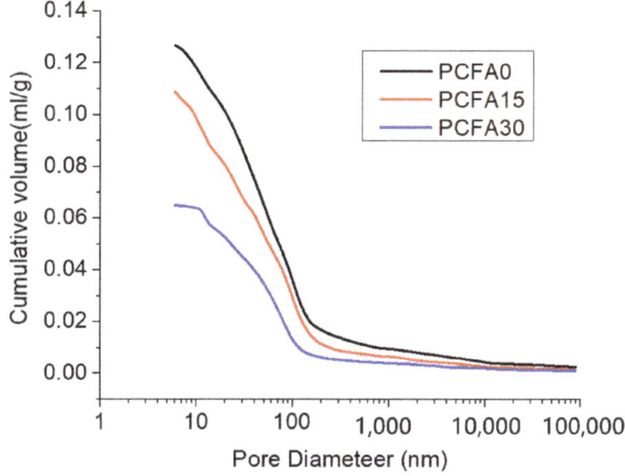

Figure 9. Cumulative porosity of concrete sample with different fly ash additions after 90 days of curing.

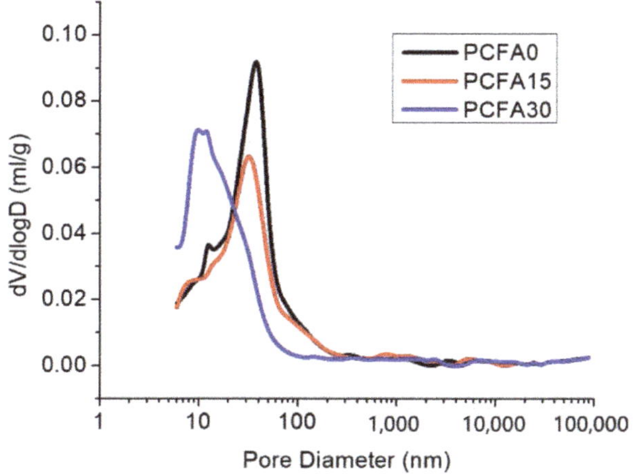

Figure 10. Pore size distribution after 90 days of curing.

Mehta and Monteriropj [28] classified pore size in concrete into four classes: Class I of <4.5 nm, Class II of 4.5–50 nm, Class III of 50–100 nm, and Class IV of > 100 nm. A pore size of <100 nm has little or no effect, while >100 nm has a negative effect on the concrete strength and permeability. Specifically, the curves in Figure 11 show that the pore volume fraction of Class I and II increased while that of Class III and IV decreased. For example, the pore volume fraction with 4.5–100 nm in PCFA30 increased by about 62.5% and 36.8% compared to PCFA0 and PCFA15, respectively, while that of 100–200 nm and larger than 200 mm for PCFA15 and PCFA30 considerably reduced to 20.83% and 15.39%, respectively, compared to that of PCFA0. This indicated that pore structures were well-improved as fly ash was added into concrete. This was caused by the second hydration of fly ash with C–H after 90 days of curing generated more C–S–H gel, causing the promotion of concrete compactness and the optimization of pore structures. Figure 12a–c shows the morphologies of concrete samples with different fly ash additions after 90 days of curing. PCFA15 and PCFA30 with fly ash had enhanced concrete density, while more pores could be found for PCFA0 without fly ash replacement. The left-unhydrated fly ash particles may have filled pores in the concrete, as shown in Figure 12d. In this case, fly ash prevented the penetration of free chloride ions and reduced the corrosion of steel rebar, leading to a higher concrete durability.

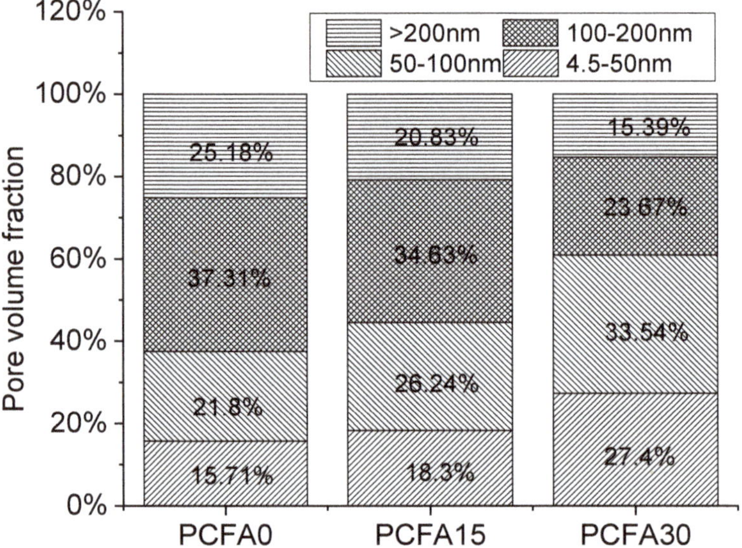

Figure 11. Pore size distribution of concrete with different fly ash additions after 90 days of curing.

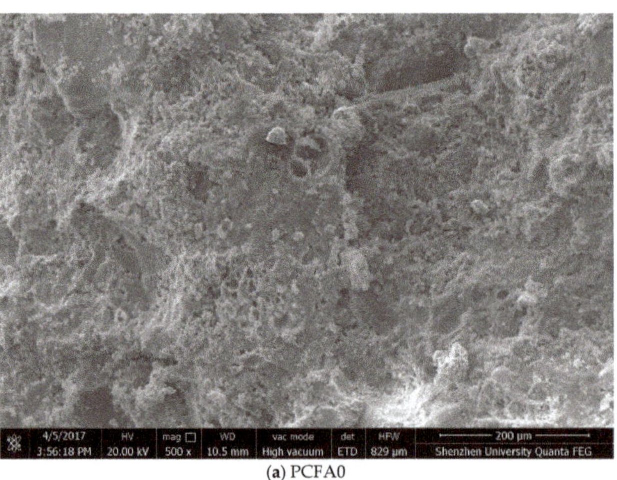

(a) PCFA0

Figure 12. *Cont.*

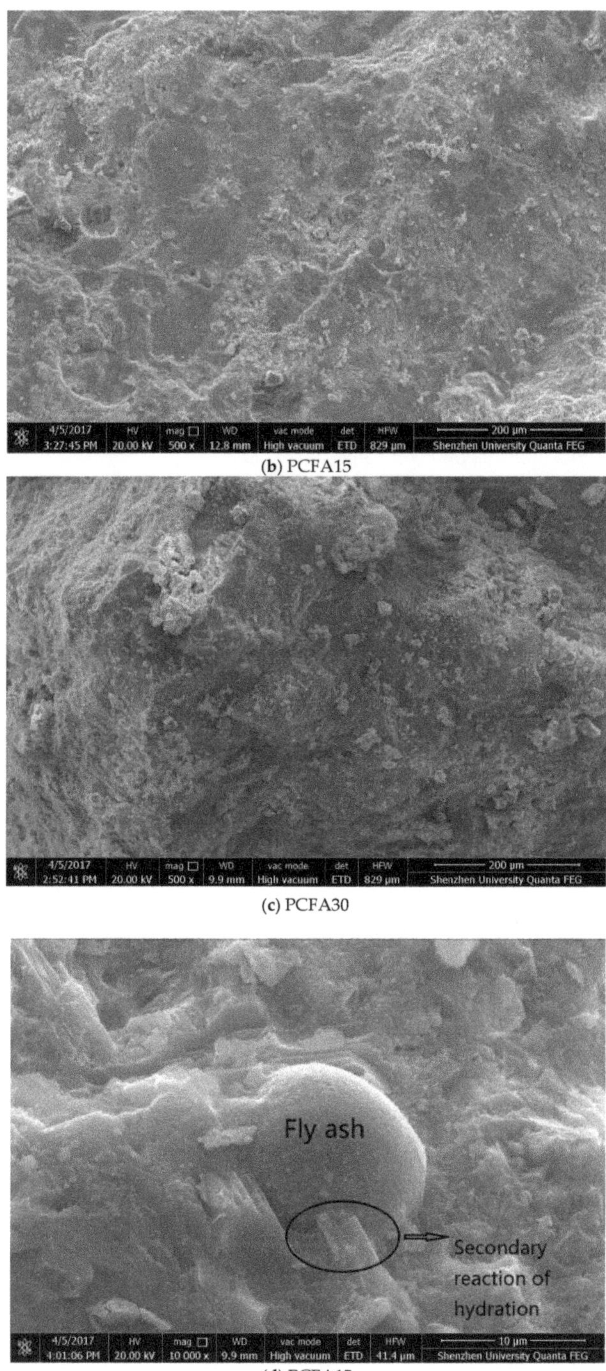

Figure 12. Morphologies of concrete samples with different fly ash additions after 90 days of curing.

4. Conclusions

This study systematically, experimentally, and analytically investigated the influence of fly ash on chloride diffusion, chloride bonding capacity, and the microstructure of normal concrete in a water soaking environment. The main findings are summarized as follows.

(1) The addition of fly ash reduced free chloride diffusion, increased the chloride binding capacity, and improved the concrete compactness, which could further improve the durability of concrete.

(2) Two individual zones of chloride deposition, namely the descending zone (0–15 mm) and the gentle zone (15–40 mm), were identified through the measurement of chloride content. The chloride concentration of samples at the 0–7.5 mm zone reduced by 54.5%, 65.9%, and 69.4%, respectively, for PCFA0, PCFA15 and PCFA30. Fly ash addition indeed decreased the free chloride concentration.

(3) In natural immersion, the chloride ion diffusion coefficients of concrete with 15% and 30% fly ash replacements for cement were reduced by 26.1% and 37.9%, respectively, compared to concrete without any fly ash addition, which indicated that the more fly ash was added in the concrete, the lower the chloride diffusion coefficient.

(4) Both chemical ion exchange and physical adsorption processes contributed to the reduction of free chloride ion concentrations. The free chloride ion content had a linear relation with the total amount of chloride ions. An equation was proposed to predict the binding capacity of chloride. Concretes with 0%, 15%, and 30% replacement for cement had binding capacity indexes of 0.1426, 0.1882, and 0.2134, respectively, which indicated that the chloride binding capacity was improved by fly ash. In this case, fly ash reduced the free chloride ion concentration, and if this concentration was less than the critical value, the rebar corrosion could be prevented, thus resulting in a longer service life of reinforced concrete. Moreover, a future study is expected to differentiate between chemical and physical adsorption process.

(5) Fly ash particles and hydration products (filling and pozzolanic effect) led to the densification of microstructures. SEM results showed that the volume of pores less than 100 nm obviously increased for concrete with 15% fly ash replacement for cement, the volume of pores sized from 4.5 to 50 nm, and 50 to 100 nm pronouncedly increased, while the volume of pores larger than 100 nm decreased significantly.

Author Contributions: Conceptualization, J.L. (Jun Liu); investigation, J.L. (Jiaying Liu); writing—original draft preparation, Z.H.; validation, J.Z.; writing—review and editing, W.L. and W.Z. All authors have read and agreed to the published version of the manuscript.

Funding: The financial support from Key-Area Research and Development Program of Guangdong Province (2019B111107002), Natural Science Foundation of China (No. 51708360 and 51978407), Shenzhen City Science and Technology Project (No. JCYJ20180305124008155, JCYJ20170302143133880, JCYJ20180305124106675), and Shenzhen International Cooperation Research Project (No.GJHZ20180928155602083) are gratefully acknowledged.

Conflicts of Interest: The authors declare no conflict of interest.

References

1. Hasanbeigi, A.; Menke, C.; Price, L. The CO_2 abatement cost curve for the Thailand cement industry. *J. Clean. Prod.* **2010**, *18*, 1509–1518. [CrossRef]
2. Chen, C.; Habert, G.; Bouzidi, Y.; Jullien, A. Environmental impact of cement production: Detail of the different processes and cement plant variability evaluation. *J. Clean. Prod.* **2010**, *18*, 478–485. [CrossRef]
3. Madhavi, T.C.; Swamy Raju, L.; Mathur, D. Durabilty and Strength Properties of High Volume Fly Ash Concrete. *J. Civil. Eng. Res.* **2014**, *4*, 7–11.
4. Abdullah, M.M.A.; Tahir, M.F.M.; Hussin, K.; Binhussain, M.; Sandu, I.G.; Yahya, Z.; Sandu, A.V. Fly ash based lightweight geopolymer concrete using foaming agent technology. *Rev. Chim.* **2015**, *66*, 1001–1003.
5. Mishra, S.R.; Kumar, S.; Park, A.; Rho, J.; Losby, J.; Hoffmeister, B.K. Ultrasonic characterization of the curing process of PCC fly ash–cement composites. *Mater. Charact.* **2003**, *50*, 317–323. [CrossRef]

6. Hemalatha, T.; Ramaswamy, A. A review on fly ash characteristics-Towards promoting high volume utilization in developing sustainable concrete. *J. Clean. Prod.* **2017**, *147*, 546–559. [CrossRef]

7. Mukherjee, A.B.; Zevenhoven, R.; Bhattacharya, P.; Sajwan, K.S.; Kikuchi, R. Mercury flow via coal and coal utilization by-products: A global perspective. *Resour. Conserv. Recycl.* **2008**, *52*, 571–591. [CrossRef]

8. Selic, E.; Herbell, J.D. Utilization of fly ash from coal-fired power plants in China. *J. Zhejiang Univ.-Sci. A* **2008**, *9*, 681–687.

9. Liu, J.; Gu, Z.; Xiang, J.; Pan, D. Permeation properties and pore structure of surface layer of fly ash concrete. *Materials* **2014**, *7*, 4282–4296. [CrossRef]

10. Liu, J.; Xing, F.; Dong, B.; Ma, H.; Pan, D. Study on water sorptivity of the surface layer of concrete. *Mater. Struct.* **2014**, *47*, 1941–1951. [CrossRef]

11. Argiz, C.; Moragues, A.; Menéndez, E. Use of ground coal bottom ash as cement constituent in concretes exposed to chloride environments. *J. Clean. Prod.* **2017**, *170*, 25–33. [CrossRef]

12. Zou, D.J.; Liu, T.J.; Qiao, G.F. Experimental investigation on the dynamic properties of RC structures affected by the reinforcement corrosion. *Adv. Struct. Eng.* **2014**, *17*, 851–860. [CrossRef]

13. Mehta, P.K. High-performance, high-volume fly ash concrete for sustainable development. In Proceedings of the International Workshop on Sustainable Development and Concrete Technology, Beijing, China, 20–21 May 2004.

14. Moghaddam, F.; Sirivivatnanon, V.; Vessalas, K. The effect of fly ash fineness on heat of hydration, microstructure, flow and compressive strength of blended cement pastes. *Case Stud. Constr. Mater.* **2019**, *10*, e00218. [CrossRef]

15. Liu, J.; Qiu, Q.; Chen, X.; Wang, X.; Xing, F.; Han, N.; He, Y. Degradation of fly ash concrete under the coupled effect of carbonation and chloride aerosol ingress. *Corros. Sci.* **2016**, *112*, 364–372. [CrossRef]

16. Thomas, M.D.A.; Matthews, J.D. Carbonation of fly ash concrete. *Mag. Concr. Res.* **1992**, *44*, 217–228. [CrossRef]

17. Liu, T.; Zou, D.; Teng, J.; Yan, G. The influence of sulfate attack on the dynamic properties of concrete column. *Constr. Build. Mater.* **2012**, *28*, 201–207. [CrossRef]

18. Naik, T.R.; Singh, S.; Ramme, B. Mechanical properties and durability of concrete made with blended fly ash. *ACI Mater. J.* **1998**, *95*, 454–462.

19. Malhotra, V.M. Role of supplementary cementing materials in reducing greenhouse gas emissions CANMET. In *Concrete Technology for a Sustainable Development in the 21st Century*; Gjorv, O.E., Sakai, K., Eds.; Spon Press: London, UK, 1999.

20. Liu, J.; Qiu, Q.; Chen, X.; Xing, F.; Han, N.; Ma, Y. Understanding the interacted mechanism between carbonation and chloride aerosol attack in ordinary Portland cement concrete. *Cem. Concr. Res.* **2017**, *95*, 217–225. [CrossRef]

21. Liu, J.; Wang, X.; Qiu, Q.; Ou, G.; Xing, F. Understanding the effect of curing age on the chloride resistance of fly ash blended concrete by rapid chloride migration test. *Mater. Chem. Phys.* **2017**, *196*, 315–323. [CrossRef]

22. Tripathi, S.R.; Ogura, H.; Inoue, H.; Hasegawa, T.; Takeya, K.; Kawase, K. Measurement of chloride ion concentration in concrete structures using terahertz time domain spectroscopy (THz-TDS). *Corros. Sci.* **2012**, *62*, 5–10. [CrossRef]

23. Al-Mattarneh, H. Determination of chloride content in concrete using near- and far-field microwave non-destructive methods. *Corros. Sci.* **2016**, *105*, 133–140. [CrossRef]

24. Liu, J.; Ou, G.; Qiu, Q.; Chen, X.; Hong, J.; Xing, F. Chloride transport and microstructure of concrete with/without fly ash under atmospheric chloride condition. *Constr. Build. Mater.* **2017**, *146*, 493–501. [CrossRef]

25. Angst, U.; Elsener, B.; Larsen, C.K.; Vennesland, Ø. Critical chloride content in reinforced concrete: A review. *Cem. Concr. Res.* **2009**, *39*, 1122–1138. [CrossRef]

26. Sun, G.; Zhang, Y.; Sun, W.; Liu, Z.; Wang, C. Multi-scale prediction of the effective chloride diffusion coefficient of concrete. *Constr. Build. Mater.* **2011**, *25*, 3820–3831. [CrossRef]

27. Wu, J.; Li, H.; Wang, Z.; Liu, J. Transport model of chloride ions in concrete under loads and drying-wetting cycles. *Constr. Build. Mater.* **2016**, *112*, 733–738. [CrossRef]

28. Mehta, P.K.; Monteriropj, M. *Concrete: Microstructure, Properties and Materials*; McGraw-Hill: New York, NY, USA, 2006; pp. 203–300.

29. Muigai, R.; Alexander, M. Durability design of reinforced concrete structures: A comparison of the use of durability indexes in the deemed-to-satisfy approach and the full-probabilistic approach. *Mater. Struct.* **2012**, *45*, 1233–1244. [CrossRef]

30. Kwon, S.J.; Na, U.J.; Park, S.S.; Jung, S.H. Service life prediction of concrete wharves with early-aged crack: Probabilistic approach for chloride diffusion. *Struct. Saf.* **2009**, *31*, 75–83. [CrossRef]

31. Huang, Z.; Padmaja, K.; Li, S.; Liew, J.R. Mechanical properties and microstructure of ultra-lightweight cement composites with fly ash cenospheres after exposure to high temperatures. *Constr. Build. Mater.* **2018**, *164*, 760–774. [CrossRef]

32. Huang, Z.Y.; Liew, J.Y.R.; Li, W. Evaluation of compressive behavior of ultra-lightweight cement composite after elevated temperature exposure. *Constr. Build. Mater.* **2017**, *148*, 579–589. [CrossRef]

33. Berry, E.E.; Hemmings, R.T.; Cornelius, B.J. Mechanisms of hydration reactions in high volume fly ash pastes and mortars. *Cem. Concr. Compos.* **1990**, *12*, 253–261. [CrossRef]

34. Kouloumbi, N.; Batis, G.; Malami, C. The anticorrosive effect of fly ash, slag and a Greek pozzolan in reinforced concrete. *Cem. Concr. Compos.* **1994**, *16*, 253–260. [CrossRef]

35. Kouloumbi, N.; Batis, G. Chloride corrosion of steel rebars in mortars with fly ash admixtures. *Cem. Concr. Compos.* **1992**, *14*, 199–207. [CrossRef]

36. Cheewaket, T.; Jaturapitakkul, C.; Chalee, W. Long term performance of chloride binding capacity in fly ash concrete in a marine environment. *Constr. Build. Mater.* **2010**, *24*, 1352–1357. [CrossRef]

37. NT (NORDTEST) Build 443. Concrete Hardened: Accelerated Chloride Penetration, Nordtest, Tekniikantie 12, FIN-02150 Espoo, Finland. 1995. Available online: http://www.nordtest.info/index.php/methods/building/item/concrete-hardened-accelerated-chloride-penetration-nt-build-443.html (accessed on 8 September 2020).

38. AASHO T260. *Standard Method of Test for Sampling and Testing for Chloride Ion in Concrete and Concrete Raw Materials*; American Association of State Highway and Transportation Officials: Washington, DC, USA, 2009.

39. Ma, H.Y.; Li, Z.J. Realistic Pore Structure of Portland Cement Paste: Experimental Study and Numerical Simulation. *Comput. Concr.* **2013**, *11*, 317–336. [CrossRef]

40. Huang, Z.Y.; Huang, Y.S.; Han, N.X.; Zhou, Y.W.; Xing, F.; Sui, T.B.; Wang, B.; Ma, H.Y. Development of limestone calcined clay cement (LC^3) concrete in South China and its bond behavior with steel bar. *J. Zhejiang Univ.-Sci. A* **2020**, *21*. [CrossRef]

41. Washburn, E.W. Note on a Method of Determining the Distribution of Pore Sizes in a Porous Material. *Proc. Natl. Acad. Sci. USA* **1921**, *7*, 115. [CrossRef] [PubMed]

42. Collepardi, M.; Marcialis, A.; Turriziani, R.C. Penetration of chloride ions into cement paste and concrete. *Am. Ceram. Soc.* **1972**, *55*, 534–535. [CrossRef]

43. Funahashi, M. Predicting corrosion free service life of a concrete structure in a chloride environment. *ACI Mater. J.* **1998**, *87*, 581–587.

44. Wang, C.; Yang, C.; Qian, J.; Zhong, M.; Zhao, S. Behavior and Mechanism of Pozzolanic Reaction Heat of Fly Ash and Ground Granulated Blast Furnace Slag at Early Age. *J. Chin. Ceram. Soc.* **2012**, *40*, 1050–1058.

45. Zibara, H.; Hooton, R.D.; Thomas, M.D.A.; Stanish, K. Influence of the C/S and C/A ratios of hydration products on the chloride ion binding capacity of lime-SF and lime-MK mixtures. *Cem. Concr. Res.* **2008**, *38*, 422–426. [CrossRef]

46. Yuan, Q.; Shi, C.; De Schutter, G.; Audenaert, K.; Deng, D. Chloride binding of cement-based materials subjected to external chloride environment—A review. *Constr. Build. Mater.* **2009**, *23*, 1–13. [CrossRef]

47. Elakneswaran, Y.; Nawa, T.; Kurumisawa, K. Electrokinetic potential of hydrated cement in relation to adsorption of chlorides. *Cem. Concr. Res.* **2009**, *39*, 340–344. [CrossRef]

48. Ben-Yair, M. The effect of chlorides on concrete in hot and arid regions. *Cem. Concr. Res.* **1974**, *4*, 405–416. [CrossRef]

49. Tang, L.P.; Nilsson, L.O. *Resistance of Concrete to Chloride Ingress: Testing and Modelling*; Spon Press: London, UK, 2012.

 *applied sciences*

Article

Properties of Foamed Lightweight High-Performance Phosphogypsum-Based Ternary System Binder

Girts Bumanis *, Jelizaveta Zorica and Diana Bajare

Department of Building Materials and Products, Riga Technical University, LV-1658 Riga, Latvia; jelizaveta.zorica@rtu.lv (J.Z.); diana.bajare@rtu.lv (D.B.)
* Correspondence: girts.bumanis@rtu.lv; Tel.: +371-26062011

Received: 21 August 2020; Accepted: 4 September 2020; Published: 8 September 2020

Featured Application: This research brings introduction to a novel type of hydraulic binder based on phosphogypsum. Ternary system binder based on phosphogypsum, Portland cement, and pozzolan was elaborated as high-performance material with the strength up to 90 MPa. This binder is characterized by low Portland cement content and low water-binder ratio. Mineralogical and microstructural investigation was performed and data represented. Furthermore, development of lightweight foamed concrete based on developed binder was created using foaming agents.

Abstract: The potential of phosphogypsum (PG) as secondary raw material in construction industry is high if compared to other raw materials from the point of view of availability, total energy consumption, and CO_2 emissions created during material processing. This work investigates a green hydraulic ternary system binder based on waste phosphogypsum (PG) for the development of sustainable high-performance construction materials. Moreover, a simple, reproducible, and low-cost manufacture is followed by reaching PG utilization up to 50 wt.% of the binder. Commercial gypsum plaster was used for comparison. High-performance binder was obtained and on a basis of it foamed lightweight material was developed. Low water-binder ratio mixture compositions were prepared. Binder paste, mortar, and foamed binder were used for sample preparation. Chemical, mineralogical composition and performance of the binder were evaluated. Results indicate that the used waste may be successfully employed to produce high-performance binder pastes and even mortars with a compression strength up to 90 MPa. With the use of foaming agent, lightweight (370–700 kg/m^3) foam concrete was produced with a thermal conductivity from 0.086 to 0.153 W/mK. Water tightness (softening coefficient) of such foamed material was 0.5–0.64. Proposed approach represents a viable solution to reduce the environmental footprint associated with waste disposal.

Keywords: phosphogypsum; ternary binder; high performance; strength; foam; lightweight material; thermal conductivity

---

1. Introduction

Gypsum binder is widely used in construction due to its ease of production, availability, and low price [1]. Physically, gypsum is infinitely recyclable; however, the recycling process requires additional energy [2]. Despite these benefits, the disadvantage of gypsum binder is its brittleness, poor resistance to cracking, and unsuitability for damp conditions. Traditional gypsum binder use has been defined in EN 12859, where the main gypsum application is associated with the production of plasters, blocks, tiles, and boards [3]. Besides natural gypsum, synthetic gypsum, produced as chemical by-product, is used widely for the production of gypsum products. There are more than 50 different types of gypsum waste [2]. However, the most common by-product is phosphogypsum (PG), flue gas desulphurization gypsum, and borogypsum [4,5]. Phosphogypsum (PG) is produced as a by-product from phosphate fertilizer production and the annual PG production reaches 280 mlj.t worldwide and only about 15% of PG is used as secondary raw material, but rest is disposed in open-type stacks [5]. Partially or completely, synthetic gypsum can be a substitute for natural gypsum as cement admixtures, gypsum-based plasters, drywalls. Researches on production of traditional gypsum binders based on PG are widely published, but they have rather limited practical application due to specific nature of PG [6]. Moreover, there are legislation limits and prejudice coming from society regarding PG, so the direct use of PG as substitution of natural gypsum is problematic. More complicated and effective way of the utilization of PG is to create an advanced and new type of binder, which has a much lower carbon footprint comparing to Portland cement, while remaining strength properties similar to Portland cement. It was reported that the addition of blast furnace slag and cement could effectively improve the mechanical strength of PG. However, fly ash played a negative role on the compressive strength of PG [7]. S. Kumar described properties of fly ash–lime–phosphogypsum ternary binder [8] or recently anhydrous gypsum was used to develop lime-pozzolan green binder [9]. Two waste-stream materials were used and only lime was defined as a primary resource, the calcination temperature (900–1100 °C) of latter is lower compared to Portland cement. In this case, the amount of PG in the binder was in the range from 10–40%. However, the disadvantage comparing to Portland cement is low compressive strength—which could be in range from 2–4 MPa [8]; nevertheless it is reasonable if it is compared to the lime binder and hydraulic lime binder. Higher strength results were obtained in ternary binder system phosphogypsum–steel slag–granulated blast-furnace slag (GGBS)–limestone cement, where the content of PG was from 25–65%, while the amount of slag was from 22–48%. The obtained strength at the age of 28 days was up to 45 MPa while the obtained binder is characterized with fast setting time (initial setting time 6–9 min, final 10–12 min) [10]. These results are more comparable to traditional binders, while the problem could be a fast setting time. The fast setting may not benefit the engineering application because there is not enough time for casting before the cement sets. In some case it was reported that the citric acid in amount from 0.03–0.15 wt.% of cement could retard setting time significantly. The use of citric acid could increase the open time from 25 to 47 min, while this admixture tended to slightly reduce compressive strength of the binder [11]. In ternary systems where Portland cement is present, superplasticizer can be used and low water-binder ratio can be achieved. Traditionally, supplementary materials that are utilized to replace ordinary Portland cement decrease the workability of the cementitious mixtures and superplasticizers such as polycarboxylate based are usually added to cement to control their fluidity [12]. The use of polycarboxylate acid-based superplasticizer could be used from 0.75–1.75 wt.%; however, reports say that it could slightly reduce early compressive strength of the material while final strength tended to increase [11]. These aspects regarding to the utilization of PG in new types of binders were considered in present research by choosing mixture composition, including use of chemical admixtures.

To continue the development of alternative waste stream binders in a production of new materials and enhance its valorization possibilities, novel lightweight foam material based on developed ternary binder was elaborated. In construction industry, there is growing interest in lightweight concretes. It combines positive properties of constructive and insulation materials and is characterized by moderate strength, low density, and improved thermal properties. Cellular concrete is composed on mortar matrix and specially created system of air cells, which occupies up to 85% of material volume [13]. High porosity limits potential of mechanical strength, but high volume of open pores is the main reason for increased water absorption and drying shrinkage. The density of traditional gypsum ranges from 600 to 1500 kg/m^3, as given in Clause 4.8.1. of EN 12859 [3]. This well-known standard covers the gypsum application range, and beyond this range, research is being conducted to make gypsum material more sustainable. Attempts to produce lightweight gypsum with foaming admixtures have yielded a material with density ranging from 300 to 600 kg/m^3 [14]. Such material has low density, superior sound, and thermal insulation and can be considered a sustainable high-performance material. High-efficiency sound-absorbing material was made also with PG, which composite structure was described by Baoguo Ma et al. [15]. Thus, aim of the work was to produce in laboratory conditions lightweight ternary system based material with density less than 600 kg/m^3, which is outside the traditional boundaries to bring the novelty of the research. Bulk density and thermal conductivity were set as target values, which should be determined together with technological properties so that gypsum material could be easily produced and handled (workability, strength). Here, research on development of highly porous ternary system gypsum-based binder material was evaluated and compared.

2. Materials and Methods

Dihydrate phosphogypsum ($CaSO_4 \cdot 2H_2O$) analyzed in this work is a waste generated by fertilizer production plant AB Lifosa (Kėdainiai, Lithuania) in wet-process phosphoric acid production, where apatite from the Kovdor mine, Kola peninsula, Russia, is decomposed by sulphuric acid. To produce 1t of orthophosphoric acid, about 3.0–4.5 t of PG is obtained [16]. At the enterprise, volcanic origin Cola apatite (Kirov and Kovdor) (containing F_2 1–2%, P_2O_5—about 38% as phosphorus) is used as a raw material in the production of phosphoric acid. Besides, phosphorites of sedimentary origin from other countries (Morocco, Jordan, Kazakhstan (Karatau), Algeria, South Africa Republic) are used. The dihydrate PG was used as secondary raw material in the present research. Chemical composition of raw materials is given in Table 1. The initial pH of PG is in the range from 2.2–2.9 while during the storage it can increase gradually. The PG was dried at 60 °C (moisture content 9–12 wt.%) and milled to powder-like particles with collision milling in semi-industrial disintegrator with the rotational speed of 50 Hz. The particle size distribution is given in Figure 1. Then, the calcium sulfate hemihydrate binder was obtained by treatment of milled gypsum powder at temperature 180 °C for 4 h. Commercial gypsum plaster (BG) was used to compare the characteristic technological properties of obtained binder.

Table 1. Chemical composition of raw materials used to prepare novel building material [weight %].

Element	Commercial Gypsum BG	Phosphogypsum PG	Metakaolin MKW	Cement CEM I 42.5N
SiO_2	3.73	1.07	51.80	22.64
Al_2O_3	1.68	0.70	34.20	5.93
Fe_2O_3	0.46	0.22	0.50	3.26
CaO	35.64	37.16	0.10	57.04
MgO	3.92	0.21	0.10	4.26
SO_3	30.90	37.38	-	3.30
Na_2O	0.31	0.48	0.60	0.10
K_2O	-	-	-	2.40
TiO_2	0.05	0.11	0.60	0.38
Cl	-	-	-	0.14
P_2O_5	-	0.57	-	0.46
LOI	22.43	19.24	11.5	-
Total	99.42	99.74	99.60	99.90

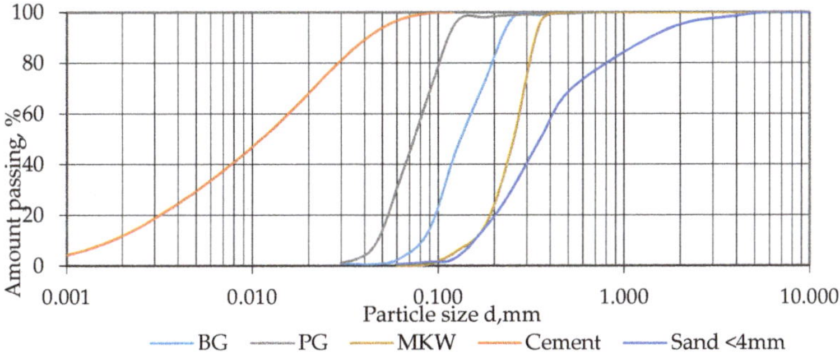

Figure 1. The particle size distribution of the raw materials used to prepare novel building material.

The other components of the binder were metakaolin as supplementary cementitious material and Portland cement CEM I 42.5N (CEM). Chemical composition of CEM is given in Table 1. The initial setting time was 182 min and the final setting was 224 min (according to LVS EN 196-3); the normal consistency was 28.2% (according to LVS EN 196-3). Blaine fineness of cement was 3787 cm^2/g (according to LVS EN 196-6). Na equivalent was 1.68. Compressive strength after 1 day was 15.4 MPa, after 2 days—32.9 MPa, after 7 days—48.8 MPa, and after 28 days—60.5 MPa. Waste metakaolin (MKW) was obtained from the porous glass granule production factory "Stikloporas" JSC (Druskininkai, Lithuania). Metakaolin is a by-product from the final stage of expanded glass granule production, where kaolinite clay powder is used as a substance for anti-agglomeration and is heated at 850 °C temperature for 40–50 min. MKW is mostly amorphous dehydration product of kaolinite, Al$_2$(OH)$_4$Si$_2$O$_5$, which exhibits strong pozzolanic activity often used in the concrete industry as a supplementary cementitious material. XRD pattern and detailed description of MKW was published before [17]. MKW contains amorphous metakaolin detected in 2θ region from 15 to 30°, and some minor crystalline phases were also detected in MKW: quartz (SiO$_2$) and Halloysite Al$_2$Si$_2$O$_5$(OH)$_4$. The specific surface area of the MKW is 15.86 m^2/g. Loss of ignition for metakaolin was 11.5%. The particle size distribution of materials used in the preparation of ternary binder is given in Figure 1.

Sika ViscoCrete G-2 is a high-performance superplasticizer/water reducer based on polycarboxylate (PCE) polymer technology. PCE is formulated for applications in systems with high calcium sulfate content or pure gypsum-based binders. Gips RETARD (TKK) is a powdered citric acid-based admixture, which was used for regulating the setting time of gypsum. Gips RETARD was added to the dry binder mixture and mixed thoroughly before water was added. Sodium dodecyl sulfate (SDS) ≥85%, pure (Acros Organics), was used as foam stabilizer for preparation of foamed samples. To produce porous, lightweight ternary system material, anion surface-active substance with stabilizing and functional agents PB-Lux was used.

Prepared mixture composition is given in Table 2. First two compositions were ternary system binder pastes (GCP-PG and GCP-BG). The difference between two of them is that in one case, commercial gypsum binder is used (abbreviation with BG), while in the other one PG binder is used (abbreviation with PG). The amount of PCE was 1.5% from the total amount of binder while set retarder R was 0.2%. Set retarder was used for samples with PG as it has a shorter set time comparing to commercial gypsum [6]. The W/B ratio was 0.34 for both mixtures. Other two mixtures were based on two mentioned binder pastes, but additional washed quartz sand (0/4 mm) was incorporated into the mixture composition with binder-sand ratio 1:1.75. The W/B ratio slightly increased with the incorporation of sand.

Table 2. Mixture composition of prepared ternary system binder paste and ternary system binder mortar.

Mixture	PG	BG	CEM	MKW	G-2	R	Sand 0/4 mm	W/B
GCP-PG	1		0.4	0.4	1.5%	0.2%	-	0.34
GCP-BG		1	0.4	0.4	1.5%	-	-	0.34
GCP-PG-S	1		0.4	0.4	1.5%	0.2	1.75	0.36
GCP-BG-S		1	0.4	0.4	1.5%	-	1.75	0.36

Mixing procedure of binder pastes was similar to traditional cement mortars. First, dry components including set retarder were homogenized for 2 min. Then, powder gradually was added into the water and superplasticizer mixture and homogenized for 2 min before casting. Mortar was mixed similar to the pastes, while the sand and extra water were added to the pre-mixed paste. Samples were cast in 20 × 20 × 20 mm or 40 × 40 × 160 mm molds for further testing.

The mixture composition of foamed ternary system binder is given in Table 3. Four mixture compositions were prepared. The amount of foam that was prepared was similar for all mixture compositions. The amount of binder paste homogenized with foams was changed for both compositions. SDS stabilizing admixture was used during the preparation of foams. Mixing procedure of binder pastes was similar to traditional cement mortars. First, dry components including set retarder were homogenized for 2 min. Then, powder gradually was added into the water and superplasticizer mixture and homogenized for 2 min before introduction in simultaneously prepared foams. Then, binder paste was cast in foams and homogenized for 3 min. Both commercial gypsum BG and phosphogypsum PG was compared. Plate samples with dimensions $35 \times 35 \times 5$ cm were prepared for thermal conductivity measurements. Samples were further cut to cubical specimens $5 \times 5 \times 5$ cm and physical properties as well as mechanical properties were tested.

Table 3. Mixture compositions of foamed ternary system binder.

Composition	Components, Weights Parts									
	Water, Foams	Water, Binder	PG	BG	PBLUX	CEMII	MKW	R	G-2	SDS
GCP-PG-1	200	410	750	-	5	312	312	2.5	21	2.5
GCP-PG-2	200	560	1050	-	5	437	437	3.5	29.5	3.5
GCP-BG-1	200	370	-	750	5	312	312	2.5	21	2.5
GCP-BG-1	200	520	-	1050	5	437	437	3.5	29.5	3.5

The setting time of the obtained binder was tested using the Vicat apparatus. Consistency of fresh mortar was tested with the Suttard's viscometer. Early age (24 h) and 28 d compressive strength was determined. After 24 h, samples were cured in moist conditions (RH 95%). At the age of 35 d, air dry samples were tested. Samples with dimension of $20 \times 20 \times 20$ mm were tested using Zwick Z100 with testing speed of 0.5 mm/min. The specific gravity and total porosity were determined by using Le Chatelier flask. Hardened samples were ground to powder with planetary ball mill Retch PM400 for 10 min with a speed of 300 rpm, and obtained powder was used to determine specific gravity. Total porosity was calculated from bulk density and specific gravity. Water absorption was measured for prismatic specimens for binder and mortar ($40 \times 40 \times 160$ mm) and cubical specimens for porous samples ($50 \times 50 \times 50$ mm). Samples were dried at 60 °C until a constant mass was obtained, then samples were immersed in water for 72 h and saturated mass was obtained. Then, water absorption was calculated and open porosity determined using the volumetric measurements of individual samples. The mineralogical composition was determined by X-ray diffraction (XRD) (PAN analytical X'Pert PRO). The macrostructure of the material was observed by a digital microscope at a magnification of 40 and 120x. Scanning Electron Microscopy (SEM) with Energy Dispersive Spectroscopy (EDX) was used (JEOL JSM 820 + IXRF systems 500 digital processing, Japan) for microscopic analysis of binder paste with 20 kV voltage. The thermal conductivity of the materials was determined using the LaserComp heat meter FOX600, according to the guidelines of Standard LVS EN 12667. Foamed ternary system binder plate specimens with dimensions of $35 \times 35 \times 5$ cm were prepared for thermal conductivity test. Temperature difference between testing plates was 20 °C (the bottom plate was +20 °C and the upper plate was 0 °C).

3. Results and Discussion

Results are represented in two main subsections where dense ternary system binder properties are described (Section 3.1) and in Section 3.2 where properties of lightweight ternary binder are represented.

3.1. Properties of High-Performance Ternary System Binder and Mortar

The ternary system binder was described through its appearance in macro and micro level, chemical and mineralogical composition through hydration processes. Fresh and hardened properties of the binder and mortar were determined and results represented.

3.1.1. Appearance of Ternary System Binder and Mortar

The macrostructure of GCP-PG mortar is given in Figure 2. The difference between binder depending on the gypsum type used (PG or BG) did not influence the appearance of the macrostructure of the material. Mortar has a homogenous structure where individual sand filler grains and the binder can be identified. As 50 wt.% of the binder is gypsum, the material has a white appearance that is characteristic for gypsum binders. The whitish appearance could be an advantage for the creation of exposed architectural or structural elements. Small black inclusions as remains of foam glass granules originated from waste metakaolin were indicated, which could be a negative factor influencing the strength of the binder and mortar.

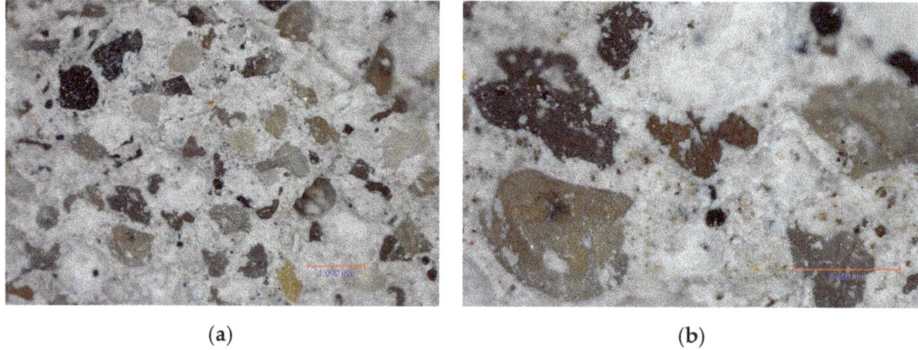

(a)   (b)

Figure 2. Macrostructure of ternary system mortar (**a**) under magnification of 40x, (**b**) under magnification of 120x.

The microstructure of the ternary binder (GCP-PG and GCP-BG) is given in Figure 3. It has a complex fine-grained structure with a large monolithic structure in smaller magnification (500x). The magnification at 2000x reveals the interaction between binder components and different regions, which can be identified.

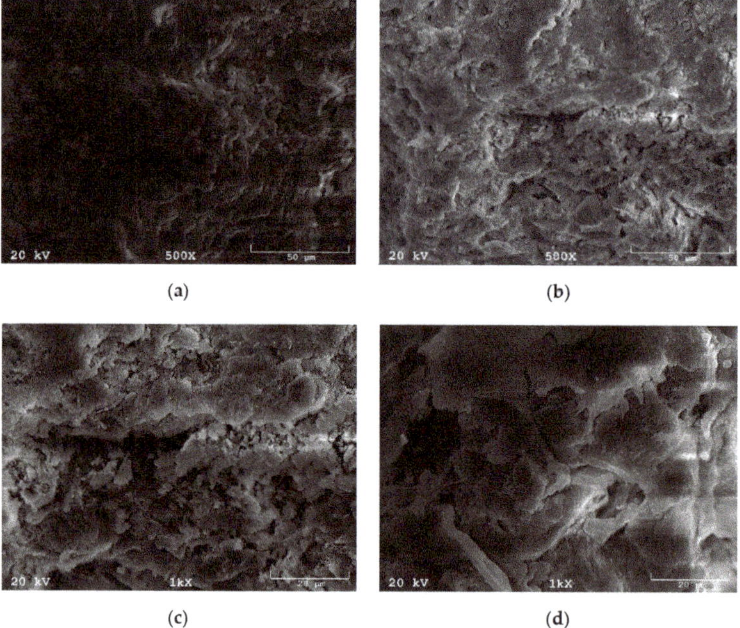

(a)   (b)

(c)   (d)

Figure 3. Microstructure of the ternary binder detected by SEM in magnification of 500x and 1000x. (**a,c**) GCP-phosphogypsum (PG), (**b,d**) GCP-BG.

3.1.2. Chemical and Mineralogical Characterization

XRD results are given in Figure 4. The mineralogical composition of a ternary binder powder was investigated before hydration (Figure 4c,d). Bassanite ($CaSO_4 \cdot 0.5H_2O$, ref 33-0310), calcium silicate oxide ($Ca_3(SiO_4)O$, ref 73-0599), and quartz (SiO_2, ref 78-1252) were identified from the source dry materials of GCP-PG. GCP-BG had the same three minerals as for GCP-PG, while additionally a dolomite ($CaMg (CO_3)_2$, ref 36-0426) and an anhydrite $Ca(SO_4)$, ref 72-0916) were also detected, which is associated with the composition of commercial BG.

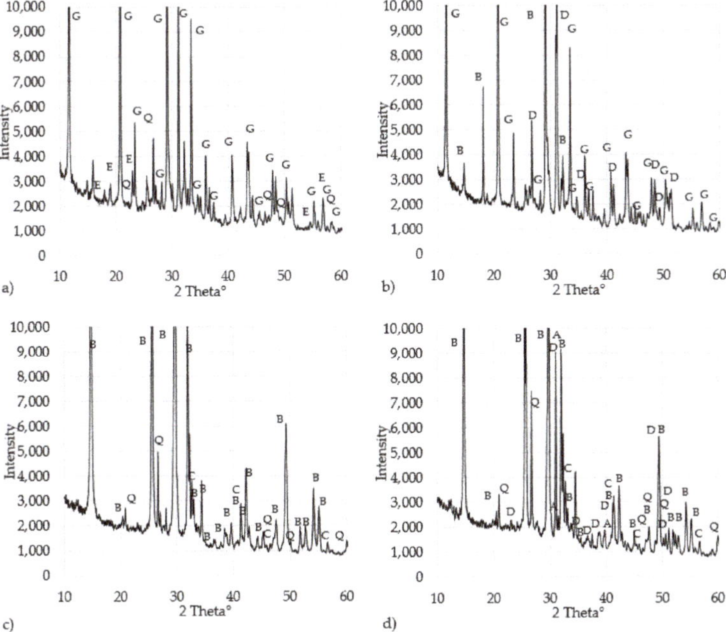

Figure 4. Mineralogical characterization of raw material mixture and hydrated paste of ternary binder: (**a**) Hydrated GCP-PG; (**b**) hydrated GCP-BG, (**c**) raw mixture of GCP-PG, (**d**) raw mixture of GCP-BG. B—Bassanite ($CaSO_4 \cdot 0.5H_2O$, ref 33-0310), C—calcium silicate oxide ($Ca_3(SiO_4)O$, ref 73-0599), Q—quartz (SiO_2, ref 78-1252), D—dolomite ($CaMg (CO_3)_2$, ref 36-0426), A—anhydrite $Ca(SO_4)$, ref 72-0916), G—gypsum ($Ca(SO_4)(H_2O)_2$ (74-1433), E—ettringite ($Ca_6(Al(OH)_6)_2(SO_4)_3(H_2O)_{26}$, ref 72-0646).

The phase change after binder hydration was detected (Figure 4a,b). Moreover, ternary binder contained gypsum, cement, and pozzolan, the only phases which were identified in a hydrated GCP-PG mixture where gypsum ($Ca(SO_4)(H_2O)_2$ (74-1433), quartz (SiO_2, ref 78-1252), and ettringite ($Ca_6(Al(OH)_6)_2(SO_4)_3(H_2O)_{26}$, ref 72-0646) were present. The formation and presence of the ettringite in such a late hydration stage are associated with high gypsum content in the mixture. In such a way, expansions can occur from excessive calcium sulfoaluminate formation after hardening and continue until the gypsum becomes depleted, that is why pozzolan was added and, according to previous studies, addition of pozzolan could even eliminate delayed ettringite formation [18]. In mixture composition, GCP-BG ettringite was not detected. Besides minerals coming from raw materials (dolomite, quartz), gypsum ($Ca(SO_4)(H_2O)_2$ (74-1433) and also gypsum hemihydrate $Ca(SO_4)(H_2O)_{0.5}$ were detected, which could be formed slowly from the hydration of the anhydrite.

The EDX analysis of the GCP-PG binder is given in Figure 5. Large amount of Ca element was identified in most EDX points analyzed, which is associated with the fact that Ca is present in all binder components (phosphogypsum, cement, and metakaolin). The gypsum source in the structure can be identified through element S, which comes from PG binder ($CaSO_4$ $0.5H_2O$). Small quantities of Al and Si were identified as elements coming from metakaolin and cement. This explains the fact that no cement minerals were identified by XRD. The XRD analysis of raw material mixture and hardened binder was investigated while no strong new mineralogical peaks were identified. Mostly transition between gypsum hemihydrate and gypsum dihydrate was observed.

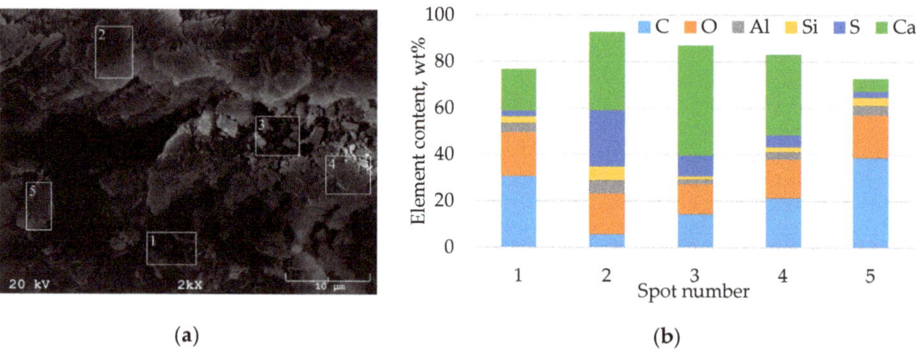

(a)   (b)

Figure 5. Energy Dispersive Spectroscopy (EDX) analysis of GCP-PG binder. (**a**) point positions analyzed by EDX, (**b**) element content in analyzed region.

3.1.3. Physical and Mechanical Properties

The bulk density of hardened binder is given in Table 4. The density of binder paste GCP-BG is slightly lower comparing to GCP-PG (1609 to 1766 kg/m^3), which can be explained by the finer particle distribution for PG and denser structure of the paste. Similar tendencies were observed for mortar. The bulk density of a mortar material increased to 1886 and 2027 kg/m^3 for GCP-BG-S and GCP-PG-S, respectively. The total porosity for GCP-PG and GCP-BG was 23 and 29 vol.%, while for mortar it was even lower −11 and 17 vol.% for GCP-PG-S and GCP-BG-S. The low porosity is associated with low W/B ratio, which is also similar to traditional concrete nature [19]. This phenomenon had an influence also on the compressive strength of the materials presented.

Table 4. Physical properties of high-performance ternary system binder.

Mixture	Consistency by Suttard Viscosimeter, mm	Setting Time, min		Dry Density, kg/m^3	Total Porosity, vol.%
		Initial	Final		
GCP-PG	370	165	280	1766 ± 9	23 ± 2.2
GCP-PG-S	230	130	170	2027 ± 14	11 ± 1.2
GCP-BG	295	70	130	1609 ± 24	29 ± 2.4
GCP-BG-S	210	75	85	1886 ± 38	17 ± 1.9

The consistency of fresh material was strongly affected by the amount of superplasticizer and low water content. The initial mixing of binder powder with water and plasticizer gives stiff mixture while during the intensive mixing, the effect of superplasticizer takes place and very viscous mixture was obtained with flow diameter −295 mm for GCP-BG and 370 mm for GCP-PG. The low W/B ratio of the paste (GCP-PG and GCP-BG) gives similar consistency as for ultra-high performance concrete [20]. The flow for GCP-PG was higher, which is associated with the finer particle nature of PG. The prepared mortar was less viscous comparing to paste while the effect of superplasticizer was also present and high workability obtained. The flow of mortar was from 210–230 mm.

The setting time was longer for GCP-PG and GCP-PG-S as set retarder was purposely used in mixture composition. The initial setting time was from 130–165 min for binder paste and mortar.

The final set time was longer for GCP-PG (280 min) than for mortar GCP-PG-S (170 min) as it was more viscous to initiate the set of the paste. GCP-BG and GCP-BG-S initial setting time was slightly influenced by the mineral additives (t_{in} 70 and 75 min, respectively), while the final set time was longer for GCP-BG (130 min) and was reduced for mortar GCP-BG-S (85 min).

The compressive strength results are given in Figure 6. The compressive strength increased during the water curing similar as for Portland cement binder. The strength after demolding at 24 h was similar among all samples (13–15 MPa). The further curing resulted in slower strength gain for mortars (GCP-PG-S and GCP-BG-S). At the age of 28 d, the strength of moist samples reached 50 MPa for GCP-PG and 30 MPa for GCP-BG. The strength increase was not observed between 7 and 28 d for GCP-BG. The mortar strength reached 26 MPa for GCP-BG-S and 43 MPa for GCP-PG-S. After sample drying, material strength increased significantly at the age of 35 d. The GCP-PG reached 88 MPa while GCP-PG-S mortar reached 50 MPa. Binder GCP-BG had 49 MPa and mortar GCP-BG-S had 39 MPa, respectively. The failure of the mortar was mostly in the transition zone of aggregate and binder. Such result is highly competitive and promising compared to traditional Portland cement mortars. The higher strength results obtained for materials based on PG could be explained by the fine nature of PG giving more reactive nature for ternary binder. Moreover, commercial gypsum BG could contain other additives that could lead to lower strength results.

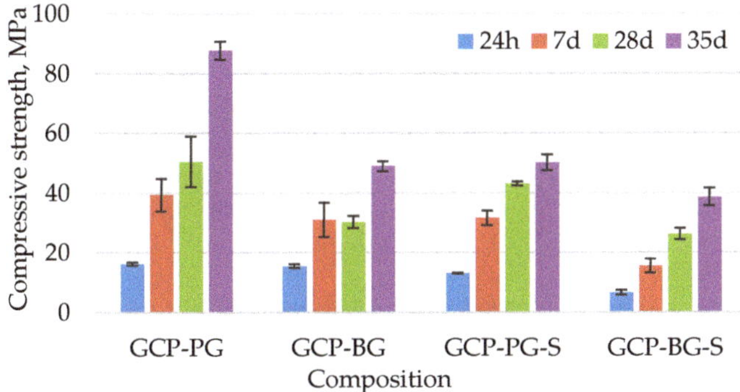

Figure 6. The compressive strength of ternary binder and mortar.

3.2. Properties of Foamed Ternary System Binder

The porous ternary system binder was described through its appearance in macro and micro levels. Physical properties such as density, thermal conductivity as well as mechanical properties such as compressive strength and compressive strength softening coefficient regarding water saturation were determined.

3.2.1. Appearance of Foamed Ternary System Binder

The macrostructure and microstructure of foamed ternary system binder are given in Figure 7. The material appearance is light grey, similar to Portland cement-based materials. The structure of all samples is monolithic and materials are easy to handle. The macrostructure is highly porous for all samples. The pores are distributed uniformly throughout the sample. The pore macrostructure for samples made with BG has larger pores comparing to samples made with PG. This was also confirmed by the pore measurements with digital microscope. Microstructure is generally homogenous while still different particles can be observed in the matrix. The samples with PG have smaller pores that could be evaluated more precisely in micro-level. Depending on the mixture composition, the increase of binder paste in the same foam amount leads to smaller pores remaining the characteristics coming from gypsum source described before. For GCP-BG-1, macro pores with size in range from 1–4 mm were formed while smaller pores were detected within the large pore walls. The smaller pores were in range from 0.2–1.0 mm. GCP-BG-2 had more uniform large pores in a range of 2 mm, while smaller pores were observed in the range from 0.2–0.6 mm. GCP-PG-1 had pores up to 0.7 mm. While most of the pore size was in range from 0.1–0.5 mm. For GCP-PG-2, the structure was similar to GCP-PG-1. The pore size was in range from 0.1–0.6 mm.

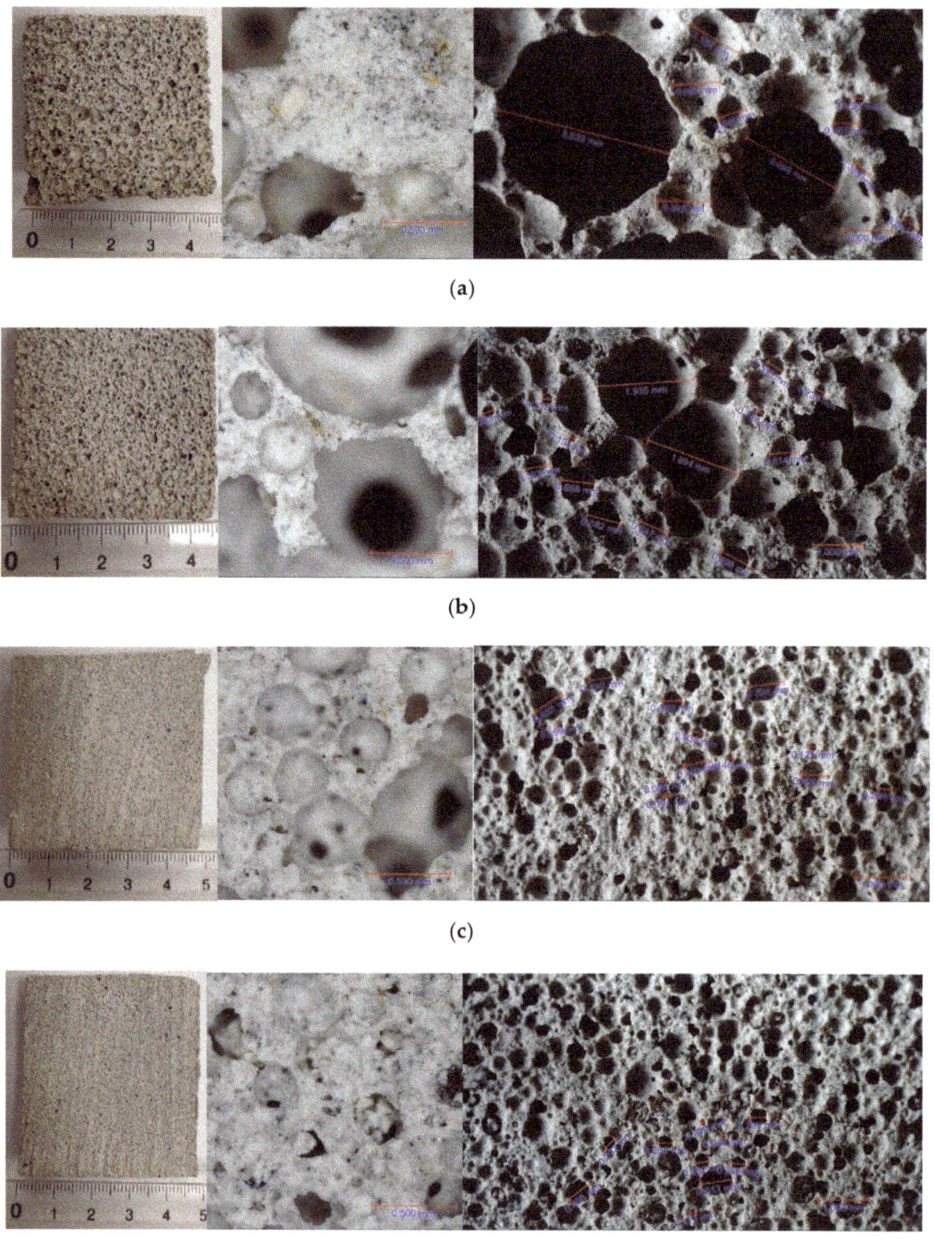

Figure 7. Macrostructure and microstructure of ternary system binder foams. (**a**) GCP-BG-1, (**b**) GCP-BG-2, (**c**) GCP-PG-1, (**d**) GCP-PG-2.

The microstructure of samples observed with SEM is given in Figure 8. The pore hierarchical structure remained the same as previously observed with digital microscope. For mixture GCP-BG, large pores are also visible in SEM. In large magnification, the structure similar to cement-based composites is visible. It has a complex fine-grained structure with the large monolithic structure in smaller magnification. The magnification at 500x reveals the interaction between binder components and different regions, which can be identified.

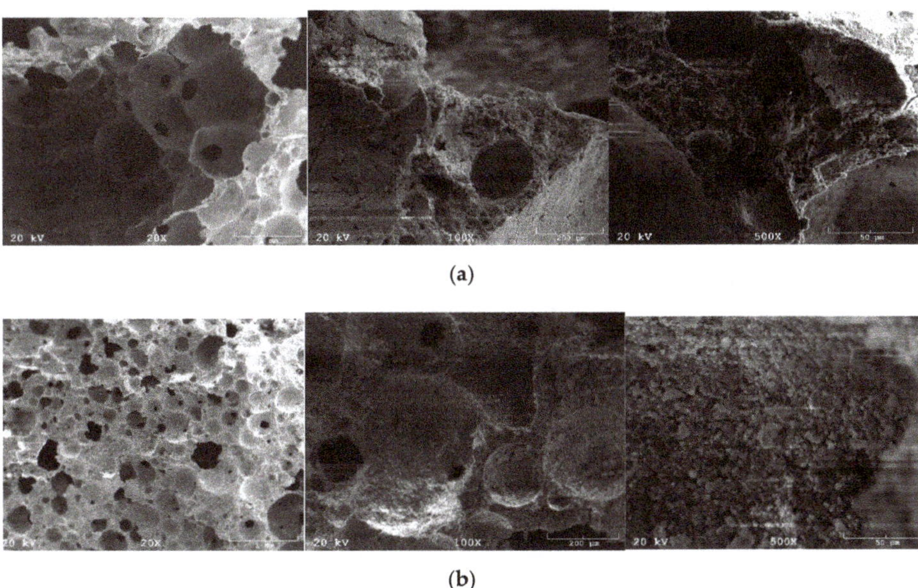

(a)

(b)

Figure 8. The micrographs of ternary system binder foam observed with SEM. (**a**) GCP-BG-1, (**b**) GCP-PG-1.

3.2.2. Physical and Mechanical Properties

The physical properties of ternary system binder foams are given in Table 5. Due to larger pores, the bulk density of GCP-BG foams is lower comparing to GCP-PG foams. The lightweight material with density 368 and 486 kg/m^3 was obtained depending on the amount of paste mixed with foams. GCP-PG foams density increased to 634 and 697 kg/m^3, respectively, which is associated with the smaller pores formed in the production process. The total porosity was higher for GCP-BG, which explains the lower bulk density. The total porosity was 84 and 79 vol.% for GCP-BG and 72 and 70 wt.% for GCP-PG. Important finding was that open porosity was formed for GCP-BG foams, while for GCP-PG more pores were closed. This is also evidenced by the fact that during the water absorption test the GCP-PG foams remained flowing after total immersion in water (total test time was 72 h). The water absorption was 105 and 91 wt.% for GCP-PG foams, while for GCP-PG it was 46 and 39 wt.%, respectively. Density for water-saturated samples increased and was in range from 758–901 kg/m^3.

Table 5. Physical properties of foamed ternary system binder.

Composition	ρ_o, kg/m^3	$\rho_o$wet, kg/m^3	Water Absorption, wt.%	Open Porosity, vol.%	Closed Porosity, vol.%	Total Porosity, vol.%	Thermal Conductivity, λ, W/m·K
GCP-BG-1	368	758	105	39	45	84	0.086
GCP-BG-2	486	901	91	43	36	79	0.123
GCP-PG-1	634	876	46	28	45	72	0.111
GCP-PG-2	697	901	39	25	44	70	0.153

The lowest thermal conductivity of produced specimens was for GCP-BG-1. With the lowest bulk density, the thermal conductivity was as low as 0.086 W/m·K. The increase of bulk density for GCP-BG increased the thermal conductivity to 0.123 W/m·K. Furthermore, GCP-PG-1 had higher bulk density, and due to finer pore size distribution, the thermal conductivity was lower comparing to GCP-BG-2 and it was 0.111. The GCP-PG-2 had thermal conductivity of 0.153 W/m·K.

The compressive strength results of prepared ternary system binder foams are given in Figure 9. The compressive strength of GCP-BG-1 was 0.33 MPa and increased to 1.0 for GCP-BG-2. The increase of compressive strength was observed for GCP-PG. For GCP-PG-1, it was 2.4 MPa and increased to 5.4 MPa for GCP-PG-2. The strength increase is associated with the increase of foam bulk density and with the size of macropores formed during production. For samples made with PG, pores were smaller. Foam samples after water absorption tests were tested for compression to determine softening coefficient of the foams. Since gypsum is not watertight material and in proposed binder it is a major part of the binders' component (50 wt.%), concerns regarding water tightness were analyzed (Table 6). It was detected that water softening coefficient was from 0.50–0.64. Similar trends were observed with concrete foams before—the compressive strength decreases with the increase of volumetric moisture content for every density ranging from 300–800 kg/m^3, but the softening rate was lower—the compressive strength in the immersed saturation state is about 0.86–0.89 times of that in the standard curing state [21].

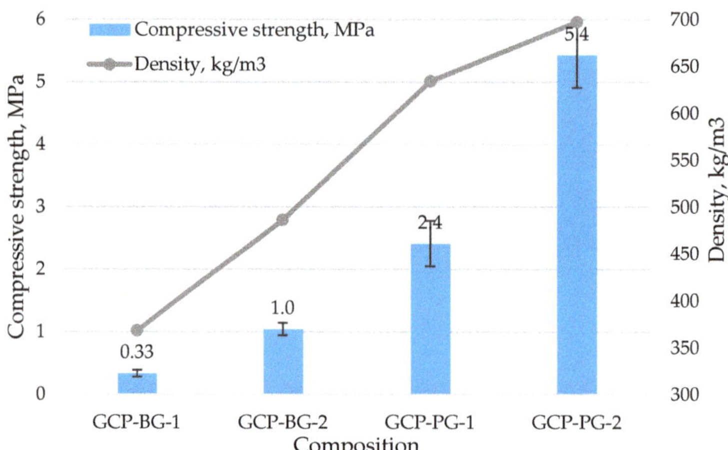

Figure 9. The relation between bulk density and thermal conductivity of foamed ternary system binder.

Table 6. Physical properties of foamed ternary system binder.

Composition	Dry Compression Strength, MPa	Saturated Compression Strength, MPa	Softening Coefficient
GCP-BG-1	0.3 ± 0.06	0.2 ± 0.02	0.56
GCP-BG-2	1.0 ± 0.10	0.6 ± 0.06	0.59
GCP-PG-1	2.4 ± 0.37	1.2 ± 0.06	0.50
GCP-PG-2	5.4 ± 0.52	3.5 ± 0.17	0.64

4. Conclusions

High-strength ternary system binder was developed containing major part of gypsum, metakaolin, and Portland cement. Valorization option for waste phosphogypsum and metakaolin was offered by incorporating these materials in high-performance cementitious composites and by producing highly porous lightweight foam materials. Proposed binder has technological properties similar to traditional cementitious binders based on Portland cement. The mixture composition with low W/B ratio of 0.34 was developed with the addition of superplasticizer. The set time of binder was adjusted by the help of the set retarder and it is comparable to Portland cement. Impressive compressive strength of up to 88 MPa was reached. The binder proved to be suitable to produce mortar with a strength of up to 50 MPa. Samples were prepared by a simple methodology that, together with the use of wastes, contributes to improve sustainability of the process. Foamed ternary system binder material with density in range from 368–697 kg/m^3 was obtained with compressive strength from 0.33 to 3.5 MPa. Besides technological properties, long-term properties such as durability, shrinkage/swelling should be evaluated, as gypsum with cement can form hazardous compounds such as ettringite, which could lead to loss of integrity of the material.

Author Contributions: Conceptualization, G.B. and D.B.; methodology, G.B. and D.B.; validation, G.B., D.B., and J.Z.; formal analysis, G.B.; investigation, G.B and J.Z.; resources, G.B. and D.B.; data curation, G.B.; writing—original draft preparation, G.B.; writing—review and editing, G.B., J.Z., and D.B.; visualization, G.B. and J.Z.; supervision, G.B. and D.B.; project administration, G.B. and D.B.; funding acquisition, G.B. All authors have read and agreed to the published version of the manuscript.

Funding: This work has been supported by the European Regional Development Fund within the Activity 1.1.1.2 "Post-doctoral Research Aid" of the Specific Aid Objective 1.1.1 "To increase the research and innovative capacity of scientific institutions of Latvia and the ability to attract external financing, investing in human resources and infrastructure" of the Operational Program "Development of sustainable and effective lightweight building materials based on secondary resources" (No. 1.1.1.2/VIAA/1/16/050).

Conflicts of Interest: The authors declare no conflict of interest.

References

1. Bicer, A.; Kar, F. Thermal and mechanical properties of gypsum plaster mixed with expanded polystyrene and tragacanth. *J. Therm. Sci. Eng. Prog.* **2017**, *1*, 59–65. [CrossRef]
2. Lushnikova, N.; Dvorkin, L. Sustainability of gypsum products as a construction material. *J. Sustain. Constr. Mater.* **2016**, 643–681. [CrossRef]
3. British Standards Institution. *BS EN 12859:2011, Gypsum Blocks—Definitions, Requirements and Test Methods*; BSI: London, UK, 30 April 2011; p. 38. ISBN 9780580710360.
4. Dragomir, A.M.; Lisnic, R.; Prisecaru, T.; Prisecaru, M.M.; Vîjan, C.A.; Nastac, D.C. Study on synthetic gypsum obtained from wet flue gas desulphurisation in thermal power plants. *J. Rev. Rom. Mater. Rom. Mater.* **2017**, *47*, 551–556.
5. Tayibi, H.; Choura, M.; López, F.A.; Alguacil, F.J.; López-Delgado, A. Environmental impact and management of phosphogypsum. *J. Environ. Manag.* **2009**, *90*, 2377–2386. [CrossRef] [PubMed]
6. Bumanis, G.; Zorica, J.; Bajare, D.; Korjakins, A. Technological properties of phosphogypsum binder obtained from fertilizer production waste. *J. Energy Procedia* **2018**, *147*, 301–308. [CrossRef]
7. Zhang, L.; Zhang, A.; Li, K.; Wang, Q.; Han, Y.; Yao, B.; Gao, X.; Feng, L. Research on the pretreatment and mechanical performance of undisturbed phosphogypsum. *J. Case Stud. Constr. Mater.* **2020**, *13*, e00400. [CrossRef]
8. Kumar, S. Fly ash-lime-phosphogympsum hollow blocks for walls and partitions. *J. Build. Env.* **2003**, *38*, 291–295. [CrossRef]
9. Morsy, M.S.; Rashad, A.M.; Shoukry, H.; Mokhtar, M.M.; El-Khodary, S.A. Development of lime-pozzolan green binder: The influence of anhydrous gypsum and high ambient temperature curing. *J. Build. Eng.* **2020**, *28*, 101026. [CrossRef]
10. Huang, Y.; Lin, Z.S. Investigation on phosphogypsum-steel slag-granulated blast-furnace slag-limestone cement. *J. Constr. Build. Mater.* **2010**, *24*, 1296–1301. [CrossRef]
11. Zhang, G.; Li, G.; Li, Y. Effects of superplasticizers and retarders on the fluidity and strength of sulphoaluminate cement. *J. Constr. Build. Mater.* **2016**, *126*, 44–54. [CrossRef]
12. Akhlaghi, O.; Aytas, T.; Tatli, B.; Sezer, D.; Hodaei, A.; Favier, A.; Scrivener, K.; Menceloglu, Y.Z.; Akbulut, O. Modified poly(carboxylate ether)-based superplasticizer for enhanced flowability of calcined clay-limestone-gypsum blended Portland cement. *J. Cem. Concr. Res.* **2017**, *101*, 114–122. [CrossRef]
13. Namsone, E.; Šahmenko, G.; Korjakins, A. Durability Properties of High Performance Foamed Concrete. *J. Procedia Eng.* **2017**, *172*, 760–767. [CrossRef]
14. Skujans, J.; Iljins, U.; Ziemelis, I.; Gross, U.; Ositis, N.; Brencis, R.; Veinbergs, A.; Kukuts, O. Experimental research of foam gypsum acoustic absorption and heat flow. *J. Chem. Eng. Trans.* **2010**, *19*, 79–84. [CrossRef]

15. Ma, B.; Jin, Z.; Su, Y.; Lu, W.; Qi, H.; Hu, P. Utilization of hemihydrate phosphogypsum for the preparation of porous sound absorbing material. *J Constr. Build. Mater.* **2020**, *234*, 117346. [CrossRef]

16. Nizevičienė, D.; Vaičiukynienė, D.; Michalik, B.; Bonczyk, M.; Vaitkevičius, V.; Jusas, V. The treatment of phosphogypsum with zeolite to use it in binding material. *J Constr. Build. Mater.* **2018**, *180*, 134–142. [CrossRef]

17. Bumanis, G.; Vitola, L.; Stipniece, L.; Locs, J.; Korjakins, A.; Bajare, D. Evaluation of Industrial by-products as pozzolans: A road map for use in concrete production. *J Case Stud. Constr. Mater.* **2020**, *13*, e00424. [CrossRef]

18. Nguyen, V.H.; Leklou, N.; Aubert, J.E.; Mounanga, P. The effect of natural pozzolan on delayed ettringite formation of the heat-cured mortars. *J Constr. Build. Mater.* **2013**, *48*, 479–484. [CrossRef]

19. Bajare, D.; Bumanis, G.; Shakhmenko, G.; Justs, J. High performance and conventional concrete properties affected by ashes obtained from different type of grasses. *J Am. Concr. Inst.* **2012**, *289*, 317–330.

20. Žvironaite, J.; Kligys, M.; Pundiene, I.; Pranckevičiene, J. Effect of limestone particles on rheological properties and hardening process of plasticized cement pastes. *J Medzg.* **2015**, *21*, 143–148. [CrossRef]

21. Zhao, W.; Su, Q.; Wang, W.; Niu, L.; Liu, T. Experimental study on the effect of water on the properties of cast in situ foamed concrete. *J Adv. Mater. Sci. Eng.* **2018**. [CrossRef]

Article

A Low-Autogenous-Shrinkage Alkali-Activated Slag and Fly Ash Concrete

Zhenming Li [1,*], Xingliang Yao [1,2], Yun Chen [1,3], Tianshi Lu [1] and Guang Ye [1,4]

[1] Department of Materials, Mechanics, Management & Design, Faculty of Civil Engineering and Geoscience, Delft University of Technology, 2628 CN Delft, The Netherlands; yaoxl1991@mail.sdu.edu.cn (X.Y.); y.chen-9@tudelft.nl (Y.C.); lutianshi2017@gmail.com (T.L.); g.ye@tudelft.nl (G.Y.)

[2] National Engineering Laboratory for Coal-Fired Pollutants Emission Reduction, Shandong University, Jinan 250061, China

[3] School of Materials Science and Engineering, South China University of Technology, Guangzhou 510640, China

[4] Magnel Laboratory for Concrete Research, Department of Structural Engineering, Ghent University, 9052 Ghent, Belgium

* Correspondence: z.li-2@tudelft.nl

Received: 17 August 2020; Accepted: 31 August 2020; Published: 2 September 2020

Abstract: Alkali-activated slag and fly ash (AASF) materials are emerging as promising alternatives to conventional Portland cement. Despite the superior mechanical properties of AASF materials, they are known to show large autogenous shrinkage, which hinders the wide application of these eco-friendly materials in infrastructure. To mitigate the autogenous shrinkage of AASF, two innovative autogenous-shrinkage-mitigating admixtures, superabsorbent polymers (SAPs) and metakaolin (MK), are applied in this study. The results show that the incorporation of SAPs and MK significantly mitigates autogenous shrinkage and cracking potential of AASF paste and concrete. Moreover, the AASF concrete with SAPs and MK shows enhanced workability and tensile strength-to-compressive strength ratios. These results indicate that SAPs and MK are promising admixtures to make AASF concrete a high-performance alternative to Portland cement concrete in structural engineering.

Keywords: alkali-activated concrete; shrinkage; cracking; internal curing; metakaolin

---

1. Introduction

An important way to reduce the CO_2 emissions from the construction sector is to use "greener" alternative binders to ordinary Portland cement (OPC). Alkali-activated materials (AAMs), or so-called geopolymers, which can be made of industrial by-products, have been reported to entail much lower CO_2 emission and embodied energy than OPC systems [1].

Blast-furnace slag (indicated as "slag" hereafter) and fly ash, as by-products from steelmaking and coal-fired electricity plants, respectively, are the two most commonly utilized precursors to produce AAMs. The literature has illustrated that alkali-activated slag and fly ash (AASF) shows superior strength, excellent durability and good fire resistance compared to OPC systems [2–4]. However, a wider application of this material has not been reached yet, partly due to the large autogenous shrinkage and the potential risk of cracking of this material, especially when NaOH and Na_2SiO_3 are used as activators [5,6].

A number of studies have been conducted to reduce the shrinkage of AASF. However, it has been found that the shrinkage-reducing agents and expansive additives that are widely adopted in OPC may be ineffective or cause side effects (e.g., strength loss) in AAMs due to the differences in microstructures and chemical environments between AAMs and OPC [7,8]. Elevated temperature curing can mitigate the shrinkage of AASF [9], but this strategy has high requirements on the curing equipment and can accelerate the setting of AASF, which is already more rapid than usually needed. Feasible strategies to mitigate the autogenous shrinkage of AASF are desired to widen the commercial acceptance of this material.

The results of previous studies [10–13] indicate that internal curing and incorporation of a small amount of metakaolin (MK) are helpful to reduce the autogenous shrinkage of alkali-activated slag (AAS) and AASF. However, both of these two strategies have their limitations. Internal curing can significantly mitigate the autogenous shrinkage caused by self-desiccation, but on the first day when autogenous shrinkage rapidly develops, the effect of internal curing is quite limited due to the possible involvement of other shrinkage mechanisms [13]. By contrast, the incorporation of MK can effectively mitigate the early-age autogenous shrinkage of AAS and AASF by reducing the high reaction rate in the acceleration period and coarsening the gel pores [10]. However, the effect of MK on later-age autogenous shrinkage is not evident. These results indicate that these two admixtures might be a good complement to each other to further lower the autogenous shrinkage of AAMs. However, the combined effect of them on the autogenous shrinkage and cracking properties of AASF systems have not been studied yet.

In this study, superabsorbent polymers (SAPs) and MK are applied to reduce the autogenous shrinkage of AASF. Experiments are conducted at both paste and concrete scales. The cracking potential of paste and concrete is evaluated by the ring test and Temperature Stress Testing Machine (TSTM), respectively. The workability and the mechanical properties of the concrete are also investigated. Eventually, with the addition of SAPs and MK, a high-performance eco-efficient alkali-activated concrete is produced.

2. Materials and Methods

2.1. Raw Materials

The main precursors used in this study were slag and Class F fly ash. The material parameters of these two precursors are shown in Tables 1 and 2.

Table 1. Chemical compositions of slag, fly ash and MK.

Oxide (wt. %)	SiO_2	Al_2O_3	CaO	MgO	Fe_2O_3	SO_3	K_2O	TiO_2	Other
Slag	31.8	13.3	40.5	9.3	0.5	1.5	0.3	1.0	1.9
Fly ash	56.8	23.8	4.8	1.5	7.2	0.3	1.6	1.2	2.8
MK	55.1	38.4	0.6	-	2.6	-	0.2	1.1	2.1

Table 2. Particle size and density of slag, fly ash and MK.

	Particle Size (μm)			Density (g/cm^3)
	D10	**D50**	**D90**	
Slag	4.6	18.3	33.2	2.9
Fly ash	10.6	48.1	83.4	2.4
MK	27.0	69.4	113.5	2.7

Bulk-polymerized SAPs with particle sizes up to about 200 μm in the dry state were used. The SAPs were a cross-linked copolymer based on acrylamide and acrylate. The composition and physical properties of MK are also shown in Tables 1 and 2, respectively. It should be noted that part of the high percentage of silica content in MK was due to the presence of quartz (43% in weight), which remains inert during the reaction process [14]. No admixture besides SAPs and MK was used.

Pellets of NaOH, deionized water and commercial water glass solution were used to synthesize the alkaline activator. For 1000 g of precursor, an activator containing 384 g of water, 1.146 mol of SiO_2 and 0.76 mol of Na_2O was applied.

2.2. Mixtures

The mixture design of the plain AASF paste and AASF paste with SAPs and/or MK is shown in Table 3.

Table 3. Mixture proportions of AASF paste with SAPs and/or MK.

Composition (wt. %)	AASF Paste	AASFIC	AASFMK	AASFICMK
Slag	1	1	0.9	0.9
Fly ash	1	1	1	1
Activator	1	1	1	1
SAPs	-	0.0032	-	0.0032
Extra activator for internal curing	-	0.064	-	0.064
MK	-	-	0.1	0.1

The dosage of SAPs was designed based on the absorption capacity of the SAPs (20 g/g activator [11]) and the chemical shrinkage of AASF paste. The ultimate chemical shrinkage of AASF paste was determined by taking the chemical shrinkage measured by dilatometry untill the age of 28 days [15,16], which was 0.026 mL/g. Taking also into account the density of the activator as well, 1.23 g/cm^3, the extra liquid provided by the SAPs should be 0.032 g per gram of binder. Therefore, the adding amount of 0.16 wt. % SAP was applied for the internal curing of AASF.

According to the results from previous studies [10,17], 5 wt. % of MK in the binder can already show a significant mitigating effect on the autogenous shrinkage of AASF paste. Higher amounts of MK may lead to considerable strength loss in the matrix. Therefore, the dosage of MK in this study was chosen as 5 wt. % of the binder as a substitution for 10 wt. % of slag.

Based on the results at the paste scale (details will be given in Sections 3.1 and 3.2), two mixtures, AASF and AASFICMK, were chosen for experiments at the concrete scale. The mixture proportions of the concrete mixtures are shown in Table 4.

Table 4. Mixture proportions of AASF and AASFICMK concrete.

Composition (kg/m^3 of Concrete)	AASF Concrete	AASFICMK Concrete
Slag	200	180
Fly ash	200	200
Activator	200	200
SAPs	-	0.64
Extra activator for internal curing	-	12.8
MK	-	20
Aggregate (0–4 mm)	789	789
Aggregate (4–8 mm)	440	440
Aggregate (8–16 mm)	525	525

2.3. Experimental Methods

2.3.1. Autogenous Shrinkage of Paste

The autogenous shrinkage of the paste was measured by the corrugated tubes method according to ASTM C1968 [18]. The detailed procedure can be found in [19].

2.3.2. Cracking Potential of Paste

The cracking potential of the paste induced by restrained autogenous shrinkage was indicated by the dual ring test [20,21]. The geometry of the rings is shown in Figure 1. The strain gauges attached to the inner surface of the inner steel ring started to record the strains of the inner ring when the paste was cast in between the two steel rings. The paste ring was sealed by aluminium foil fixed with asphalt tape during the test to avoid moisture loss. The apparatus was put in a temperature-controlled room with the temperature fixed at 20 °C.

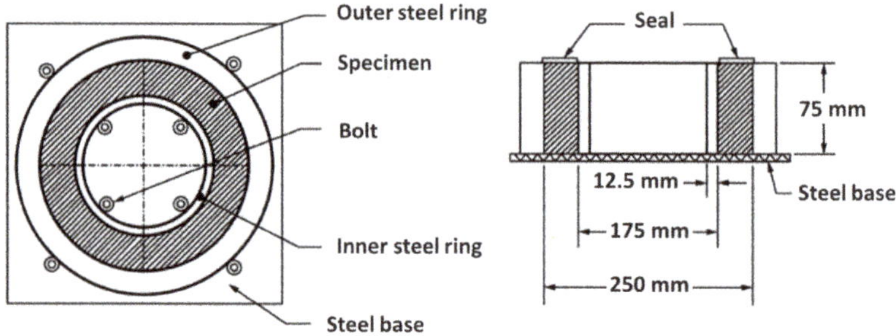

Figure 1. Dimensions of the steel rings, after [22].

The maximal stress in the paste was calculated according to Equation (1) based on the strain and dimensions of the rings [23].

$$\sigma_{max} = -\varepsilon \cdot E_{steel}\left(\frac{R_{IP}^2 - R_{II}^2}{2R_{IP}^2}\right)\left(\frac{R_{OP}^2 + R_{IP}^2}{R_{OP}^2 - R_{IP}^2}\right) \tag{1}$$

where σ_{max} is the maximum stress in the paste ring; ε is the measured strain of the inner steel ring; E_{steel} is the elastic modulus of the ring; R_{II}, R_{IP} and R_{OP} are the inter radius of the inner steel ring (75 mm), the inner radius of the paste (87.5 mm) and its outer radius (125 mm), respectively.

2.3.3. Workability of Concrete

The workability of AASFICMK concrete was measured. The slump was measured according to NEN 12350 [24]. The largest diameter of the flow spread of the concrete and the diameter of the spread at right angles to it were measured immediately after the cone was lifted up. The setting time of AASFICMK concrete was determined by the Vicat method [25] on the corresponding paste.

2.3.4. Autogenous Shrinkage of Concrete

The autogenous shrinkage of the concrete was measured with an Autogenous Deformation Testing Machine (ADTM) [26]. The prismatic mold for the concrete was made of thin steel plates and external insulating materials. The size of the mold is $1000 \times 150 \times 100$ mm^3 (see the top left corner of Figure 2). The mold was connected with cryostats by a series of circulation canals located between the plates and the insulating material.

The length change of the concrete was measured with two external quartz rods located next to the side mold. Linear Variable Differential Transformers (LVDTs) were installed at both ends of the rods. The LVDTs measured the movement of the steel bars, which were cast in the concrete. The distance between the two embedded steel bars was 750 mm. The measurement of the deformation of AASF and AASFICMK concrete started at 11h and 12 h after casting, respectively, when the concrete was stiff enough to hold the measuring bars and the LVDTs that were connected with them. Attention was paid to the sealing of the molds in order to avoid moisture loss to the environment.

2.3.5. Cracking Potential of Concrete

The cracking initiation in the concrete was monitored by a TSTM. The TSTM was equipped with a horizontal steel frame in which compressive and tensile force could be applied on the concrete specimen. A temperature-controlled mold was used for the concrete casting in order to obtain any prescribed thermal condition. The mold was similar to the ADTM mold described in Section 2.3.4. The whole specimen was of a dog-bone shape and the testing area of interest was of a prismatic shape

($1000 \times 150 \times 100$ mm^3), see the bottom left corner of Figure 2. The deformation of the concrete was kept at zero (nominally, in reality within ± 1 μm range) so that a full restraint condition could be reached. When the total deformation of the concrete went beyond the threshold, a load was applied to pull or push the concrete back to the original position. The load was recorded with the load cell with a loading capacity of 100 kN and a resolution of 0.049 kN. A sudden drop in the load to around zero indicated the occurrence of cracking in the concrete.

2.3.6. Strength of Concrete

Concrete cubes ($150 \times 150 \times 150$ mm^3) for the compressive and splitting strength tests were cast and cured in sealed and temperature-controlled steel molds (see the top middle of Figure 2). The moulds were connected with cryostats by parallel circulation tubes and the upper surface was sealed by plastic film.

The compressive strength and splitting strength of the concrete were measured according to NEN-EN 12390 [27]. The measurements were conducted at the age of 1, 3, 7 and 28 days and the day when the concrete beam in the TSTM cracked. One cube was tested for compressive strength and two for splitting strength.

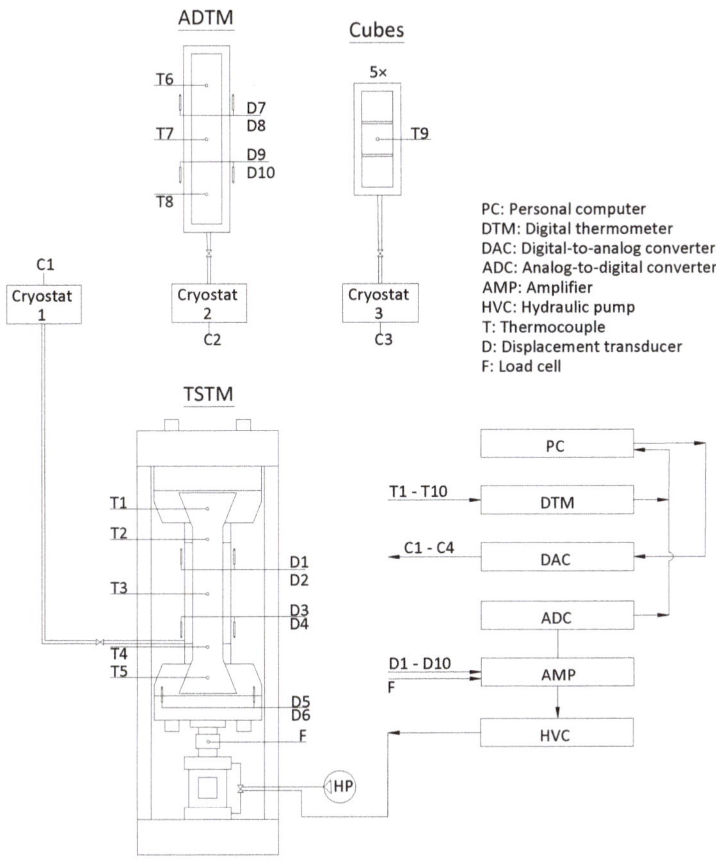

Figure 2. Overview of the set-up for the concrete properties, after [28]. Top left: ADTM. Top middle: Cubes. Bottom left: TSTM. Bottom right: controlling systems.

To keep consistency of the materials, the concrete samples for strength, autogenous shrinkage and cracking potential measurements were from the same batch of casting. All the samples were cured in a sealed condition. The whole set-up, including the TSTM, ADTM, cubes and controlling systems is schematically shown in Figure 2. Various thermocouples were used to monitor the temperatures of the samples. To minimize the influence of thermal deformation on the autogenous shrinkage, the temperatures of the middle parts of the specimen, i.e., T3 for the TSTM, T7 for the ADTM, and T9 for the cubes, were controlled at 20 °C.

3. Results and Discussion

3.1. Autogenous Shrinkage of Paste

The autogenous shrinkage curves of the paste mixtures are shown in Figure 3. The autogenous shrinkage of AASF paste reached more than 2000 µm/m at 1 day and around 4000 µm/m at the age of 7 days. This magnitude is higher than the autogenous shrinkage of common OPC-based systems irrespective of the presence of supplementary materials [29,30]. The shrinkage mechanism was discussed in a previous study [19]. It can be seen from Figure 3 that both the additions of SAPs and MK resulted in lower autogenous shrinkage of AASF paste. In particular, the addition of SAPs greatly mitigate the autogenous shrinkage of AASF paste after the first day. By contrast, MK was more effective on the first day; afterward, the effect of MK became less evident.

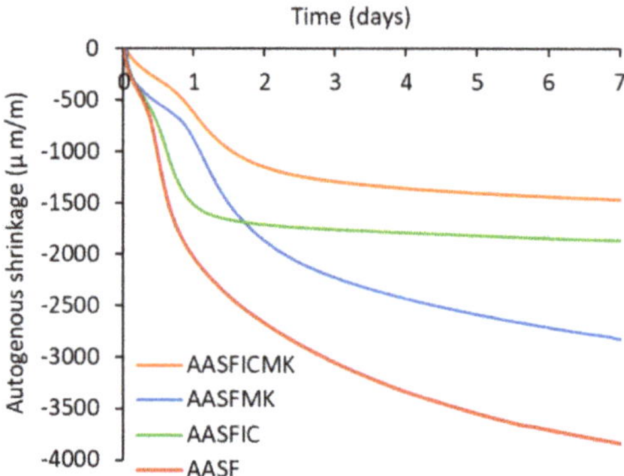

Figure 3. Autogenous shrinkage of AASF paste with SAPs and/or MK.

When SAPs and MK were added together into AASF, the autogenous shrinkage of the paste was the lowest among all the four mixtures in the whole week. The mitigating effect of the combination of SAPs and MK was more evident than when they were applied individually. Both the early-age and later-age autogenous shrinkage were significantly mitigated compared to those of the plain AASF paste. For example, the 1-day and 7-day autogenous shrinkage of AASFICMK paste was only 30% and 40% of that of AASF paste, respectively. This result indicates that SAPs and MK complement each other in mitigating the autogenous shrinkage of AASF.

3.2. Cracking Potential of Paste

It should be noted that low autogenous shrinkage does not necessarily mean low cracking potential. If the mitigating effect on the autogenous shrinkage is at the cost of dramatic loss in strength loss, the material may be subject to higher cracking risk [31]. To investigate the effect of SAPs and MK on the cracking potential of the paste, the ring test was used to measure the stress in the paste mixtures under a restrained condition. The results are shown in Figure 4. The sudden drop in the stress to around zero indicated the occurrence of cracking.

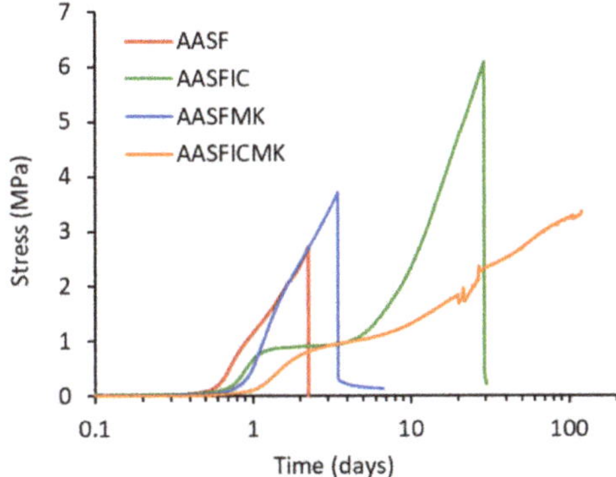

Figure 4. Autogenous-shrinkage-induced stress in AASF paste with SAPs and/or MK. A logarithmic scale is used on the x-axis in order to distinguish individual curves. The small fluctuation of the stress in AASFICMK paste on around 20 days and 30 days was due to the temperature fluctuation in the curing room.

Figure 4 shows that the plain AASF paste cracked on the third day after casting when the internal stress reached around 2.7 MPa. Substituting 10 wt. % slag by MK prolonged the cracking time by about 1 day and the paste broke at a stress of 3.7 MPa. The cracking potentials of AASF and AASFMK pastes were both "high" according to ASTM C1581 [22]. With internal curing by SAPs, the paste did not crack until 29 days of curing when the stress reached 6 MPa. Since the cracking time of AASFIC was close to 28 days, and the stress rate at the cracking time was 0.14 MPa/day, the cracking potential of AASFIC could be classified as "medium-low" according to ASTM C1581 [22].

The results in Figure 4 indicate that both SAPs and MK were helpful in reducing the cracking potential of the paste. Meanwhile, the addition of SAPs or MK did not lead to low strength of the matrix, as indicated by the high failure stress of the pastes. When SAPs and MK were applied together into AASF, the paste showed no cracking within 3 months of curing, which could not be realized by using only SAPs or MK. According to the low stress rate (<0.1 MPa/day), the cracking risk of AASFICMK paste was rather low [22].

Since the combined incorporation of SAPs and MK led to the lowest autogenous shrinkage and the lowest cracking potential, the mixture AASFICMK was further studied at the concrete level to develop low-shrinkage and low-cracking-potential AAMs concrete. The plain AASF concrete was studied as a reference mixture.

3.3. Workability and Consistence of Fresh Concrete

During the casting of AASFICMK concrete, a good flowability was observed. The slump of AASFICMK concrete was measured to be 280 mm (Figure 5a). The concrete quickly spread over the whole flow table (700 × 700 mm) after the cone was lifted up (Figure 5b). This slump flow value corresponded to the class SF2 for self-compacting concrete [32]. The initial and final setting times of AASFICMK measured by the Vicat method were 58 min and 117 min, respectively. The long setting time and the large slump flow indicated very good workability of AASFICMK concrete.

Figure 5. Slump (**a**) and flowability (**b**) of AASFICMK concrete.

3.4. Strength of Concrete

The strength development of AASFICMK concrete is shown in Figure 6 with the plain AASF concrete for comparison. It can be seen that with the incorporation of SAPs and MK, the compressive and splitting tensile strength of AASFICMK concrete was generally lower than that of AASF concrete. The reduced strength was contributed by both SAPs and MK. To provide internal curing to the concrete, extra liquid was added during mixing to be absorbed by the SAPs (see Table 4). The SAPs after absorption would act as liquid reservoirs during reaction and also as defects due to the large voids left when the liquid was released. The increased porosity of the concrete led to reduced strength [11]. Besides, the incorporation of MK was found to hinder the reaction rate in the acceleration period and could therefore reduce the strength in the very early age [17], although its impact on the 28-day strength was minor. When SAPs and MK were added together, their reducing effects were combined. Nonetheless, the 1-day compressive strength of AASFICMK concrete reached 2.1 MPa, which enabled a successful demolding at that age. The 28-day compressive strength of AASFICMK concrete reached 51 MPa, which was already sufficient for most structural uses as specified, for example, in the standard ACI 318 [33].

Besides strength values, the splitting tensile strength-to-compressive strength (f_t/f_c) ratio is also an important parameter that allows for the estimation of f_t by knowing f_c or vice versa [34]. The ratio also provides insight into the stress type (compression or tension) to which the concrete is more prone. The f_t/f_c ratio of AASFICMK concrete is compared with that of AASF concrete in Figure 7. On the first day, the f_t/f_c ratio of AASFICMK concrete was lower than that of AASF concrete which was probably because that the bonding between the aggregate and the paste in AASFICMK was still weak due to the retarding effect of MK and SAPs on the early-age reaction rates of AASF [10,11]. After the first day, however, the f_t/f_c ratios of AASFICMK concrete were always higher than those of AASF concrete. The higher f_t/f_c ratio of AASFICMK indicates that the incorporation of MK and SAPs could improve the tensile resistance of AASF concrete.

According to [31,35], a low f_t/f_c ratio is related to the development of microcracking in the concrete, for example in the paste surrounding aggregates, which harms the tensile strength more than the compressive strength of concrete. As shown in Figures 3 and 4, the incorporation of SAPs and MK reduced the autogenous shrinkage and the cracking potential of AASF paste. Therefore, the development of microcracking in AASFICMK concrete was supposed to be less severe than in AASF concrete. This may be the main reason why AASFICMK concrete showed a higher f_t/f_c ratio than the plain AASF concrete. Whether the combination of SAPs and MK can reduce the autogenous shrinkage and the potential for cracking of concrete at the macro level will be verified in the next sections.

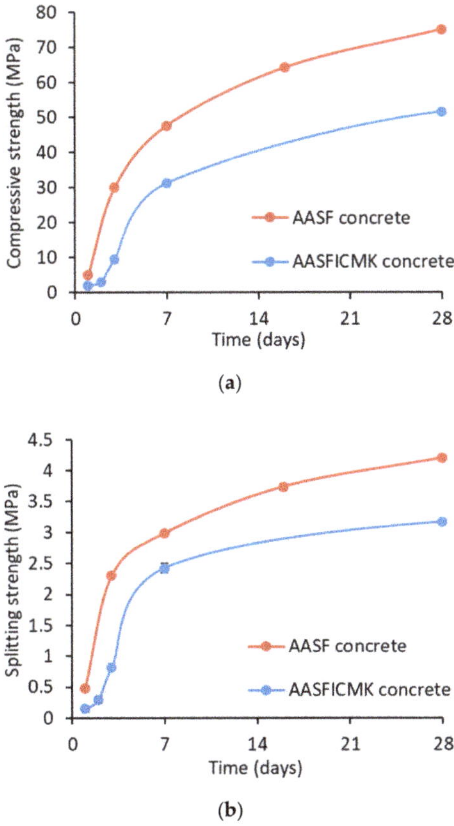

Figure 6. Compressive (**a**) and splitting strength (**b**) of AASFICMK concrete in comparison with AASF concrete. For splitting strength, the error bar is shown in the diagram, but it is too small to be clearly distinguished from the marker.

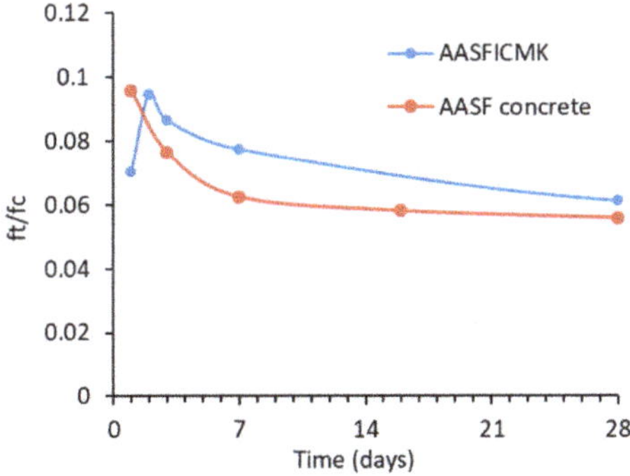

Figure 7. Splitting tensile strength-to-compressive strength (f_t/f_c) ratios of AASFICMK concrete, in comparison with AASF concrete.

3.5. Autogenous Shrinkage of Concrete

Figure 8 shows the autogenous shrinkage of the concrete. The plain AASF concrete showed large autogenous shrinkage, reaching more than 340 μm/m at the age of 28 days. In comparison, the autogenous shrinkage of AASFICMK concrete was less than 120 μm/m after a month of curing. This indicates that the utilization of SAPs and MK could effectively mitigate the autogenous shrinkage of AASF concrete. The autogenous shrinkage of AASFICMK was even lower than that of OPC concrete (see the results in [29,36]). The slight expansion of the concrete at an early age as shown in Figure 8 might be due to artifacts rather than a material behavior, since AASFICMK paste did not show expansion (see Figure 3). When stiffness of the concrete was low, the small pushing force from the LVDTs could move the embedded measuring bars a little bit, which enlarged the distance between the two measuring bars, even if the concrete itself did not expand [31]. After the first 3 days, the "expansion" was compensated by the shrinkage of the concrete.

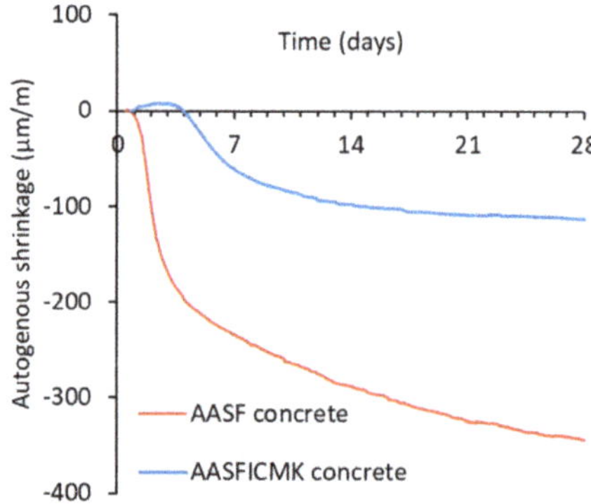

Figure 8. Autogenous shrinkage of AASF and AASFICMK concrete.

3.6. Cracking Potential of Concrete

The stress evolutions in the plain AASF concrete and AASFICMK concrete are shown in Figure 9. A sudden drop in the stress to around zero indicated the failure of the concrete due to tensile stress. It can be seen that the stress generated in AASFICMK was much lower than that in AASF. In the first 4 days, a small compressive stress was detected in AASFICMK due to the slight "expansion" of the concrete (see Figure 8). Afterwards, a tensile stress started to develop. The stress in AASFICMK was only 50% and 30% of the stress in the plain AASF concrete at the age of 7 days and 14 days, respectively.

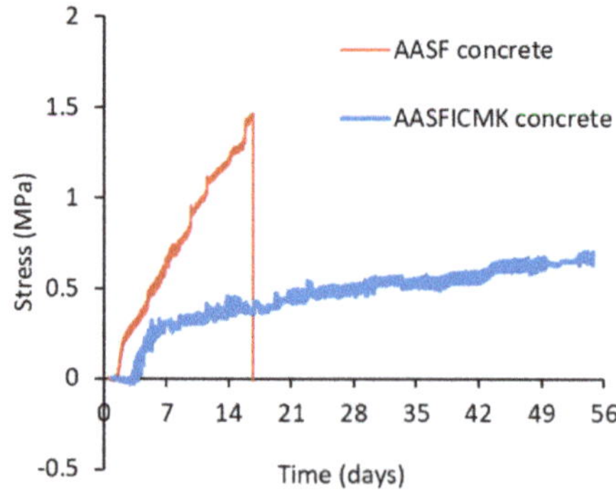

Figure 9. Self-induced stress in AASF and AASFICMK concrete.

According to the cracking time and stress rate, the cracking potential of AASF concrete was classified as "moderate" [22]. With the incorporation of SAPs and MK, AASFICMK concrete did not crack within 56 days. The stress rate after the first week reached below 0.01 MPa/day, indicating a very "low" cracking potential of the concrete [22].

The superior workability, the high 28-day strength, and the low cracking potential indicate that AASFICMK concrete could be considered as a highly commercially competitive construction material. Furthermore, due to the very low cracking potential of AASFICMK concrete, there is a lot of room for further tailoring the current mixture design in order to reach optimal overall performance of the concrete for different applications. For example, for the cases where the autogenous shrinkage is not very critical, lower liquid/binder ratios, lower dosages of SAPs/MK or higher amounts of slag could be used, by which higher strength of the concrete can be easily achieved.

4. Conclusions

In this paper, internal curing by SAPs and incorporation of MK were used to mitigate the autogenous shrinkage of slag-and-fly-ash-based AAMs activated by NaOH/Na$_2$SiO$_3$. The ring test and TSTM were used to track the shrinkage-induced stress and cracking potential of the paste and concrete, respectively.

It was found that both SAPs and MK were effective in mitigating the autogenous shrinkage and the self-induced stress of AASF paste and concrete. The dosages of 0.16 wt. % of SAPs and 5 wt. % of MK are recommended, which yielded an alkali-activated concrete (AASFICMK) with very low autogenous shrinkage and cracking potential and high enough strength. AASFICMK concrete also showed satisfactory workability. These results indicate that SAPs and MK are promising admixtures to produce high-performance AASF concrete with low shrinkage.

Author Contributions: Z.L.: conceptualization, methodology, investigation, writing—original draft. X.Y.: investigation. Y.C.: investigation. T.L.: methodology. G.Y.: project administration, writing—review & editing. All authors have read and agreed to the published version of the manuscript.

Funding: This research is funded by the Netherlands Organisation for Scientific Research (NWO) under grant number of 15803 and China Scholarship Council (CSC) under grant number of 201506120072 and 201906050022.

Acknowledgments: Maiko van Leeuwen is also gratefully acknowledged for his help with the experiments.

Conflicts of Interest: The authors declare that they have no conflict of interest.

References

1. Duxson, P.; Provis, J.L.; Lukey, G.C.; van Deventer, J.S.J. The role of inorganic polymer technology in the development of 'green concrete'. *Cem. Concr. Res.* **2007**, *37*, 1590–1597. [CrossRef]
2. Provis, J.L.; Van Deventer, J.S.J. *Geopolymers: Structures, Processing, Properties and Industrial Applications*; Woodhead: Cambridge, UK, 2009.
3. Arbi, K.; Nedeljković, M.; Zuo, Y.; Ye, G. A Review on the Durability of Alkali-Activated Fly Ash/Slag Systems: Advances, Issues, and Perspectives. *Ind. Eng. Chem. Res.* **2016**, *55*, 5439–5453. [CrossRef]
4. Fang, G.; Ho, W.K.; Tu, W.; Zhang, M. Workability and mechanical properties of alkali-activated fly ash-slag concrete cured at ambient temperature. *Constr. Build. Mater.* **2018**, *172*, 476–487. [CrossRef]
5. Cartwright, C.; Rajabipour, F.; Radli, A. Shrinkage Characteristics of Alkali-Activated Slag Cements. *J. Mater. Civ. Eng.* **2014**, *27*, 1–9. [CrossRef]
6. Nedeljković, M.; Li, Z.; Ye, G. Setting, Strength, and Autogenous Shrinkage of Alkali-Activated Fly Ash and Slag Pastes: Effect of Slag Content. *Materials* **2018**, *11*, 2121. [CrossRef]
7. Bilek, V., Jr.; Kalina, L.; Fojtík, O. Shrinkage-Reducing Admixture Efficiency in Alkali-Activated Slag across the Different Doses of Activator. In *Key Engineering Materials*; Trans Tech Publications Ltd.: Zurich, Switzerland, 2018; Volume 761, pp. 19–22.
8. Ye, H.; Radlińska, A. Shrinkage mitigation strategies in alkali-activated slag. *Cem. Concr. Res.* **2017**, *101*, 131–143. [CrossRef]

9. Bakharev, T.; Sanjayan, J.G.; Cheng, Y. Effect of elevated temperature curing on properties of alkali-activated slag concrete. *Cem. Concr. Res.* **1999**, *29*, 1619–1625. [CrossRef]

10. Li, Z.; Nedeljković, M.; Chen, B.; Ye, G. Mitigating the autogenous shrinkage of alkali-activated slag by metakaolin. *Cem. Concr. Res.* **2019**, *122*, 30–41. [CrossRef]

11. Li, Z.; Wyrzykowski, M.; Dong, H.; Granja, J.; Azenha, M.; Lura, P.; Ye, G. Internal curing by superabsorbent polymers in alkali-activated slag. *Cem. Concr. Res.* **2020**, *135*, 106123. [CrossRef]

12. Tu, W.; Zhu, Y.; Fang, G.; Wang, X.; Zhang, M. Internal curing of alkali-activated fly ash-slag pastes using superabsorbent polymer. *Cem. Concr. Res.* **2019**, *116*, 179–190. [CrossRef]

13. Li, Z.; Zhang, S.; Liang, X.; Ye, G. Internal curing of alkali-activated slag-fly ash paste with superabsorbent polymers. *Constr. Build. Mater.* **2020**. under peer review.

14. Li, Z.; Ye, G. Experimental study of the chemical deformation of metakaolin based geopolymer. In Proceedings of the SynerCrete'18 International Conference on Interdisciplinary Approaches for Cement-based Materials and Structural Concrete, Funchal, Portugal, 24–26 October 2018; pp. 443–448.

15. ASTM C1608. *Standard Test Method for Chemical Shrinkage of Hydraulic Cement Paste*; ASTM: West Conshohocken, PA, USA, 2007; pp. 667–670.

16. Li, Z.; Zhang, S.; Zuo, Y.; Chen, W.; Ye, G. Chemical deformation of metakaolin based geopolymer. *Cem. Concr. Res.* **2019**, *120*, 108–118. [CrossRef]

17. Li, Z.; Liang, X.; Chen, Y.; Ye, G. Effect of metakaolin on the autogenous shrinkage of alkali-activated slag-fly ash paste. *Constr. Build. Mater.* **2020**. under peer review.

18. ASTM C1968. *Standard Test Method for Autogenous Strain of Cement Paste and Mortar*; ASTM: West Conshohocken, PA, USA, 2013; pp. 1–8. [CrossRef]

19. Li, Z.; Lu, T.; Liang, X.; Dong, H.; Ye, G. Mechanisms of autogenous shrinkage of alkali-activated slag and fly ash pastes. *Cem. Concr. Res.* **2020**, *135*, 106107. [CrossRef]

20. Eren, Ö.; Abdalkader, A.H.M. Plastic shrinkage cracking of fiber reinforced concrete. In Proceedings of the 10th East Asia-Pacific Conference on Structural Engineering and Construction (EASEC), Bankok, Thailand, 3–5 August 2006; pp. 473–480. [CrossRef]

21. Hossain, A.B.; Weiss, J. Assessing residual stress development and stress relaxation in restrained concrete ring specimens. *Cem. Concr. Compos.* **2004**, *26*, 531–540. [CrossRef]

22. ASTM C1581. *Standard Test Method for Determining Age at Cracking and Induced Tensile Stress Characteristics of Mortar and Concrete under Restrained Shrinkage*; ASTM: West Conshohocken, PA, USA, 2009; pp. 1–7. [CrossRef]

23. Schlitter, J.L.; Senter, A.H.; Bentz, D.P.; Nantung, T.; Weiss, W.J. A dual concentric ring test for evaluating residual stress development due to restrained volume change. *J. ASTM Int.* **2010**, *7*, 1–13.

24. NEN-EN 12350-2. *Testing Fresh Concrete—Part 2: Slump-Test*; BSI: London, UK, 2009.

25. ASTM C191. *Standard Test Methods for Time of Setting of Hydraulic Cement by Vicat Needle BT—Standard Test Methods for Time of Setting of Hydraulic Cement by Vicat Needle*; ASTM: West Conshohocken, PA, USA, 2013.

26. Lokhorst, S.J. *Deformational Behaviour of Concrete Influenced by Hydration Related Changes of the Microstructure*; Delft University of Technology: Delft, The Netherlands, 2001.

27. NEN-EN 12390-3. *Testing hardened concrete—Part 3: Compressive Strength of Test Specimens*; BSI: London, UK, 2009.

28. Sule, M.S. *Effect of Reinforcement on Early-Age Cracking in High Strength Concrete*; Delft University of Technology: Delft, The Netherlands, 2003.

29. Lura, P. *Autogenous Deformation and Internal Curing of Concrete*; Delft University of Technology: Delft, The Netherlands, 2003.

30. Lu, T.; Li, Z.; Huang, H. Effect of Supplementary Materials on the Autogenous Shrinkage of Cement Paste. *Materials* **2020**, *13*, 3367. [CrossRef]

31. Li, Z.; Zhang, S.; Liang, X.; Ye, G. Cracking potential of alkali-activated slag and fly ash concrete subjected to restrained autogenous shrinkage. *Cem. Concr. Compos.* **2020**, 103767. [CrossRef]

32. NEN-EN 12350-8. *Testing Fresh Concrete—Part 8: Self-Compacting Concrete—Slump-Flow Test*; BSI: London, UK, 2019.

33. ACI (American Concrete Institute). *Building Code Requirements for Structural Concrete and Commentary*; ACI: Brussels, Belgium, 2011.

34. Arioglu, N.; Canan Girgin, Z.; Arioglu, E. Evaluation of ratio between splitting tensile strength and compressive strength for concretes up to 120 MPa and its application in strength criterion. *ACI Mater. J.* **2006**, *103*, 18–24.
35. Li, Z.; Lu, T.; Chen, Y.; Wu, B.; Ye, G. Prediction of the autogenous shrinkage of alkali-activated slag and fly ash concrete. *Mater. Struct.* **2020**. under peer review.
36. Darquennes, A.; Staquet, S.; Delplancke-Ogletree, M.P.; Espion, B. Effect of autogenous deformation on the cracking risk of slag cement concretes. *Cem. Concr. Compos.* **2011**, *33*, 368–379. [CrossRef]

Article

Mechanical Properties and Freeze–Thaw Durability of Basalt Fiber Reactive Powder Concrete

Wenjun Li [1], Hanbing Liu [1], Bing Zhu [1], Xiang Lyu [1], Xin Gao [2] and Chunyu Liang [1,*]

[1] College of Transportation, Jilin University, Changchun 130025, China; wenjun18@mails.jlu.edu.cn (W.L.); lhb@jlu.edu.cn (H.L.); zhubing18@mails.jlu.edu.cn (B.Z.); lvxiang18@mails.jlu.edu.cn (X.L.)
[2] China Resources Land Limited, Changchun 130025, China; xingao17@mails.jlu.edu.cn
* Correspondence: liangcy@jlu.edu.cn; Tel.: +86-159-4305-9951

Received: 20 July 2020; Accepted: 14 August 2020; Published: 16 August 2020

Abstract: Basalt fiber has a great advantage on the mechanical properties and durability of reactive powder concrete (RPC) because of its superior mechanical properties and chemical corrosion resistance. In this paper, basalt fiber was adopted to modified RPC, and plain reactive powder concrete (PRPC), basalt fiber reactive powder concrete (BFRPC) and steel fiber reactive powder concrete (SFRPC) were prepared. The mechanical properties and freeze–thaw durability of BFRPC with different basalt fiber contents were tested and compared with PRPC and SFRPC to investigate the effects of basalt fiber contents and fiber type on the mechanical properties and freeze–thaw durability of RPC. Besides, the mass loss rate and compressive strength loss rate of RPC under two freeze–thaw conditions (fresh-water freeze–thaw and chloride-salt freeze–thaw) were tested to evaluate the effects of freeze–thaw conditions on the freeze–thaw durability of RPC. The experiment results showed that the mechanical properties and freeze–thaw resistance of RPC increased as the basalt fiber content increase. Compared with the fresh-water freeze–thaw cycle, the damage of the chloride-salt freeze–thaw cycle on RPC was great. Based on the freeze–thaw experiment results, it was found that SFRPC was sensitive to the corrosion of chloride salts and compared with the steel fiber, the improvement of basalt fiber on the freeze–thaw resistance of RPC was great.

Keywords: reactive powder concrete; basalt fiber; chloride-salt corrosion; freeze–thaw durability; mechanical properties

---

1. Introduction

In the practical application of concrete structural engineering, the deterioration of concrete and the shortening of its service life are related to freeze–thaw damage, chloride ion penetration, sulfate attack and carbonization [1,2]. Accordingly, it is important to improve the durability of concrete structures in practical applications, which is a crucial factor to reduce the cost of infrastructure construction and maintenance. In cold areas, most concrete structures appear in the natural environment, and the freeze–thaw cycle reduces the service life and becomes one of the main sources of damage to the concrete structure [3]. Especially in cold areas, where de-icing salt has been adopted, the concrete damage caused by the coupling action of chloride corrosion and freeze–thaw cycle is serious [4]. The destruction of concrete structures caused by freeze–thaw cycle has aroused widespread concern. The durability of concrete is closely related to the internal pore structure. Concrete structure with low-porosity could effectively reduce the pore solution inside the concrete, thus reducing the internal water pressure caused by the freezing of the pore solution and improving the freeze–thaw resisrance of concrete [5]. Besides, the compact concrete structure could hinder the transmission of chloride and greatly reduce the damage caused by chloride ion penetration [6,7].

Reactive powder concrete (RPC), proposed in the 1990s, is an ultra-high-performance concrete characterized by dense microstructure and uniform concrete matrix [8]. The traditional RPC is developed through microstructure enhancement technology and the main improvement is related to matrix uniformity, porosity and microstructure, including eliminating coarse aggregates and reducing the water–binder ratio to improve matrix uniformity [9], optimizing particle gradation to decrease matrix porosity [10] and increasing the silica component by adding silica fume to improve the microstructure of the matrix structure [11]. RPC is seen as a potential material for the field of special prestressed and precast concrete components due to its low permeability, superior mechanical properties and durability [12,13]. Although the manufacturing cost of RPC is generally high, the thickness of concrete component is possible to be reduced due to its ultra-high mechanical properties, resulting in materials and costs saving. Therefore, RPC has certain economic advantages in practical applications [14]. However, RPC still has the characteristics of high brittleness of ordinary concrete and the brittleness of RPC increases as the strength increase. Therefore, the addition of fiber is usually adopted to increase the toughness of RPC [15].

Adding fiber into brittle concrete is a widely used technology. The bridging and pulling effects of fiber could improve the brittleness and impact resistance of the mixture [16,17], and the three-dimensional randomly scattered fiber could effectively inhibit the generation and propagation of cracks [18]. Previous reports have found that polypropylene fiber [19], carbon fiber [20] and steel fiber [21] can be used as reinforcing materials for RPC, and the fiber commonly used to reinforce concrete is steel fiber [22]. To improve the flexural strength, compressive strength, elastic modulus and ductility of RPC, the recommended content of steel fiber is 2% volume fraction [23]. Moreover, adding steel fiber into concrete is judged as an effective method to prevent spalling. However, steel fiber with high content will converge into a spherical shape and the steel fiber is sensitive to chemical corrosion, which results in a decrease in properties of concrete. Besides, adding steel fiber into the concrete will reduce the workability and increase the cost [24].

Basalt fiber is an environmentally friendly and high-performance fiber made of natural volcanic basalt rock, which can be used to reinforce concrete. Due to its excellent acid and alkali resistance, great mechanical properties and high temperature stability, the application prospect of basalt fiber as fiber-reinforced material is broad [25]. Basalt fiber can not only effectively increase the strength and durability of concrete but also has a positive effect on the toughness and crack resistance [26,27]. The excellent reinforcement effect of basalt fiber promotes its application in RPC. Wang et al. [28] researched the effect of basalt fiber on the mechanical properties of high-performance concrete. The results showed that the strength of high-performance concrete increased with the increase in basalt fiber content, and the compressive improved slightly, while flexural strength increased significantly. Grzeszczyk et al. [29] evaluated the effect of basalt fiber content on the mechanical properties of RPC; their results showed that the flexural strength of RPC increased with the increase in basalt fiber content. Liu et al. [30] reported that the basalt fiber could increase the toughness and bending strength of basalt fiber RPC beams, and improve the resistance to steel–concrete interface damage. Most research is about the mechanical properties of basalt fiber-reinforced reactive powder concrete. There is not enough research about the freeze–thaw durability of fiber-reinforced RPC, and the most common fiber in the freeze–thaw durability research of fiber-reinforced RPC is steel fiber [2,31]. Therefore, there are few studies were reported on the effect of basalt fiber on freeze–thaw durability of RPC. Moreover, in cold regions, where de-icing salt has been adopted, the coupling effect of chloride corrosion and freeze–thaw poses a challenge to the application of RPC and, because of the excellent corrosion resistance of basalt fiber, the improvement of basalt fiber on the freeze–thaw durability of RPC is worth exploring. Therefore, it is necessary to investigate the enhancement effect of basalt fiber on the performance of RPC.

In this paper, the mechanical properties and freeze–thaw durability of basalt fiber reactive powder concrete (BFRPC) with different basalt fiber contents (4, 8 and 12 kg/m^3) were tested and compared with plain reactive powder concrete (PRPC) and steel fiber reactive powder concrete (SFRPC) to investigate the impacts of basalt fiber contents and fiber type on the mechanical properties and freeze–thaw durability of RPC. Moreover, the freeze–thaw resistance of RPC under the fresh-water freeze–thaw cycle and the chloride-salt freeze–thaw cycle was tested to investigate the effects of freeze–thaw conditions on the freeze–thaw durability of RPC. The purpose of this paper is to provide a reference for the practical application of BFRPC in cold regions.

2. Materials and Methods

2.1. Materials

Type P.O Portland cement with a strength of 42.5 MPa, manufactured by Yatai Cement Co., Ltd., Jilin, China, was used in this paper. The properties of this cement are given in Table 1. The SF93 silica fume with a specific surface area of 18,100 m^2/kg obtained from Si'ao Technology Co., Ltd., Changchun, China, was used in this study. Table 2 shows the chemical composition of silica fume and cement. Three types of quartz sand, with sizes of 20–40, 40–80 and 80–120 mesh, manufactured by Zhenxing quartz sand factory, Luoyang, were used. The proportion of three types of quartz sand was 2:2:1. The quartz powder (400 mesh) was used to fill micro pores. The chemical composition of quartz sand is given in Table 3. HPWR-Q8011 polycarboxylic superplasticizer (water reduction rate 25%) obtained from Qinfen Building Materials Co., Ltd., Shanxi, China, was used to prepare all tested concrete. Chopped basalt fiber, produced by Anjie Composite Material Co., Ltd., Haining, China, was used to reinforce RPC in this study and the properties of basalt fiber are shown in Table 4. Steel fiber with a length of 13 mm and a diameter of 200 μm was adopted in this paper, obtained from Daxing Matel Fiber Co., Ltd., Ganzhou, China, which had a tensile strength of 2850 MPa. The appearance of basalt fiber and steel fiber is given in Figure 1. The tap water was used as mixed water.

Table 1. Basic properties of cement.

Density (kg/m^3)	Specific Surface Area (m^2/kg)	Setting Time (min)		Compressive Strength (MPa)	Flexural Strength (MPa)
		Initial Setting	Final Setting		
3160	385	91	145	62.2	9.1

Table 2. The chemical composition of cement and silica fume.

Material	Chemical Composition (%)					
	SiO$_2$	Al$_2$O$_3$	Fe$_2$O$_3$	CaO	MgO	SO$_3$
Cement	22.60	5.60	4.30	62.70	1.70	2.50
Silica fume	93.3	0.73	0.49	0.85	1.21	1.02

Table 3. The chemical composition of quartz sand.

Material	Chemical Composition (%)					
	SiO$_2$	Fe	Al$_2$O$_3$	K$_2$O	Na$_2$O	H$_2$O
Quartz Sand	99.68	0.0062	0.0122	0.0011	0.002	0.02

Table 4. The performance indicator of basalt fiber.

Type	Length (mm)	Diameter (μm)	Linear Density (tex)	Tensile Strength (MPa)	Elastic Modulus (GPa)	Breaking Strength (N/tex)	Elongation (%)
Basalt fiber	22	23	2392	2836	62	0.69	3

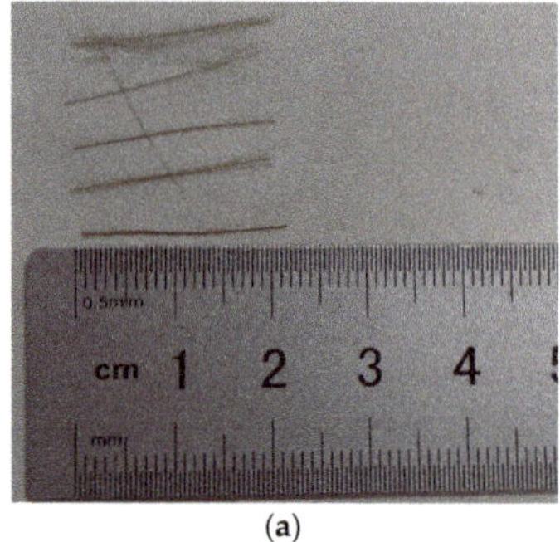

(a)

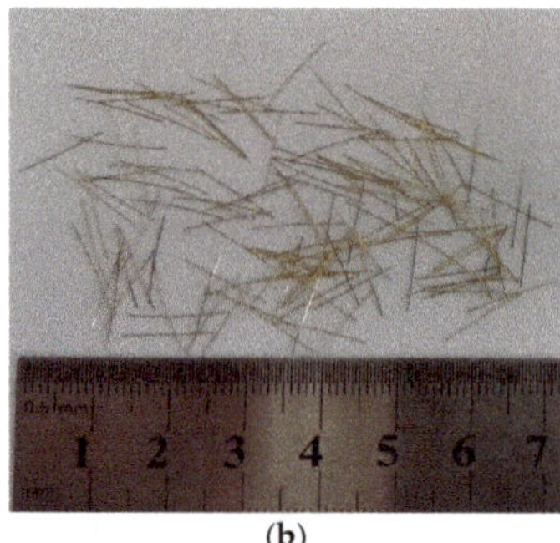

(b)

Figure 1. Fiber: (**a**) Basalt fiber (**b**) Steel fiber.

2.2. Mixture Proportion and Specimen Preparation

The best mixture proportion of RPC was obtained from the response surface method in our group's previous study [32], and based on the optimal mixture proportion; basalt fiber and steel fiber were added into RPC with an external mixing method. The content of basalt fiber was 4, 8 and 12 kg/m^3, respectively. The content of steel fiber was 2% volume fraction [23]. Moreover, PRPC specimens without fiber were prepared as the control group to investigate the effects of basalt fiber and steel fiber on the mechanical properties and freeze–thaw durability of RPC. The mixture proportion of BFRPC and SFRPC is shown in Table 5.

Table 5. The mixture proportion of basalt fiber reactive powder concrete (BFRPC) and steel fiber reactive powder concrete (SFRPC).

Mix ID	Basalt Fiber (kg/m^3)	Steel Fiber (%)	Cement (kg/m^3)	Quartz Sand (kg/m^3)	Silica Fume (kg/m^3)	Quartz Powder (kg/m^3)	Water (kg/m^3)	Water Reducer (kg/m^3)
R	-	-	834.73	939.03	208.68	308.85	166.95	52.17
BF4	4	-	834.73	939.03	208.68	308.85	166.95	52.17
BF8	8	-	834.73	939.03	208.68	308.85	166.95	52.17
BF12	12	-	834.73	939.03	208.68	308.85	166.95	52.17
SF	-	2	834.73	939.03	208.68	308.85	166.95	52.17

All RPC mixtures were prepared by using an ISO 679 mixer with a capacity of 5 L. The degree of dispersion of basalt fiber and steel fiber in the mixture has a significant impact on the properties of RPC. During the process of mixing, the fibers were put into the mixer step by step to ensure the dispersion of fibers. The specific of mixing were as follows. (1) Put the quartz sand and fiber into the mixer and mix for 2 min. The friction between the quartz sand particles makes the fiber evenly dispersed. (2) Put the cement, quartz powder and silica fume into the mixer and mix for 3 min. (3) Dissolve the polycarboxylic superplasticizer into the water and add the solution into the mixer in two lots, mixing for 3 min each time. After stirring, the mixture was filled into a metal mold with a size of 40 × 40 × 160 mm, immediately, and compressed by vibrating on the ZT-96 vibrator table. The specimens were placed in an environment of 95% humidity and 20 °C for 24 h and then de-moulded. The specimens were cured for 48 h under the condition of 90 °C steam curing, the heating speed was 12 °C/h and the cooling speed was 15 °C/h.

2.3. Testing Methods

2.3.1. Flexural Strength

According to the three-point flexural test described in Chinese national standard GB/T 17671-1999 [33] to evaluate the flexural strength of RPC. The distance between the two fulcrum points was 100 mm. The specimens were placed with the side face up in the testing machine and align the center line of the specimens with the upper fixture. The loading speed was set to 50 N/s. Three specimens with volumes of 40 × 40 × 160 mm were maded in each group for testing and the mean of the three measured results was used as the flexural strength. The specific calculation process is shown in Formula (1). The specific experimental device of flexural strength is shown in Figure 2.

$$f_f = \frac{1.5 F_f \mathrm{L}}{\mathrm{b}^3} \tag{1}$$

where f_f refers to the flexural strength of specimens (MPa), F_f refers to the failure load (N), L refers to the distance between the two fulcrum and L= 100 mm, b refers to the cross-sectional width of the specimens and b = 40 mm.

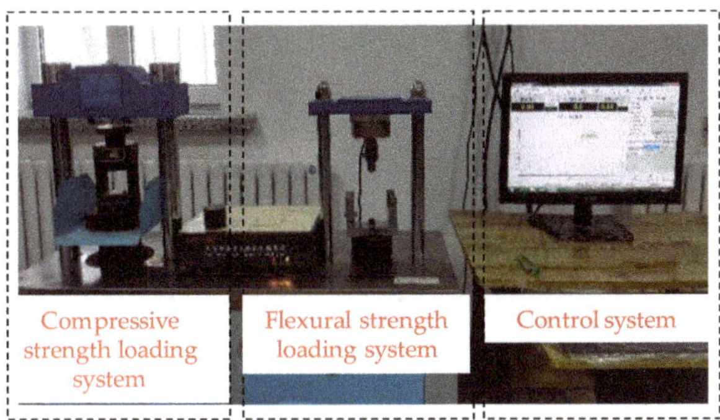

Figure 2. The specific experimental device.

2.3.2. Compressive Strength

According to the Chinese national standard GB/T 17671-1999 [33] to evaluate the compressive strength of RPC. The six fracture blocks obtained after the flexural strength test were adopted for the compressive strength test. Removed the debris on the surface of the fracture blocks and placed it in the fixture. The two sides of the fracture blocks were used as the compression surface. Adjusted the position of the specimen so that the compression surface and the fixture were in full contact. The loading rate was set to 2.4 kN/s. The compressive strength of RPC was the mean of the six measured results. The specific calculation process is shown in Formula (2). The specific experimental device of compressive strength is shown in Figure 2.

$$f_C = \frac{F}{\mathrm{A}} \tag{2}$$

where f_C refers to the compressive strength of RPC (MPa), F refers to the failure load (N), A refers to the compression surface and A = 1600 mm^2.

2.3.3. Freeze–Thaw Cycle

In this paper, the freeze–thaw durability of PRPC, SFRPC and BFRPC under two freeze–thaw conditions was investigated. The specific freeze–thaw cycle test grouping is shown in Table 6. According to the rapid freeze–thaw method described in Chinese national standard GB/T50082-2009 [34] to investigate the freeze–thaw durability of PRPC, SFRPC and BFRPC. The freezing temperature is −18 ± 2 °C and the freezing time is 3 h. The melting temperature is 5 ± 2 °C and the melting time is greater than one quarter of the entire freezing and melting cycle time. The specific arrangement of the freeze–thaw cycle test is shown in Figure 3.

Table 6. Freeze–thaw cycle test grouping.

Mix ID	WR	WBF12	WSF	NR	NBF4	NBF8	NBF12	NSF
Freeze–thaw medium	fresh water	fresh water	fresh water	5 wt% NaCl	5 wt% NaCl	5 wt% NaCl	5 wt% NaCl	5 wt% NaCl

Note: W refers to fresh water, N refers to 5 wt% NaCl solution.

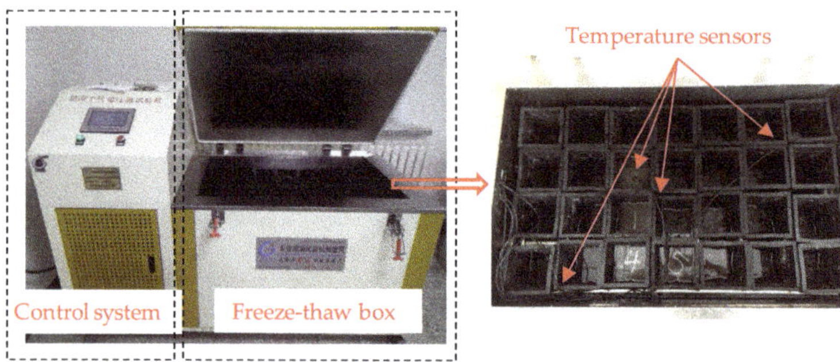

Figure 3. Freeze–thaw cycle test.

The cured specimens were immersed in fresh water and 5 wt% NaCl solution for 48 h. The groups of WR, WB12 and WS were immersed in fresh water and the groups of NR, NB4, NB8, NB12 and NS were immersed in 5 wt% NaCl solution. After immersion for 48 h, the the moisture on the surface of the specimens wiped off; the initial mass of the specimens was weighed and the initial compressive strength was measured. Then, the remaining specimens were put into the freeze–thaw test machine and injected fresh water and 5 wt% NaCl solution until the water level was 5 cm above the top surface of the specimens. The number of freeze–thaw cycle was set to 800, and the mass and compressive strength of the specimens were measured every 100 cycles. The mass loss rate and compressive strength loss rate were adopted to investigate the freeze–thaw durability of BFRPC and SFRPC. The specific calculation formulas are as follows:

$$MR = \frac{M_i - M_0}{M_0} \times 100\% \tag{3}$$

$$CR = \frac{f_{Ci} - f_{C0}}{f_{C0}} \times 100\% \tag{4}$$

where *MR* refers to the mass loss rate, M_i refers to the mass at the Nth cycles, M_0 refers to the initial mass, *CR* refers to the compressive strength loss rate, f_{Ci} refers to the compressive strength at the Nth cycles, f_{C0} refers to the initial compressive strength.

3. Results and Discussion

3.1. Mechanical Properties

In this paper, the mechanical properties of BFRPC with different basalt fiber contents were tested and compared with PRPC and SFRPC to evaluate the effects of basalt fiber contents and fiber type on the mechanical properties of RPC.

3.1.1. Compressive Strength

Figure 4 shows that the effects of basalt fiber contents and fiber type on the compressive strength of RPC. It can be known from Figure 4 that the addition of basalt fiber could improve the compressive strength of RPC and the compressive strength of RPC increases with the increase in basalt fiber content, which is consistent with the conclusion of Wang et al. [28]. When the basalt fiber content is 12 kg/m^3, the compressive strength of BFRPC reaches the maximum value of 149.40 MPa, which is 6% higher than that of PRPC. Liu et al. [35] pointed out that evenly distributed basalt fiber inside the RPC could inhibit the initiation of cracks and bear part of the load, thus improving the compressive strength of RPC. Besides, basalt fiber is composed of oxides such as silica, alumina and magnesia, and its chemical composition is similar to cement. The bond strength between basalt fiber and cement paste is great [36], thus improving the compressive strength of RPC. Compared with SFRPC, the compressive strength of BFRPC with a basalt fiber content of 12 kg/m^3 is 4.3% lower, which indicates that basalt fiber is not as effective as steel fiber in improving the compressive strength of RPC.

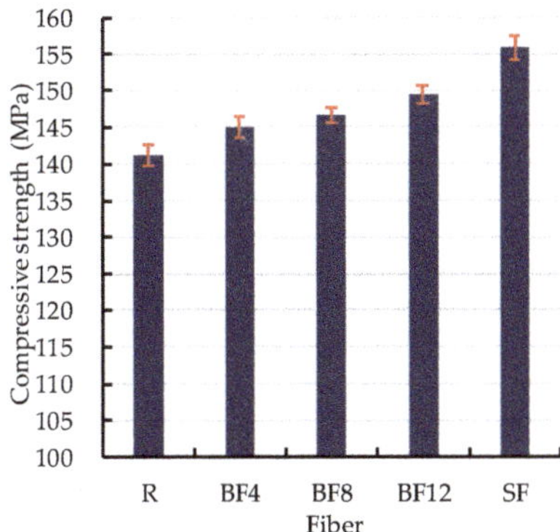

Figure 4. Effect of basalt fiber content and fiber type on the compressive strength of RPC.

3.1.2. Flexural Strength

The experiment result of the flexural strength test is shown in Figure 5. Figure 5 shows that the flexural strength of RPC increases as the basalt fiber content increases, which is the same as the improvement of basalt fiber on the compressive strength. The increased trend of flexural strength is a slightly different from that of compressive strength. Compared with compressive strength, when the basalt fiber is from 8 to 12 kg/m^3, a significant increase in flexural strength of BFRPC occurs and the flexural strength of BFRPC reaches 16.23 MPa, which is 18.5% higher than that of PRPC. Compared with compressive strength, the improvement of basalt fiber on the flexural strength of RPC is greater, aggreging with the conclusion conducted by Wang et al. [28]. The improvement of the flexural strength of BFRPC is related to the uniform distribution of basalt fiber inside the RPC. The uniform distributed basalt fiber could inhibit the generation and propagation of cracks and bear part of the stress, reducing the stress concentration near the crack and redistributing the stress [16], thus improving the flexural strength of RPC. Besides, the basalt fiber inside the RPC submits a three-dimensional random distribution; this distribution system can limit the deformation of RPC under loading conditions, which leads to an increase in the flexural strength of RPC. The three-dimensional distribution system of basalt fiber is gradually improved as the basalt fiber content increase, which is reflected in the significant increase in flexural strength of BFRPC with a basalt fiber content of 12 kg/m^3. Moreover, comparing the flexural strength of SFRPC with that of BFRPC (12 kg/m^3 content), it can be known that the flexural strength of BFRPC and SFRPC is similar, which indicates that when the basalt fiber content is 12 kg/m^3, the reinforcement effect of basalt fiber on the flexural strength of RPC is about same as steel fiber; Branston et al. [25] reported a similar conclusion.

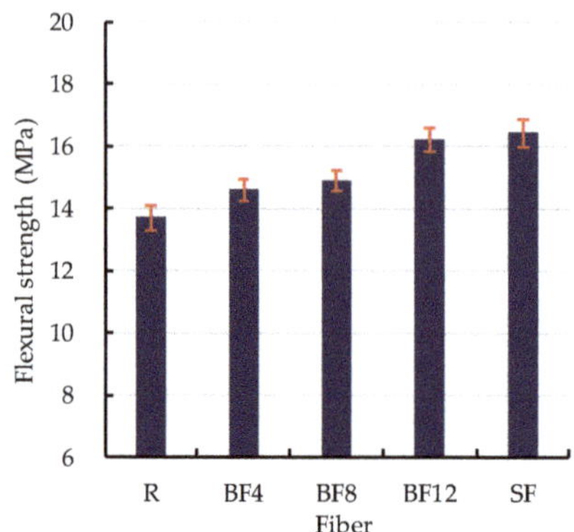

Figure 5. Effect of basalt fiber content and fiber type on the flexural strength of RPC.

3.2. Effect of Fiber on Freeze–Thaw Durability of RPC

3.2.1. Mass Loss

The mass loss of BFRPC with different basalt fiber contents was tested under the chloride-salt freeze–thaw cycle and compared with the PRPC and SFRPC to evaluate the effects of basalt fiber contents and fiber type on the freeze–thaw durability of RPC. Figure 6 shows that the effects of basalt fiber contents and fiber type on the mass loss of RPC. It can be known from Figure 6 that the mass loss rate of RPC increases with the increase in the number of freeze–thaw cycles. The mass loss of RPC is mainly caused by surface scaling and the edges of the specimen surface will fall off as the freeze–thaw cycle test continue, resulting in fiber exposure [2]. It is worth noting that the mass loss rate of BFRPC (8 and 12 kg/m^3 content) is negative in the initial part of the freeze–thaw test, which is related to the water absorption of basalt fiber. In the initial part of the freeze–thaw test, the mass of water absorbed by the basalt fiber is greater than that of the surface scaling of RPC caused by freeze–thaw damage. When freeze–thaw cycle levels are the same, the mass loss rate of RPC decreases as the basalt fiber content increase, which indicates that the basalt fiber could improve the freeze–thaw resistance. When the number of freeze–thaw cycles is less than 600, the mass loss rate of SFRPC is in a stable increase stage and the mass loss is mainly caused by surface scaling. When it exceeds 600, a significant increase in the mass loss rate occurs, and the reason is that the steel fiber exposed to the chloride ion environment corrodes rapidly, thus leading to more peeling off the edge of the specimen surface. The mass loss rate of SFRPC reaches 3.03% after 800 freeze–thaw cycles, which is higher than that of BFRPC. This indicates that steel fiber is not as effective as basalt fiber in improving the freeze–thaw durability of RPC.

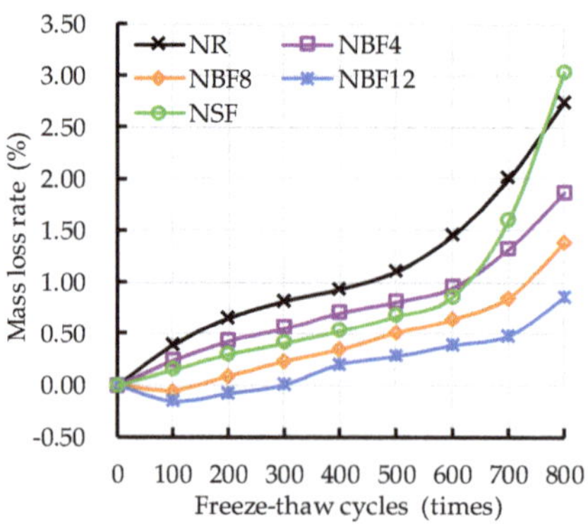

Figure 6. Mass loss rate versus freeze–thaw cycles.

3.2.2. Compressive Strength Loss

The compressive strength of BFRPC with different basalt fiber contents was tested under the chloride-salt freeze–thaw cycle and the compressive strength loss rate was calculated by Formula (4), and compared with the PRPC and SFRPC to evaluate the impacts of basalt fiber contents and fiber type on the freeze–thaw durability of RPC.

The compressive strength and compressive strength loss rate vs. freeze–thaw cycles are shown in Figure 7. It can be seen from Figure 7a that, for all RPC specimens, the compressive strength of RPC decreases as the freeze–thaw test continues, which indicates that the freeze–thaw durability of RPC gradually deteriorates under the chloride-salt freeze–thaw cycle. During the freeze–thaw test, the frost expansion of the pore solution inside the RPC leads to the initiation of micro cracks, thus reducing the compressive strength of RPC, which is consistent with the conclusion reported by An et al. [31]. Comparing the compressive strength of SFRPC with that of BFRPC (12 kg/m^3 content), it can be known that the decline in the compressive strength of SFRPC is higher and the compressive strength of SFRPC and BFRPC is almost the same after 800 freeze–thaw cycles, which indicates that the corrosion of steel fiber not only increases the mass loss of RPC, but also has a negative effect on the compressive strength of RPC. It can be seen from Figure 7b that, for PRPC, BFRPC and SFRPC specimens, the compressive strength loss rate increases as the freeze–thaw test continues. At all freeze–thaw cycle levels, the compressive strength loss rate of BFRPC is lower than that of PRPC, and the higher the basalt fiber content, the lower the compressive strength loss rate of BFRPC, which indicates that adding basalt fiber could improve the freeze–thaw durability of RPC. This is because the addition of basalt fiber increases the number of harmless pores which could effectively reduce the freezing pressure of the pore solution, thus improving the freeze–thaw durability of RPC [37]. Moreover, comparing the compressive strength loss rate of SFRPC with that of BFRPC, the compressive strength loss rate of BFRPC (8 and 12 kg/m^3 content) is lower than that of SFRPC throughout the freeze–thaw cycle test. It indicates that steel fiber is not as effective at enhancing the freeze–thaw resistance of RPC as basalt fiber.

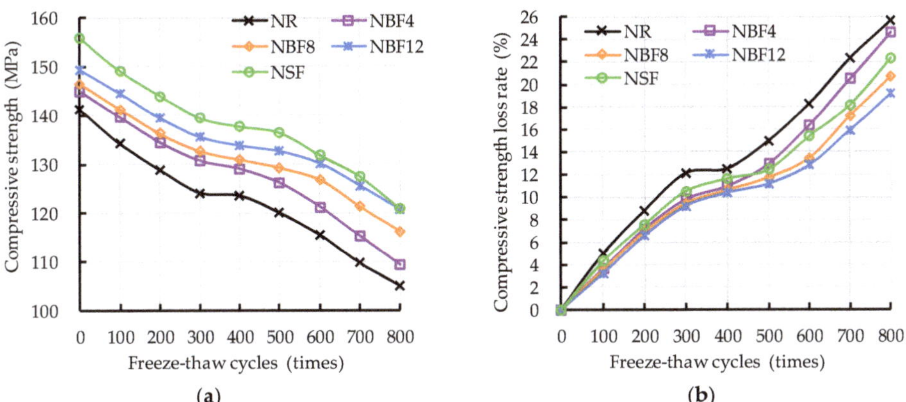

Figure 7. Compressive strength and compressive strength loss rate versus freeze–thaw cycles: (**a**) Compressive strength; (**b**) Compressive strength loss rate.

3.3. Effect of Freeze–Thaw Condition on the Freeze–Thaw Durability of RPC

3.3.1. Mass Loss

The mass loss of RPC was tested under two freeze–thaw conditions and the mass loss rate was calculated by Formula (3) to evaluate the effects of freeze–thaw conditions on the freeze–thaw durability of RPC. The mass loss rate vs. freeze–thaw cycles is shown in Figure 8. This figure shows that the mass loss rate of RPC increases as the freeze–thaw cycle test continues, and the presence of basalt fiber water absorption leads to a negative value of the mass loss rate of BFRPC in the early stage of the freeze–thaw test; as the density of the 5-wt% NaCl solution is greater than that of fresh water, when the number of freeze–thaw cycles is less than 300, the mass loss rate of BFRPC (chloride-salt freeze–thaw) is lower than that of BFRPC (fresh-water freeze–thaw). Compared with the fresh-water freeze–thaw cycle, the mass loss rate of RPC is higher under the chloride-salt freeze–thaw cycle, it can be known that chloride-salt freeze–thaw cycles causes more surface scaling, which indicates that the chloride-salt freeze–thaw cycle accelerates the deterioration of the freeze–thaw durability of RPC. After 800 freeze–thaw cycles, the mass loss rate of BFRPC under two freeze–thaw conditions is less than 1%. Besides, at all freeze–thaw cycle levels, the difference in mass loss rate of BFRPC under two freeze–thaw conditions is less than 0.25%. This indicates that the basalt fiber could significantly improve the freeze–thaw durability and chloride ion resistance of RPC.

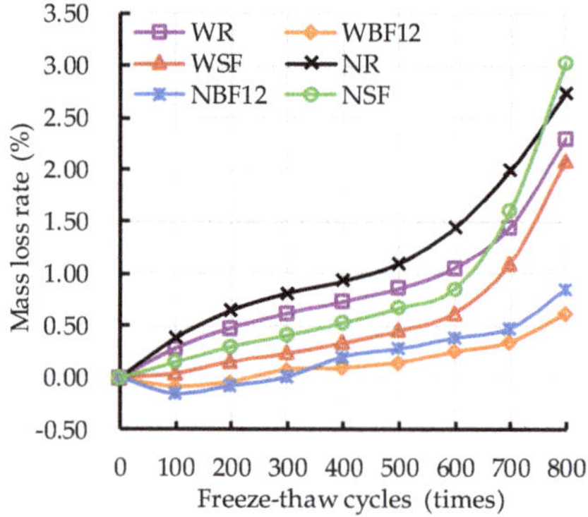

Figure 8. Mass loss versus freeze–thaw cycles.

3.3.2. Compressive Strength Loss

The compressive strength of RPC was tested under two freeze–thaw conditions and the compressive strength loss rate was calculated by Formula (4) to evaluate the effects of freeze–thaw conditions on the freeze–thaw durability of RPC.

The compressive strength and its loss rate vs. freeze–thaw cycles are shown in Figure 9. Figure 9a shows that, for all RPC mixtures, the compressive strength of RPC decreases with the freeze–thaw test continues. Before the freeze–thaw test, the compressive strength of RPC soaked in 5 wt% NaCl solution for 48 h is lower than that of RPC soaked in water, which shows that the immersion of 5 wt% NaCl solution has a certain negative effect on the compressive strength of RPC. When the number of freeze–thaw cycles is the same, the compressive strength of RPC under chloride-salt freeze–thaw cycle is lower than that of RPC under fresh-water freeze–thaw cycle, which indicates that compared with the fresh-water freeze–thaw cycle, chloride-salt freeze–thaw cycle causes greater damage to RPC, similar conclusion is reported by Vaitkevičius et al. [38]. It can be seen from Figure 9b that the compressive strength loss rate of RPC increases as the freeze–thaw cycles increase. At all freeze–thaw cycle levels, compared with the RPC under fresh-water freeze–thaw cycle, the compressive strength loss rate of RPC under chloride-salt freeze–thaw cycle is higher. Due to freeze–thaw cycle, the pore cracked and the microcracks generated provide channels for further penetration of chloride ions. Therefore, the coupling effect of freeze–thaw cycle and chloride ion erosion has more aggressive impact on the durability of RPC [2,39]. After 800 freeze–thaw cycles, the compressive strength loss rate of BFRPC under two freeze–thaw conditions is less than 20% and the difference in compressive strength loss rate of BFRPC under two freeze–thaw conditions is less than 0.9%. It indicates that the basalt fiber could significantly improve the freeze–thaw durability and chloride ion resistance of RPC and the conclusion is consistent with the mass loss rate test.

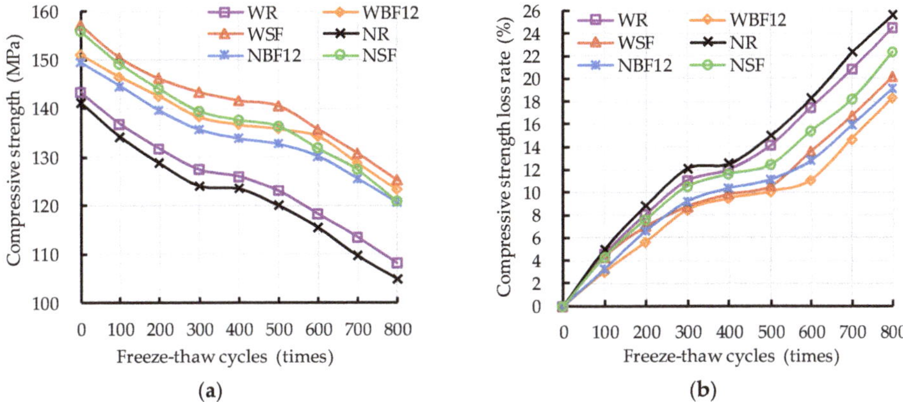

Figure 9. Compressive strength and Compressive strength loss rate versus freeze–thaw cycles: (**a**) Compressive strength; (**b**) Compressive strength loss rate.

4. Conclusions

In this paper, the mechanical properties and freeze–thaw durability of BFRPC with different basalt fiber contents were tested and compared with PRPC and SFRPC to investigate the effects of basalt fiber contents and fiber type on the mechanical properties and freeze–thaw durability of RPC. Besides, the freeze–thaw durability of RPC under two freeze–thaw conditions was tested to investigate the effects of freeze–thaw conditions on the freeze–thaw durability of RPC. Based on the results of the experiment, the conclusions are as follows:

1. The compressive strength and flexural strength can be improved by adding basalt fiber into RPC, and the compressive and flexural strength of BFRPC increase as the basalt fiber content increases. Compared with PRPC, the compressive strength and flexural strength of BFRPC (12 kg/m^3 content) are increased by 6% and 18.5%, respectively; the improvement of basalt fiber on the flexural strength of RPC is greater than compressive strength.

2. The freeze–thaw durability of BFRPC increases as the basalt fiber content increases. After 800 freeze–thaw cycles, the mass loss rate and compressive strength loss rates of BFRPC with a basalt fiber content of 12 kg/m^3 are 0.85% and 19.17%, respectively, which are lower than that of PRPC. The addition of basalt fiber could significantly improve the freeze–thaw durability of RPC.

3. Compared with the fresh-water freeze–thaw cycle, the mass loss rate and compressive strength loss rate of RPC are higher under the chloride-salt freeze–thaw cycle, and the damage of the chloride-salt freeze–thaw cycle on RPC is great.

4. After 800 freeze–thaw cycles, the mass loss rate and compressive strength loss rates of SFRPC are greater than 2% and 20%, respectively, which are higher than that of BFRPC. Steel fiber is not as effective at enhancing the freeze–thaw durability of RPC as basalt fiber.

Author Contributions: Conceptualization, H.L. and X.G.; methodology, C.L. and X.G.; formal analysis, W.L. and X.L.; investigation, W.L., B.Z. and X.L.; writing—original draft preparation, W.L.; writing—review and editing, C.L.; funding acquisition, H.L. All authors have read and agreed to the published version of the manuscript.

Funding: This research was funded by the Industrial Technology Research & Development Special Project of Jilin Province (2018C042-1); the Transportation Science and Technology Program of Jilin Province (2018-1-9); and the Science Technology Development Program of Jilin Province (20200403157SF).

Acknowledgments: The authors would like to express their appreciation to the anonymous reviewers for their constructive suggestions and comments on improving the quality of the paper.

Conflicts of Interest: The authors declare no conflict of interest.

References

1. Glasser, F.P.; Marchand, J.; Samson, E. Durability of concrete—Degradation phenomena involving detrimental chemical reactions. *Cem. Concr. Res.* **2008**, *38*, 226–246. [CrossRef]
2. Wang, Y.; An, M.Z.; Yu, Z.R.; Han, S.; Ji, W.Y. Durability of reactive powder concrete under chloride-salt freeze–thaw cycling. *Mater. Struct.* **2017**, *50*, 18. [CrossRef]
3. Wang, Y.R.; Cao, Y.B.; Zhang, P.; Ma, Y.W.; Zhao, T.J.; Wang, H.; Zhang, Z.H. Water absorption and chloride diffusivity of concrete under the coupling effect of uniaxial compressive load and freeze-thaw cycles. *Constr. Build. Mater.* **2019**, *209*, 566–576. [CrossRef]
4. Sun, W.; Mu, R.; Luo, X.; Miao, C.W. Effect of chloride salt, freeze–thaw cycling and externally applied load on the performance of the concrete. *Cem. Concr. Res.* **2002**, *32*, 1859–1864. [CrossRef]
5. Cai, H.; Liu, X. Freeze-thaw durability of concrete: Ice formation process in pores. *Cem. Concr. Res.* **1998**, *28*, 1281–1287. [CrossRef]
6. Liu, Z.Y.; Zhang, Y.S.; Jiang, Q. Continuous tracking of the relationship between resistivity and pore structure of cement pastes. *Constr. Build. Mater.* **2014**, *53*, 26–31. [CrossRef]
7. Wang, L.C.; Ueda, T. Mesoscale Modeling of Chloride Penetration in Unsaturated Concrete Damaged by Freeze-Thaw Cycling. *J. Mater. Civ. Eng.* **2014**, *26*, 955–965. [CrossRef]
8. Richard, P.; Cheyrezy, M. Composition of reactive powder concretes. *Cem. Concr. Res.* **1995**, *25*, 1501–1511. [CrossRef]
9. Chan, Y.W.; Chu, S.H. Effect of silica fume on steel fiber bond characteristics in reactive powder concrete. *Cem. Concr. Res.* **2004**, *34*, 1167–1172. [CrossRef]
10. Bonneau, O.; Vernet, C.; Moranville, M.; Aitcin, P.C. Characterization of the granular packing and percolation threshold of reactive powder concrete. *Cem. Concr. Res.* **2000**, *30*, 1861–1867. [CrossRef]
11. Mostofinejad, D.; Nikoo, M.R.; Hosseini, S.A. Determination of optimized mix design and curing conditions of reactive powder concrete (RPC). *Constr. Build. Mater.* **2016**, *123*, 754–767. [CrossRef]
12. Yazıcı, H.; Yardımcı, M.Y.; Aydın, S.; Karabulut, A.S. Mechanical properties of reactive powder concrete containing mineral admixtures under different curing regimes. *Constr. Build. Mater.* **2009**, *23*, 1223–1231. [CrossRef]
13. Long, G.C.; Wang, X.Y.; Xie, Y.J. Very-high-performance concrete with ultrafine powders. *Cem. Concr. Res.* **2002**, *32*, 601–605. [CrossRef]
14. Yazıcı, H.; Yardımcı, M.Y.; Yiğiter, H.; Aydın, S.; Türkel, S. Mechanical properties of reactive powder concrete containing high volumes of ground granulated blast furnace slag. *Cem. Concr. Compos.* **2010**, *32*, 639–648. [CrossRef]
15. Afroughsabet, V.; Biolzi, L.; Ozbakkaloglu, T. High-performance fiber-reinforced concrete: A review. *J. Mater. Sci.* **2016**, *51*, 6517–6551. [CrossRef]
16. Al-Masoodi, A.H.H.; Kawan, A.; Kasmuri, M.; Hamid, R.; Khan, M.N.N. Static and dynamic properties of concrete with different types and shapes of fibrous reinforcement. *Constr. Build. Mater.* **2016**, *104*, 247–262. [CrossRef]
17. Zhou, W.; Hu, H.B.; Zheng, W.Z. Bearing capacity of reactive powder concrete reinforced by steel fibers. *Constr. Build. Mater.* **2013**, *48*, 1179–1186. [CrossRef]
18. Al-Tikrite, A.; Hadi, M.N.S. Stress–Strain Relationship of Unconfined RPC Reinforced with Steel Fibers under Compression. *J. Mater. Civ. Eng.* **2018**, *30*, 04018234. [CrossRef]
19. Hiremath, P.N.; Yaragal, S.C. Performance evaluation of reactive powder concrete with polypropylene fibers at elevated temperatures. *Constr. Build. Mater.* **2018**, *169*, 499–512. [CrossRef]
20. Wang, H.; Gao, X.J.; Liu, J.Z.; Ren, M.; Lu, A.X. Multi-functional properties of carbon nanofiber reinforced reactive powder concrete. *Constr. Build. Mater.* **2018**, *187*, 699–707. [CrossRef]
21. Tai, Y.S.; Pan, H.H.; Kung, Y.N. Mechanical properties of steel fiber reinforced reactive powder concrete following exposure to high temperature reaching 800 °C. *Nucl. Eng. Des.* **2011**, *241*, 2416–2424. [CrossRef]
22. Scheinherrová, L.; Vejmelková, E.; Keppert, M.; Bezdička, P.; Doleželová, M.; Krejsová, J.; Grzeszczyk, S.; Matuszek-Chmurowska, A.; Černý, R. Effect of Cu-Zn coated steel fibers on high temperature resistance of reactive powder concrete. *Cem. Concr. Res.* **2019**, *117*, 45–57. [CrossRef]
23. Scrivener, K.L.; Kirkpatrick, R.J. Innovation in use and research on cementitious material. *Cem. Concr. Res.* **2008**, *38*, 128–136. [CrossRef]

24. Olivito, R.S.; Zuccarello, F.A. An experimental study on the tensile strength of steel fiber reinforced concrete. *Compos. Part B Eng.* **2010**, *41*, 246–255. [CrossRef]
25. Branston, J.; Das, S.; Kenno, S.Y.; Taylor, C. Mechanical behaviour of basalt fibre reinforced concrete. *Constr. Build. Mater.* **2016**, *124*, 878–886. [CrossRef]
26. Algin, Z.; Ozen, M. The properties of chopped basalt fibre reinforced self-compacting concrete. *Constr. Build. Mater.* **2018**, *186*, 678–685. [CrossRef]
27. Kizilkanat, A.B.; Kabay, N.; Akyüncü, V.; Chowdhury, S.; Akça, A.H. Mechanical properties and fracture behavior of basalt and glass fiber reinforced concrete: An experimental study. *Constr. Build. Mater.* **2015**, *100*, 218–224. [CrossRef]
28. Wang, D.H.; Ju, Y.Z.; Shen, H.; Xu, L.B. Mechanical properties of high performance concrete reinforced with basalt fiber and polypropylene fiber. *Constr. Build. Mater.* **2019**, *197*, 464–473. [CrossRef]
29. Grzeszczyk, S.; Matuszek-Chmurowska, A.; Vejmelková, E.; Černý, R. Reactive Powder Concrete Containing Basalt Fibers: Strength, Abrasion and Porosity. *Materials* **2020**, *13*, 2948. [CrossRef]
30. Liu, H.; Lyu, X.; Zhang, Y.; Luo, G.; Li, W. Bending Resistance and Failure Type Evaluation of Basalt Fiber RPC Beam Affected by Notch and Interfacial Damage Using Acoustic Emission. *Appl. Sci.* **2020**, *10*, 1138. [CrossRef]
31. An, M.Z.; Wang, Y.; Yu, Z.R. Damage mechanisms of ultra-high-performance concrete under freeze–thaw cycling in salt solution considering the effect of rehydration. *Constr. Build. Mater.* **2019**, *198*, 546–552. [CrossRef]
32. Liu, H.; Liu, S.; Wang, S.; Gao, X.; Gong, Y. Effect of Mix Proportion Parameters on Behaviors of Basalt Fiber RPC Based on Box-Behnken Model. *Appl. Sci.* **2019**, *9*, 2031. [CrossRef]
33. Quality and Technology Supervision Bureau of the People's Republic of China. *Method of Testing Cements–Determination of Strength*; Quality and Technology Supervision Bureau of the People's Republic of China: Beijing, China, 1999. (In Chinese)
34. Ministry of Housing and Urban-Rural Construction of the People's Republic of China. *Standard for Test Methods of Long-Term Performance and Durability of Ordinary Concrete*; Ministry of Housing and Urban-Ural Construction of the People's Republic of China: Beijing, China, 2009. (In Chinese)
35. Liu, H.; Liu, S.; Zhou, P.; Zhang, Y.; Jiao, Y. Mechanical Properties and Crack Classification of Basalt Fiber RPC Based on Acoustic Emission Parameters. *Appl. Sci.* **2019**, *9*, 3931. [CrossRef]
36. Rybin, V.A.; Utkin, A.V.; Baklanova, N.I. Alkali resistance, microstructural and mechanical performance of zirconia-coated basalt fibers. *Cem. Concr. Res.* **2013**, *53*, 1–8. [CrossRef]
37. Zhao, Y.R.; Wang, L.; Lei, Z.K.; Han, X.F.; Shi, J.N. Study on bending damage and failure of basalt fiber reinforced concrete under freeze-thaw cycles. *Constr. Build. Mater.* **2018**, *163*, 460–470. [CrossRef]
38. Vaitkevičius, V.; Šerelis, E.; Rudžionis, Ž. Nondestructive testing of ultra-high performance concrete to evaluate freeze-thaw resistance. *Mechanika* **2012**, *18*, 164–169. [CrossRef]
39. Zhou, Z.D.; Qiao, P.Z. Durability of ultra-high performance concrete in tension under cold weather conditions. *Cem. Concr. Compos.* **2018**, *94*, 94–106. [CrossRef]

Article

Alkali Activated Paste and Concrete Based on of Biomass Bottom Ash with Phosphogypsum

Danutė Vaičiukynienė [1,*], Dalia Nizevičienė [2], Aras Kantautas [3], Vytautas Bocullo [1] and Andrius Kielė [1]

[1]   Faculty of Civil Engineering and Architecture, Kaunas University of Technology, LT-44249 Kaunas, Lithuania; vytautas.bocullo@ktu.lt (V.B.); andrius.kiele@ktu.lt (A.K.)
[2]   Faculty of Electrical and Electronics Engineering, Kaunas University of Technology, Studentu St. 48, LT-51367 Kaunas, Lithuania; dalia.nizeviciene@ktu.lt
[3]   Faculty of Chemical Technology, Kaunas University of Technology, Radvilėnų Pl. 19, LT-50254 Kaunas, Lithuania; aras.kantautas@ktu.lt
*   Correspondence: danute.vaiciukyniene@ktu.lt or danute.palubinskaite@ktu.lt

Received: 9 July 2020; Accepted: 24 July 2020; Published: 28 July 2020

Abstract: There is a growing interest in the development of new cementitious binders for building construction activities. In this study, biomass bottom ash (BBA) was used as aluminosilicate precursor and phosphogypsum (PG) was used as a calcium source. The mixtures of BBA and PG were activated with the sodium hydroxide solution or the mixture of sodium hydroxide solution and sodium silicate hydrate solution. Alkali activated binders were investigated using X-ray powder diffraction (XRD), X-ray fluorescence (XRF) and scanning electron microscopy (SEM) test methods. The compressive strength of hardened paste and fine-grained concrete was also evaluated. After 28 days, the highest compressive strength reached 30.0 MPa—when the BBA was substituted with 15% PG and activated with NaOH solution—which is 14 MPa more than control sample. In addition, BBA fine-grained concrete samples based on BBA with 15% PG substitute activated with $NaOH/Na_2SiO_3$ solution showed higher compressive strength compered to when NaOH activator was used −15.4 MPa and 12.9 MPa respectfully. The $NaOH/Na_2SiO_3$ activator solution resulted reduced open porosity, so potentially the fine-grained concrete resistance to freeze and thaw increased.

Keywords: biomass bottom ash; phosphogypsum; alkali activated fine-grained concrete

---

1. Introduction

In recent years, the amount of biomass bottom ash (BBA) originating from Lithuanian combustion plants is constantly increasing. This type of ash is classified as nonhazardous wastes, so BBA is deposited in local landfills. Consequently, it is very important to reuse ash and to reduce discarding at landfill site. According to Carrasco-Hurtado et al. [1] environmental study showed that the amount of heavy metals in BBA is usually lower than that in fly ash so for that reason it is possible to recycle it in construction materials.

Giergiczny et al. [2] investigated composite cement and concrete containing low-calcium and high-calcium fly ash and granulated blast furnace slag. When large quantities of ash or/and slag were incorporated in the cement system, the properties (e.g., long setting time, low early strength, etc.) of samples were improved. One utilization method for BBA could be the incorporation into construction materials. The chemical and mineral composition of BBA is appropriate for reusing in the production of new, low-carbon building materials. In this way, the replacement of traditional initial material [3,4] such as fly ash or slag in alkali activated materials (AAM) by BBA leads to important environmental benefits [5–7]. As demand for ecological alternatives to Portland cement like alkali activated materials (AAM) is growing, there is interest to utilize phosphogypsum (PG) in AAM. AAM binders are aluminosilicate materials like fly ash, slag, red clay that can be activated with an activator solution—NaOH, Na_2SiO_3, KOH, etc. Concrete produced with these raw materials has shown potential results: the compressive strength of alkali activated fly ash paste reaching over 25 MPa [8,9].

PG can expand the base of the AAM raw materials. Approximately 4.5–5.5 tons of PG is generated per ton of phosphoric acid production using wet process [10]. It is estimated annually 100–280 million metric tons of PG are generated globally. The PG waste is usually stockpiled in landfills. Landfilling stocks results leaching, and hazardous constituents get into groundwater and underlying soils [11]. Pérez-López et al. investigated PG deposited over Tinto river saltmarshes for 40 years until 2010. Study have shown the high potential of contamination of the whole PG stack, including those stack zones that were restored and supposedly should have stop leaching of toxic solutions [12].

Previous studies [13–17] show interest in using PG or gypsum in AMMs. Multiple studies have investigated the optimum amount of gypsum compounds and it was determined that the optimal amount of $CaSO_4$ in alkali activated systems is close to 10% wt [13,14]. Gypsum takes significant part in the activation processes–it completely dissolves and participates in solid product formation. In the alkali's activation reactions PG is a supplier of SO_4^{2-} and Ca^{2+} ions enhancing the formation of secondary reaction products. When PG is present, portlandite and ettringite initially forms after dissolution form in AAM system. The hardened AAM consists mainly of amorphous hydration products, intermixed with thenardite and minor amounts of secondary gypsum. The incorporation of PG results in shorter initial setting time, but longer final setting time. There is significant increase of compressive strength when activator is NaOH [13]. The compressive strength development can be attributed to lower porosity [13,14]. PG decreases of Ca/Si ratios in the C–A–S–H gels and it could be the reason of a higher polymerized network [13]. The AAM samples with PG inclusion exhibited an average of 1.2 times greater residual strength than samples without PG, after being treated at 400, 600, 800 and 1000 °C temperatures [15]. Boonserm et al. [16] had found that the additive of flue gas desulfurization gypsum significantly improved the geopolymerization of the mixtures of bottom ash and fly ash. The compressive strength of samples increased too in that samples were up to 10% of gypsum was used. This increase is explained with the formation of additional amount of CSH. Similar results gave Khater et al. [17]. A 10% PG additive improved samples mechanical properties and microstructure. Samples were formed from the fly ash, PG and cement kiln dust mixtures. Rashad et al. [14] investigated the alkali activated fly ash and PG. When 5% or 10% of semi hydrate PG was incorporated in the system, the mechanical properties, improvement was detected. Chang et al. [18] investigated the influence of phosphoric acid and gypsum on the sodium silicate-based alkali-activated slag pastes. It was determined that the addition of phosphoric acid acted as a retarder.

In previous already published papers, alkali activated systems based on fly ash or fly ash and bottom ash with phosphogypsum were investigated. In this work, only biomass bottom ash was used as an aluminosilicate source. Some amount of BBA was substituted with PG. Both (BBA and PG) are local availability byproducts. The aim of this work is to investigate alkali activated paste and fine-grained concrete with BBA and PG, and to describe the effect of PG on the properties of newly formed AAM systems. Two types of alkali activators were used: NaOH solutions and the mixtures made from NaOH solution and sodium silicate hydrate (WG).

2. Methodology

2.1. The Characterization of Raw Materials and the Mixing Composition of Alkali Activated Biomass Bottom Ash Pastes and Concretes

In this study, the precursor was made from biomass bottom ash (BBA) which was obtained from the combustion plant located in Lithuania. First, this BBA was dried at 100 °C temperature for 24 h and then it was milled in ball mill.

The PG used in this work had α hemihydrate type. It is the waste product of orthophosphoric acid production. PG is formed by the reaction of sulfuric acid from natural apatite according this Equation (1) [19]:

$$Ca_5(PO_4)_3\,F + 5H_2SO_4 + 2.5H_2O = 3H_3PO_4 + 5CaSO_4 \cdot 0.5H_2O + HF; \tag{1}$$

The powder of PG was taken from the conveyor belt of waste removal and dried at $100 \pm 5\,°C$ temperature.

According to SEM analysis the BBA particles have irregular shape with angular morphologies (Figure 1a). Microscopic analysis showed that semihydrate phosphogypsum crystals are of dense structure and irregularly shaped parallelepipeds (Figure 1b).

Figure 1. Microstructure of (**a**) biomass bottom ash and (**b**) phosphogypsum.

The particle size distribution for BBA is presented in Figure 2a. The particles size is in wide range from 0.9 µm to 460 µm. The PG particles are finer (Figure 2b) and they are in the range from 0.8 µm to 38 µm. The specific surfaces areas according to Blaine for PG is 201 m^2/kg and for BBA is 396 m^2/kg.

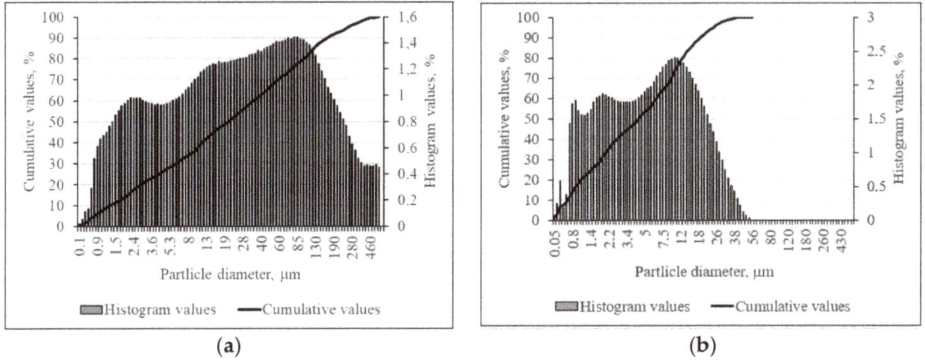

Figure 2. Granulometric composition of biomass bottom ash (**a**) and phosphogypsum (**b**).

The mean diameter of PG particles is 74.7 µm and for the particles of BBA is 58.6 µm.

The XRD analysis showed that in BBA dominated quartz, anorthoclase, gehlenite and calcium hydroxide with small amounts of calcium carbonate calcium oxide and magnesium oxide (Figure 3a). BBA has semi-amorphous semi-crystalline structure with a broad peak close to silica [20].

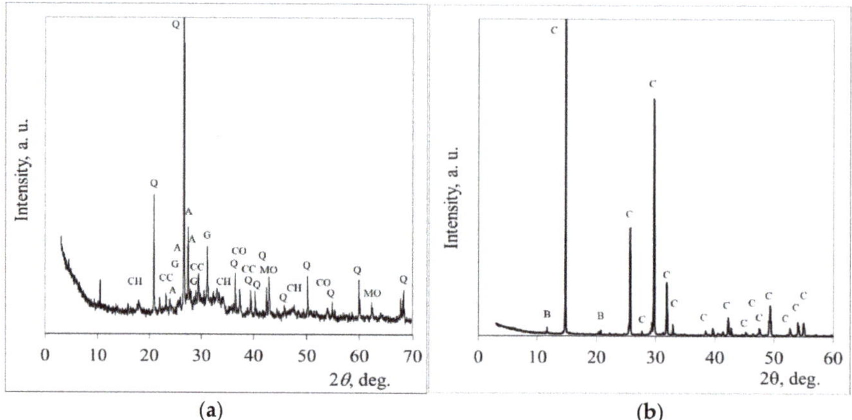

Figure 3. The mineral composition (XRD analysis) of (**a**) biomass bottom ash and (**b**) phosphogypsum. Notes: Q—quartz, SiO_2 (77–1070); CH—calcium hydroxide, $Ca(OH)_2$ (84–1271); CC—calcium carbonate, $Ca(CO)_3$ (72–1652); A—anorthoclase, $(Na,K)(Si_3Al)O_8$ (75–1631); G—gehlenite, $Ca_2\ Al(AlSiO_7)$ (79–2421); CO—calcium oxide, CaO (4–777); MO—magnesium oxide, MgO (78–430); C—bassanite, $CaSO_4 \cdot 0.5H_2O$ (33–310); B—brushite, $CaPO_3(OH) \cdot 2H_2O$ (11–293).

In the mineral composition of PG, according to XRD analysis (Figure 3b), dominated bassanite ($CaSO_4 \cdot 0.5H_2O$) and a small amount of brushite ($CaPO_3(OH) \cdot 2H_2O$). PG exhibit well definite crystalline structure [21].

This byproduct of biomass combustion in power plants (BBA) has the relatively high calcium, silicon and alkali contents (Table 1).

Table 1. Chemical composition of initial materials, wt%.

	CaO	SiO$_2$	Na$_2$O	Al$_2$O$_3$	MnO	MgO	K$_2$O	Fe$_2$O$_3$	P$_2$O$_5$	TiO$_2$	SO$_3$	F	Other
BBA	49.0	22.4	0.28	2.51	0.30	8.29	8.69	2.18	5.05	0.33	0.58	–	0.39
PG	38.60	0.37	–	0.13	–	0.04	–	0.03	0.81	–	53.48	0.14	6.4

According to the chemical composition of the PG, CaO and SO_3 are the major components of this material (Table 1). There is some amount of acidic impurities such as P_2O_5–0.81%; including water-soluble—0.10% and F—0.14%, which make PG difficult to reuse. Loss on ignition was 6.4%. The pH of the water suspension–4.7.

Two types of alkali activators were used (Table 2). The first type was sodium hydroxide solution made with commercial NaOH pellets (analytical grades). The second one was made from the mixture of NaOH solution and the sodium silicate hydrate (WG) solution (silicate modulus 3.0, concentration 36%).

Table 2. The mixing composition of alkali activated biomass bottom ash pastes.

Samples	BBA, g	PG, g	WG, g	H_2O, g	NaOH, g	SiO_2/Al_2O_3	SiO_2/Na_2O
PG 0-2	100	0	–	23	20.1	12.0	2
PG 5-2	95	5	–	28	19.9	12.0	2
PG 10-2	90	10	–	29	18.8	12.0	2
PG 15-2	85	15	–	30	17.8	12.0	2
PG 20-2	80	20	–	30	16.7	12.0	2
PG 25-2	75	25	–	30	15.7	12.0	2
PGWG 0-2	100	0	20.0	23	24.5	14.8	2
PGWG 5-2	95	5	15.8	28	22.6	14.3	2
PGWG 10-2	90	10	14.0	29	21.2	14.2	2
PGWG 15-2	85	15	15.7	30	20.5	14.6	2
PGWG 20-2	80	20	16.5	30	19.7	14.9	2
PGWG 25-2	75	25	19.5	30	19.4	15.6	2
PG 0-3	100	0	–	23	13.0	12.0	3
PG 5-3	95	5	–	29	12.4	12.0	3
PG 10-3	90	10	–	30	11.7	12.0	3
PG 15-3	85	15	–	30	11.1	12.0	3
PG 20-3	80	20	–	30	10.4	12.0	3
PG 25-3	75	25	–	30	9.8	12.0	3
PGWG 0-3	100	0	6.4	23	13.0	13.3	3
PGWG 5-3	95	5	7.1	27	12.4	13.5	3
PGWG 10-3	90	10	6.0	28	11.7	13.3	3
PGWG 15-3	85	15	7.3	30	11.1	13.7	3
PGWG 20-3	80	20	7.6	30	10.4	13.9	3
PGWG 25-3	75	25	9.7	30	9.8	18.4	3

Paste samples size was $20 \times 20 \times 20$ mm; their composition is given in Table 2. The BBA was substituted for PG at various amounts: 5%, 10%, 15%, 20% and 25%. The ratio of water and solid materials (BBA + PG) in the mixtures was regulated and ranged from 0.23 to 0.30. First, dry components were thoroughly mixed. Then, the mixtures were filled with the solutions of alkali activator (Table 2). These solutions were prepared by dissolving NaOH in water. When the complex alkali activator was used WG solution was filled to sodium hydroxide solution.

The total hydration duration was 28 days. The first day samples hydrated in room temperature, the second day at 60 °C temperature—and for the remaining 26 days, in room temperature again. All this time samples were covered with polyethylene covering materials which protect the samples from dehydration.

The sand from Kvesai quarry (Lithuania) was used as fine aggregate to produce alkali activated concrete samples the 0/4 fraction sand. The particle density of sand was 2.65 Mg/m^3. The amount of initial materials for concrete samples is shown in Table 3.

Table 3. Mixing composition of alkali activated biomass bottom ash concrete, g.

Ingredients	The Activation with NaOH			The Activation with NaOH + Na_2OnSiO_2-mH_2O		
	CPG 0-3	CPG 15-3	CPG 20-3	CPWG 0-3	CPWG 15-3	CPWG 20-3
BBA	450	383	360	450	383	360
PG	0	67.5	90	0	67.5	90
Sand 0/4	1350	1350	1350	1350	1350	1350
NaOH	58.5	49.7	46.8	58.5	49.7	46.8
Water glass	0	0	0	28.8	32.9	34.2
Water	210	210	210	200	200	200

The compressive strength of hardened AAM paste was evaluated after 7 and after 28 days. To perform the test a hydraulic press ToniTechnik 2020 was used. The compressive strength of samples was determined in accordance with EN 196-1:2005. At least of three samples were tested of each type and the Sample Standard Deviation (SSD) was calculated according to Equation (2):

$$SSD = \sqrt{\frac{\sum (x - \bar{x})^2}{(n - 1)}};$$

(2)

where, x takes on each value in the set; $\bar{x}$ is the average (statistical mean) of the set of values; n is the number of values.

2.2. The Experimental Techniques

The mineral composition of initial materials and AAM hardened pastes was carried out by using X-ray powder diffraction analysis. Data were collected by DRON-6 X-ray diffractometer with Bragg–Brentano geometry using Ni-filtered Cu Kα radiation and graphite monochromator, operating with the voltage of 30 kV and emission current of 20 mA. The step-scan covered angular range of 2–70° in steps of 2 = 0.02°. The powder X-Ray diffraction patterns were identified with references available in PDF-2 data base [22].

The chemical composition of BBA and PG was evaluated by using XRF analysis. For this purpose, a Bruker X-ray S8 Tiger WD using a rhodium (Rh) tube, an anode voltage Ua up to 60 kV and an electric current I up to 130 mA were used. The compressed samples were measured in a helium atmosphere [23]. The hydration water (loss on ignition, %) in phosphogypsum was calculated after heating the material at the temperature 400 °C. The pH phosphogypsum was measured with the AD8000 professional multi-parameter pH-ORP-Conductivity-TDS-TEMP bench meter, with a measuring range of −2.00 to 16.00 pH, a resolution of 0.01 pH and an accuracy of ±0.01 pH. The pH measurements of water suspensions were conducted when the ratio of water (W) and solid material (S) W/S was 10.

Microstructure investigation of BBA, PG and hardened pates was performed using a high-resolution scanning electron microscope ZEISS EVO MA10 [24]. The resolution of the images (of secondary electrons in a high vacuum) of this microscope is at least 3 nm with 30 kV and at least 10 nm with 3 kV. In the performed analysis, the acceleration voltage was 5 kV.

A laser particle size analyzer (CILAS 1090 LD) was used for the evaluation of the particle size of the BBA and PG. The distribution of solid particles in the air stream was 12 wt%–15 wt%. Compressed air (2500 mbar) was used as a dispersing phase [25].

The compressive and flexural strength of hardened AAM paste and fine-grained concrete samples were determined by using hydraulic press Toni Technik 2020 according to the EN 196–1. The size of hardened AAM paste samples was 20 × 20 × 20 mm. Fine-grained concrete samples were 40 × 40 × 160 mm prisms. For each data point at least tree samples were tested. The compressive strength of hardened AAM paste was evaluated after 2 and 28 days of hydration. The mechanical properties of fine-grained concrete samples were tested after 7 and 28 days.

The total and open porosity of alkali activated fine-grained concrete samples was evaluated by water absorption according to Skripkiunas et al. [26].

3. Results and Discussion

The compressive strength of alkali-activated BBA samples is shown in Figure 4. There are two types of samples: one type of hardened AAM pastes was alkali activated by using NaOH solutions and the mixture of NaOH and sodium silicate hydrate solutions was used in the second type of samples. When SiO_2/Na_2O molar ratio was 2 the samples containing 20% PG substitute had the highest compressive strength. In this case compressive strength reached 24.3 MPa and precursors were alkali activated with NaOH solutions (Figure 4a). Similar values of compressive strength (23.0 MPa) were obtained for that samples which were activated with the mixtures of NaOH and sodium silicate hydrate solutions. In this case the optimal content of PG substituting was 15% (by mass of BBA). In all investigated cases (Figure 4a) the substitution of BBA to PG had gains in compressive strength.

This substitution is recommended not to exceed 25%. Similar compressive strength (25.83 MPa) had geopolymer samples formed with circulating fluidized bed combustion coal bottom ash according to Topçu et al. [27].

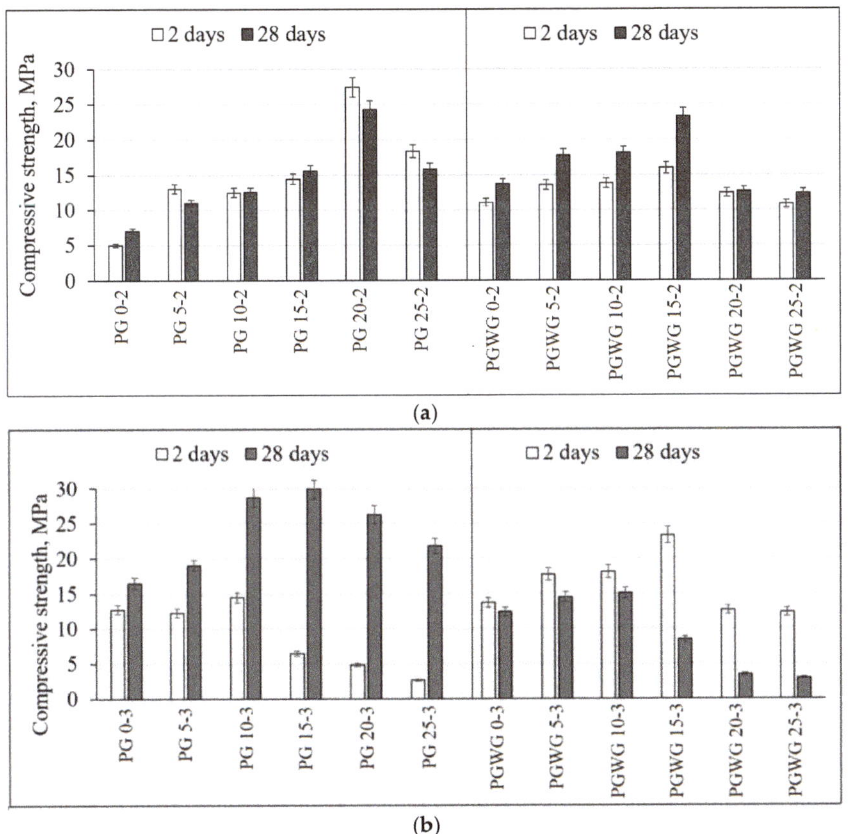

Figure 4. Compressive strength of alkali-activated biomass bottom ash pastes when SiO_2/Na_2O molar ratio is (**a**) 2 and (**b**) 3 (Table 2).

Figure 4b shows the compressive strength of the alkali activated BAA paste with SiO_2/Na_2O molar ratio 3. The positive effect was detected in this case. In the PG 15-3 samples with the alkali activator of NaOH solutions the highest compressive strength reached 30.0 MPa after 28 days. By using the same molar ratio but as alkali activator the mixture of NaOH and sodium silicate hydrate solutions was the compressive strength was reached 23.0 MPa after 7 days of hardening. After longer duration (28 days) of hardening, the reduction was observed of more than 3 times of compressive strength (8.0 MPa). This reduction of compressive strength may be explained by the fast alkali reactions resulted in quick strength gain after 2 days. This gain should be due the increase of gel like matrix. After 28 days the structure samples showed cracks on the surface which could be caused by the drying shrinkage (Figure 4b) [28]. In all investigated cases the use of calcium promoter such as PG which substituted BBA had positive effect to compressive strength gain. After 28 days the compressive strength was higher than compressive strength of reference samples. Similar results related with positive effect of calcium promoters in bottom ash geopolymer fine-grained concrete report Hanjitsuwanet al [29].

The mineral composition of alkali activated biomass bottom ash is shown in Figure 5. The X-ray diffraction study is carried out only on the 8 pastes because they are the ones that have shown the highest compressive strength values. The reference compositions were investigated as well. In all X-ray diffraction patterns it is possible to detect quartz and calcium hydroxide which left unreacted from BBA. During alkali reactions calcium silicate hydrate, calcium aluminum oxide hydroxide hydrate, sodium aluminum silicate hydrate formed. When PG was incorporated in the system, additional mineral calcium aluminum hydroxide hydrate formed (PG 20-2, PGWG 15-2, PG 15-3 and PGWG 15-3). The crystal phases remained the same in all samples and it did not depend on the molar SiO_2/Na_2O ratios which were used in this work. By using lower SiO_2/Na_2O molar ratio ($SiO_2/Na_2O = 2$) the higher amount of alkali had an impact on the formation of $Na_2CO_3(H_2O)$ (without PG) and Na_2SO_4 (with PG). The formation of $Na_2CO_3(H_2O)$ had negative affect on the development of compressive strength [30]. When PG was inserted in the alkali activated BBA, PG reacted with NaOH and this reaction products Na_2SO_4 with $Ca(OH)_2$ were (Figure 5a). During hydration process, Na_2SO_4 is an effective activator for alkali activated binders [31]. Sodium sulfate could motivate the formation of calcium aluminum silicate hydrate and calcium silicate hydrate. As seen in Figure 5a,b, the main peak of calcium silicate hydrate is more intensive in the samples where PG was incorporated.

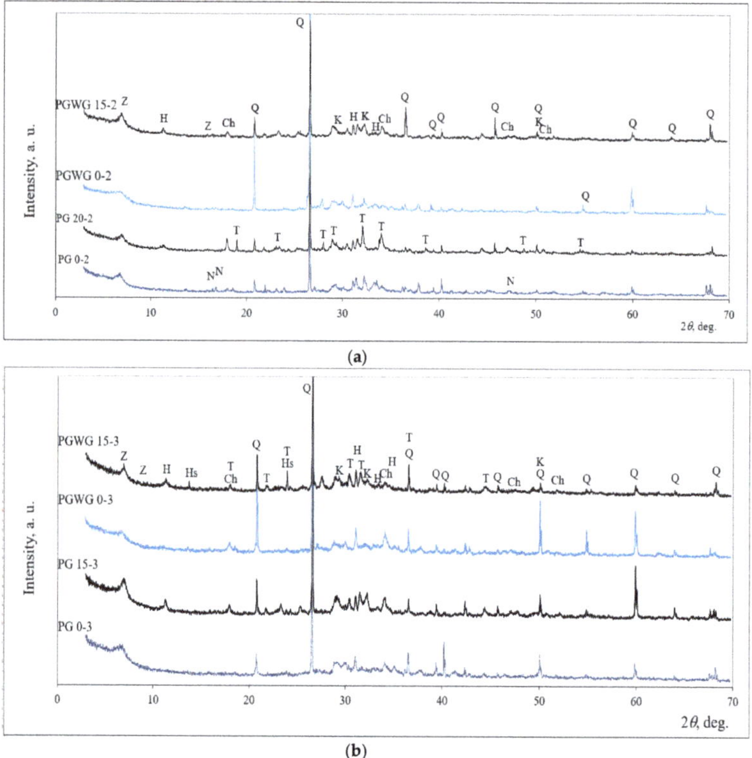

Figure 5. X-ray diffraction patterns of alkali activated biomass bottom ash when SiO_2/Na_2O molar ratio is (**a**) 2 and (**b**) 3. Notes: Q—quartz, SiO_2 (83–2465); Ch—calcium hydroxide, $Ca(OH)_2$ (84–1268); K—calcium silicate hydrate, $Ca_{1.5}SiO_{3.5}x\,H_2O$ (33–306); T—thenardite Na_2SO_4 (74–2036); Z—sodium aluminum silicate hydrate $Na_{96}Al_{96}Si_{96}O_{384}216H_2O$ (39–222); N—sodium carbonate hydrate $Na_2CO_3(H_2O)$ (70–845); Hs—hydroxy–sodalite $Na_6(AlSiO_4)_6\,8H_2O$ (72–2329); H—alcium aluminum hydroxide hydrate $Ca_2Al(OH)_7\cdot3H_2O$ (33–255).

In the samples with higher amount of SiO_2 the molar SiO_2/Na_2O ratio was 3. The peaks of new formed hydrates appear more intensive (Figure 5b). This could be related with formation of higher amount of polymerization products in alkali activated system. The hydroxy–sodalite was detected in the sample PGWG 15-3 [32].

Figure 6 shows the morphology of alkali activated biomass bottom ash after 28 days of hardening. PG 15-3 and PGWG 15-3 samples exhibited different microstructures. In the microstructure PG 15-3 sample varied honeycomb-like C–S–H and honeycomb type amorphous gel structures (Figure 6a) [33].

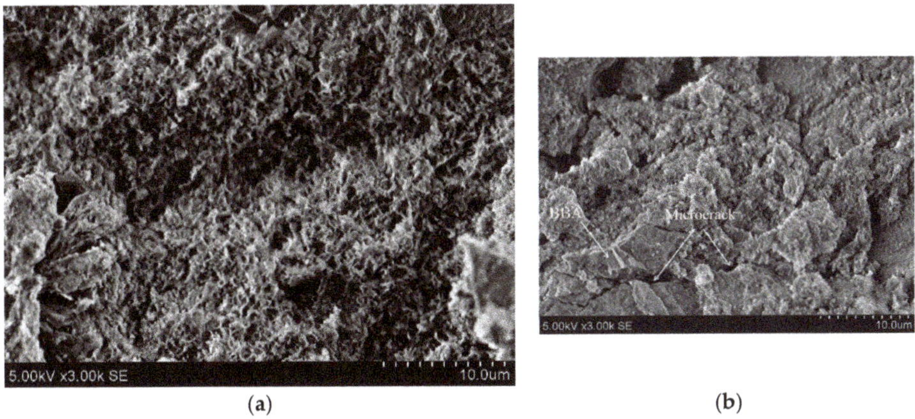

(a) (b)

Figure 6. Microstructure of alkali activated biomass bottom ash samples (**a**) PG 15-3 and (**b**) PGWG 15-3.

It can be observed that in the PGWG 15-3 sample showed a higher degree of microcracking and unreacted the particle of BBA were detected as well (Figure 6b) [34]. This PGWG 15-3 sample had a more compact microstructure by comparing with the microstructure of PG 15-3 sample. This compact microstructure is closely related to the increased amount of hydration products which increased the amount of microcracks [35].

The compressive and flexural strength of alkali activated fine-grained concretes are shown in Table 4. As the aluminosilicate precursor the mixture of BBA with PG was used. The two types of alkali activator solutions were chosen: sodium hydroxide solution and the mixture of sodium hydroxide solution and sodium silicate hydrate. The proportions of PG and BBA were chosen according to the values of paste samples compressive strength (Figure 4). According to Ding et al. [36] the compressive strength values of the alkali-activated pastes, fine-grained concretes and concretes with the same pastes were unequal. Fine-grained concretes had significantly lower compressive strength compared with the paste samples.

Table 4. Compressive strength, flexural strength and density of alkali activated fine-grained concretes samples.

Samples	Compressive Strength, MPa	Flexural Strength, MPa	Compressive Strength, MPa	Flexural Strength, MPa	Density (After 28 Days), kg/m^3
	After 7 Days of Hardening		After 28 Days of Hardening		
CPG 0-3	4.51 ± 0.26	1.52 ± 0.05	10.94 ± 0.48	2.30 ± 0.12	1965 ± 14
CPG 15-3	4.79 ± 0.23	2.22 ± 0.06	12.90 ± 0.45	2.31 ± 0.15	1971 ± 14
CPG 20-3	3.94 ± 0.22	1.39 ± 0.06	9.33 ± 0.41	2.00 ± 0.17	1853 ± 15
CPWG 0-3	6.42 ± 0.28	1.32 ± 0.05	10.73 ± 0.42	1.84 ± 0.16	1998 ± 15
CPWG 15-3	8.40 ± 0.24	2.18 ± 0.08	15.42 ± 0.43	2.41 ± 0.11	2019 ± 14
CPWG 20-3	6.20 ± 0.27	1.20 ± 0.07	11.94 ± 0.31	1.54 ± 0.13	1885 ± 13

Chindaprasirt et al. [37] investigated and compared fly ash and bottom ash fine-grained concretes. The values of compressive strength are different for alkali activated fly ash and for bottom ash. Fly ash fine-grained concrete reached 35 MPa while bottom ash fine-grained concrete had compressive strength in the range of 10–18 MPa. Such a difference is explained by the degree of polymerization. The polymerization of bottom ash is lower than the fly ash during alkali activation. All these samples were cured at 65 °C for 48 h. In this work, samples were cured at lower 60 °C temperature and duration was shorter-24 h. The fine-grained concretes samples had similar compressive strength 12.9 MPa and 15.4 MPa when activated with NaOH solution and the mixture of NaOH/Na$_2$SiO$_3$ solution, respectively (Table 4). The higher compressive strength could be related with the higher amount of active silicon (sodium silicate hydrate solution) [38]. The flexural strength was similar for both types of fine-grained concretes. A little bit higher value of flexural strength (2.4 MPa) were obtained for the sample with mixture of NaOH solution and Na$_2$SiO$_3$ solution (CPWG 15-3) compared with CPG 15-3 sample.

The porosity study is carried out on the two-alkali activated fine-grained concrete samples shown in Figure 7. It is considered to be because they are the ones that have shown the highest compressive strength values. The X-ray diffraction study is carried out only on the 8 pastes because they are the ones that have shown the highest compressive strength values. The reference compositions were investigated as well. It is possible to predict the durability (freeze–thaw resistance) of alkali activated fine-grained concrete according to these parameters of porosity. The total porosity (P) is almost the same for both types of fine-grained concretes. The open porosity (Pa) which determined by water absorption of alkali activated fine-grained concrete was less (10.9%) for CPG 15-3 samples compared with CPWG 15-3 samples 13.4%. Different situation is with close porosity (Pu). In this case CPG 15-3 samples had higher 16.8% close porosity compared with CPWG 15-3 samples which had 14.1%. Therefore, the alkali activator of NaOH and Na$_2$SiO$_3$ solutions had influence on the formation higher amount of close porosity and lower amount of open porosity while the total porosity remained the almost the same in activated fine-grained concrete samples. According to Nagrockienė et al. [39] concrete with higher closed porosity have better freeze–thaw resistance. Hence, the fine-grained concrete activated with the NaOH and Na$_2$SiO$_3$ solution should have higher freeze–thaw resistance compared with fine-grained concrete activated with just NaOH solution.

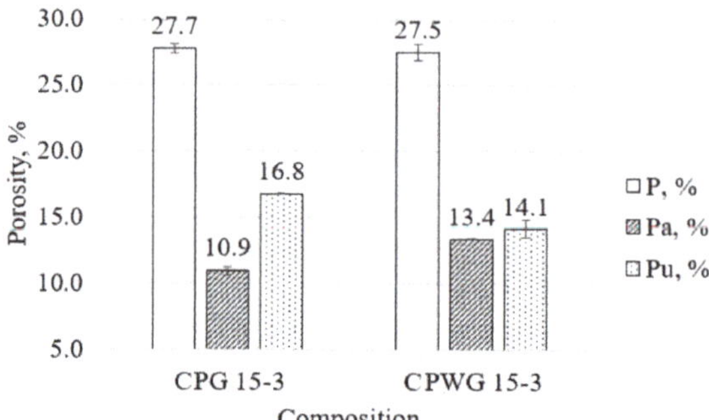

Figure 7. The porosity of alkali activated fine-grained concrete samples. P—total porosity; Pa—open porosity; Pu—closed porosity.

4. Conclusions

In this study, the compressive strength of hardened alkali activated past and fine-grained concretes was determined. To improve the reaction degree of BBA calcium promoter such as PG was used. The following observation were made:

- The higher values of compressive strength were detected when SiO_2/Na_2O molar ratio was three compared with the samples where samples had $SiO_2/Na_2O = 2$. When SiO_2/Na_2O molar ratio was two, the samples containing 20% PG substitute had the highest compressive strength (24.3 MPa) after 28 days, while compressive strength of the samples with SiO_2/Na_2O molar ratio three peaked after 28 days at 30.0 MPa;
- Overall, paste samples activated with $NaOH/Na_2SiO_3$ solution were weaker compared activated with NaOH. The compressive strength peaked at 23.0 MPa after seven days with 15% PG substitute, but after 28 days was reduced to 8.0 MPa. The reduction could be attributed to the cracking caused by the drying shrinkage. In addition, SEM images showed a higher degree of microcracking and unreacted the particle of BBA were detected when $NaOH/Na_2SiO_3$ solution was activator;
- The amount of PG substitute was 15%–20% of BBA mass and it is recommended not to exceed 25%. The higher amount of PG significantly reduced the compressive strength;
- In contrary to paste samples, fine-grained concrete samples have shown different tendencies. When BBA with 15% PG was activated with $NaOH/Na_2SiO_3$ solution compressive strength reached 15.4 MPa after 28 days, but when the same precursors were activated with NaOH compressive strength was reduced by 16% (12.9 MPa). The higher compressive strength could be related with the higher amount of active silicon;
- Even though total porosity of fine-grained concrete activated by NaOH and $NaOH/Na_2SiO_3$ solution was nearly the same -27.5% and 27.7% respectfully, there is difference in open and closed porosity. When sodium silicate was present the open porosity was reduced from 16.8% to 14.1%. Hence, the fine-grained concrete activated with the NaOH and Na_2SiO_3 solution should have higher freeze–thaw resistance;
- After 28 days of hydration the highest compressive strength reached 30.0 MPa the samples activated NaOH solution and by using 15% of BBA substitution to PG. The possibility of potential use of BBA (silicon and aluminum sources) and PG (calcium source) as binder precursor for production of AAMs was confirmed. This can suggest a solution for both alkaline binders industry to provide a new precursor and fertilizers industry to solve the storage problems of PG.

Author Contributions: We all authors: D.V., D.N., A.K. (Aras Kantautas), V.B. and A.K. (Andrius Kielė) declare that we contributed to this article in equal parts about 20% everybody: D.V. and V.B. prepared the introduction part. Characterization of initial materials and the part of experimental procedures was prepared by D.N. and A.K. (Andrius Kielė). The parts of "Results and discussion" and "Conclusion" were written and evaluated by all authors. All authors have read and agreed to the published version of the manuscript.

Funding: This research received no external funding.

Acknowledgments: This research work was supported by the Lithuanian Science Council project "The utilization of industrial waste in alkali-activated concrete", project code P-MIP-17-363.

Conflicts of Interest: The authors declare no conflicts of interest.

References

1. Carrasco-Hurtado, B.; Corpas-Iglesias, F.A.; Cruz-Pérez, N.; Terrados-Cepeda, J.; Pérez-Villarejo, L. Addition of bottom ash from biomass in calcium silicate masonry units for use as construction material with thermal insulating properties. *Constr. Build. Mater.* **2014**, *52*, 155–165. [CrossRef]
2. Giergiczny, Z. Fly ash and slag. *Cem. Concr. Res.* **2019**, *124*, 105826. [CrossRef]
3. Nowoświat, A.; Gołaszewski, J. Influence of the variability of calcareous fly ash properties on rheological properties of fresh mortar with its addition. *Materials* **2019**, *12*, 1942. [CrossRef] [PubMed]
4. Wei, X.; Ming, F.; Li, D.; Chen, L.; Liu, Y. Influence of water content on mechanical strength and microstructure of alkali-activated fly ash/GGBFS mortars cured at cold and polar regions. *Materials* **2020**, *13*, 138. [CrossRef]

5. Pérez-Villarejo, L.; Bonet-Martínez, E.; Eliche-Quesada, D.; Sánchez-Soto, P.J. Biomass bottom ash and aluminium industry slags-based geopolymers. In *Vitrification and Geopolimerization of Wastes for Immobilization or Recycling, Proceedings of the Vitrogeowastes, Elche, Spain, 14–15 September 2017*; Rincón, J.M., Vidal, M.J., Rincón, J.M., Vidal, M.J., Eds.; Universidad Miguel Hernández: Elche, Spain, 2017; Available online: http://innovacionumh.es/editorial/Vitrogeowastes.pdf (accessed on 15 September 2017).

6. Onori, R.; Will, J.; Hoppe, A.; Polettini, A.; Pomi, R.; Boccaccini, A.R. Bottom ash-based geopolymer materials: Mechanical and environmental properties. In *Ceramic Engineering and Science Proceedings*; Wiley: Hoboken, NJ, USA, 2011; Volume 32, pp. 71–82. [CrossRef]

7. ul Haq, E.; Padmanabhan, S.K.; Licciulli, A. Synthesis and characteristics of fly ash and bottom ash based geopolymers—A comparative study. *Ceram. Int.* **2014**, *40*, 2965–2971. [CrossRef]

8. Borg, R.P.; Briguglio, C.; Bocullo, V.; Vaičiukynienė, D. Preliminary investigation of geopolymer binder from waste materials. *Rom. J. Mater.* **2017**, *47*, 370–378.

9. Bocullo, V.; Vaičiukynienė, D.; Gečys, R.; Daukšys, M. Effect of ordinary portland cement and water glass on the properties of alkali activated fly ash concrete. *J. Miner.* **2020**, *10*, 1–10. [CrossRef]

10. JSC Lifosa, Ekstrakcinė fosforo rūgštis. Available online: https://www.lifosa.com/lt/ekstrakcine-fosforo-rugstis (accessed on 13 March 2020).

11. Mashifana, T.P. Chemical treatment of phosphogypsum and its potential application for building and construction. *Proced. Manufact.* **2019**, *35*, 641–648. [CrossRef]

12. Pérez-López, R.; Macías, F.; Cánovas, C.R.; Sarmiento, A.M.; Pérez-Moreno, S.M. Pollutant flows from a phosphogypsum disposal area to an estuarine environment: An insight from geochemical signatures. *Sci. Total Environ.* **2016**, *553*, 42–51. [CrossRef]

13. Gijbels, K.; Iacobescu, R.I.; Pontikes, Y.; Schreurs, S.; Schroeyers, W. Alkali-activated binders based on ground granulated blast furnace slag and phosphogypsum. *Constr. Build. Mater.* **2019**, *215*, 371–380. [CrossRef]

14. Rashad, A.M. Potential use of phosphogypsum in alkali-activated fly ash under the effects of elevated temperatures and thermal shock cycles. *J. Clean. Prod.* **2015**, *87*, 717–725. [CrossRef]

15. Vaičiukynienė, D.; Nizevičienė, D.; Kielė, A.; Janavičius, E.; Pupeikis, D. Effect of phosphogypsum on the stability upon firing treatment of alkali-activated slag. *Constr. Build. Mater.* **2018**, *184*, 485–491. [CrossRef]

16. Boonserm, K.; Sata, V.; Pimraksa, K.; Chindaprasirt, P. Improved geopolymerization of bottom ash by incorporating fly ash and using waste gypsum as additive. *Cem. Concr. Comp.* **2012**, *34*, 819–824. [CrossRef]

17. Khater, H.M.; Zedane, S.R. Geopolymerization of industrial by-products and study of their stability upon firing treatment. *Int. J. Engineer. Technol.* **2012**, *2*, 308–316.

18. Chang, J.J.; Yeih, W.; Hung, C.C. Effects of gypsum and phosphoric acid on the properties of sodium silicate-based alkali-activated slag pastes. *Cem. Concr. Compos.* **2005**, *27*, 85–91. [CrossRef]

19. Vaičiukynienė, D.; Nizevičienė, D.; Šeduikytė, L. Sustainable approach of the utilization of production waste: The use of phosphogypsum and AlF3 production waste in building materials. In *Energy Efficient, Sustainable Building Materials and Products*; Hager, I., Ed.; Wydawnictwo Politechniki Krakowskiej: Kraków, Poland, 2017; pp. 183–197.

20. Darsanasiri, A.G.N.D.; Matalkah, F.; Ramli, S.; Al-Jalode, K.; Balachandra, A.; Soroushian, P. Ternary alkali aluminosilicate cement based on rice husk ash, slag and coal fly ash. *J. Build. Eng.* **2018**, *19*, 36–41. [CrossRef]

21. Rajković, M.B.; Tošković, D.V. Phosphogypsum surface characterization using scanning electron microscopy. *Acta Period. Technol.* **2003**, *34*, 61–70. [CrossRef]

22. Bruker, D8 Advance Diffractometer (Bruker AXS) Technical Details. Available online: https://www.bruker.com/products/x-ray-diffractionand-elemental-analysis/x-ray-diffraction/d8-advance.html (accessed on 13 March 2020).

23. Bruker, X-ray S8 Tiger WD Series 2 Technical Details. Available online: https://www.bruker.com/products/x-ray-diffraction-andelemental-analysis/x-ray-fuorescence/s8-tiger.html (accessed on 13 March 2020).

24. Zeiss. *EVO MA and LS Series Scanning Electron Microscopes for Materials Analysis and Life Science. Operator User Guide, Version 1.0*; Zeiss: Cambridge, MA, USA, 2008.

25. CILAS, 1090 Particle Size Analyzer. Available online: https://www.pharmaceuticalonline.com/doc/cilas-1090-particle-size-analyzer-0002 (accessed on 13 March 2020).

26. Girskas, G.; Skripkiūnas, G. The effect of synthetic zeolite on hardened cement paste microstructure and freeze-thaw durability of concrete. *Constr. Build. Mater.* **2017**, *142*, 117–127. [CrossRef]

27. Topçu, İ.B.; Toprak, M.U.; Uygunoğlu, T. Durability and microstructure characteristics of alkali activated coal bottom ash geopolymer cement. *J. Clean. Prod.* **2014**, *81*, 211–217. [CrossRef]

28. Samantasinghar, S.; Singh, S. Effects of curing environment on strength and microstructure of alkali-activated fly ash-slag binder. *Constr. Build. Mater.* **2020**, *235*, 117481. [CrossRef]

29. Hanjitsuwan, S.; Phoo-ngernkham, T.; Damrongwiriyanupap, N. Comparative study using Portland cement and calcium carbide residue as a promoter in bottom ash geopolymer mortar. *Constr. Build. Mater.* **2017**, *133*, 128–134. [CrossRef]

30. Kovalchuk, G.; Fernández-Jiménez, A.; Palomo, A. Alkali-activated fly ash. Relationship between mechanical strength gains and initial ash chemistry. *Mater. De Construcción* **2008**, *58*, 35–52. [CrossRef]

31. Douglas, E.; Brandstetr, J. A preliminary study on the alkali activation of ground granulated blast-furnace slag. *Cem. Concr. Res.* **1990**, *20*, 746–756. [CrossRef]

32. Fernandez-Jimenez, A.; Monzo, M.; Vincent, M.; Barba, A.; Palomo, A. Alkaline activation of metakaolin—Fly ash mixtures: Obtain of Zeoceramics and Zeocements. *Micropor. Mesopor. Mater.* **2008**, *108*, 41–49. [CrossRef]

33. Karim, M.R.; Hossain, M.M.; Elahi, M.M.A.; Zain, M.F.M. Effects of source materials, fineness and curing methods on the strength development of alkali-activated binder. *J. Build. Eng.* **2020**, *29*, 101147. [CrossRef]

34. Santa, R.A.A.B.; Bernardin, A.M.; Riella, H.G.; Kuhnen, N.C. Geopolymer synthetized from bottom coal ash and calcined paper sludge. *J. Clean. Prod.* **2013**, *57*, 302–307. [CrossRef]

35. Gholampour, A.; Ho, V.D.; Ozbakkaloglu, T. Ambient-cured geopolymer mortars prepared with waste-based sands: Mechanical and durability-related properties and microstructure. *Compos. Part. B Eng.* **2019**, *160*, 519–534. [CrossRef]

36. Ding, Y.C.; Cheng, T.W.; Dai, Y.S. Application of geopolymer paste for concrete repair. *Struct. Concr.* **2017**, *18*, 561–570. [CrossRef]

37. Chindaprasirt, P.; Jaturapitakkul, C.; Chalee, W.; Rattanasak, U. Comparative study on the characteristics of fly ash and bottom ash geopolymers. *Waste Manag.* **2009**, *29*, 539–543. [CrossRef]

38. Huang, G.; Ji, Y.; Li, J.; Zhang, L.; Liu, X.; Liu, B. Effect of activated silica on polymerization mechanism and strength development of MSWI bottom ash alkali-activated mortars. *Constr. Build. Mater.* **2019**, *201*, 90–99. [CrossRef]

39. Nagrockienė, D.; Skripkiūnas, G.; Girskas, G. Predicting frost resistance of concrete with different coarse aggregate concentration by porosity parameters. *Mater. Sci.* **2011**, *17*, 203–207. [CrossRef]

Article

Experimental Characterization of Prefabricated Bridge Deck Panels Prepared with Prestressed and Sustainable Ultra-High Performance Concrete

Muhammad Naveed Zafar [1], Muhammad Azhar Saleem [1], Jun Xia [2],* and Muhammad Mazhar Saleem [1]

[1] Department of Civil Engineering, University of Engineering and Technology, Lahore PK 54890, Pakistan; naveedzafar93@gmail.com (M.N.Z.); msale005@fiu.edu (M.A.S.); saleem@email.arizona.edu (M.M.S.)
[2] Civil Engineering Department, Xi'an Jiaotong-Liverpool University, Suzhou 215123, China
* Correspondence: jun.xia@xjtlu.edu.cn

Received: 29 June 2020; Accepted: 22 July 2020; Published: 26 July 2020

Abstract: Enhanced quality and reduced on-site construction time are the basic features of prefabricated bridge elements and systems. Prefabricated lightweight bridge decks have already started finding their place in accelerated bridge construction (ABC). Therefore, the development of deck panels using high strength and high performance concrete has become an active area of research. Further optimization in such deck systems is possible using prestressing or replacement of raw materials with sustainable and recyclable materials. This research involves experimental evaluation of six full-depth precast prestressed high strength fiber-reinforced concrete (HSFRC) and six partial-depth sustainable ultra-high performance concrete (sUHPC) composite bridge deck panels. The composite panels comprise UHPC prepared with ground granulated blast furnace slag (GGBS) with the replacement of 30% cement content overlaid by recycled aggregate concrete made with replacement of 30% of coarse aggregates with recycled aggregates. The experimental variables for six HSFRC panels were depth, level of prestressing, and shear reinforcement. The six sUHPC panels were prepared with different shear and flexural reinforcements and sUHPC-normal/recycled aggregate concrete interface. Experimental results exhibit the promise of both systems to serve as an alternative to conventional bridge deck systems.

Keywords: prefabricated; bridge deck; prestressed; UHPC; sustainable

---

1. Introduction

Replacement of traditional cast-in-place concrete bridge decks is time demanding and causes long delays in traffic movement [1,2]. Such constraints make the decisions of replacing the worn-out decks more difficult and cause loss of interest in the expansion of bridges [3]. This problem is more pronounced in densely populated urban areas. Therefore, there is a need to develop such deck systems, which can make the process of replacement and new construction quick and safe. Such systems should satisfy not only the strength requirements but also have the flexibility to fit in the existing bridge superstructures.

Prefabricated bridge deck panels are becoming increasingly popular because of their accelerated construction, reduced requirement of on-site labor, and shorter traffic disruption. For example, over 10 years, the application of prefabricated bridge decking in the United States increased by 25% with 49,043 bridges in 2007 to 61,416 bridges in 2016 [4]. Several researchers like Saleem et al. [5], Ghasemi et al. [6], Al-Ramahee et al. [3] have proposed precast bridge systems developed with ultra-high performance concrete (UHPC), high strength steel (HSS), and fiber-reinforced polymers (FRP). First developed in France in the 1990s [7], UHPC comprises of steel fibers, silica fumes, ground quartz, and superplasticizer, in addition to cement and sand [8,9]. It has no coarse aggregate. In comparison to regular concrete, UHPC has high flowability, high resistance to cracking, very low permeability, and low shrinkage/creep properties [10]. The minimum compressive strength is more than 150 MPa, and good tensile strength exceeds 5 MPa. The modulus of elasticity crosses 46 GPa [11,12]. A low-profile asymmetric waffle deck system was developed by Saleem et al. [5] using UHPC and HSS. The self-weight of this system was 1.55 kN/m^2, which was further optimized to 1.0 kN/m^2 by Ghasemi et al. [6]. These systems were tested with single and multiple ribs, both in simple-span and two-span configurations. Static and dynamic testing proved their potential and feasibility to replace the conventional bridge decks for fixed and moveable bridges. Ghasemi et al. [13], for the first time, combined UHPC with carbon fiber reinforced polymer (CFRP) rebars to develop waffle bridge deck panels. The idea was to offset the brittle behavior of CFRP with the ductile nature of UHPC. The study confirmed that with only 102 mm overall depth and self-weight of 0.9 kN/m^2, the deck panels meet the strength and serviceability requirements for a stringer spacing of 1.2 m. Al-Ramahee et al. [3] developed precast deck panels by combining UHPC with FRP laminates. No reinforcement in the form of rebars was used. With an overall depth of 127 mm and self-weight of 0.5–0.6 kN/m^2, the system met the strength and serviceability requirement (Δ_{all} = L/800) of AASHTO LRFD [14]. UHPC materials have also been used in other bridge-related applications, such as serving as the bonding material for precast prestressed concrete bridge girders [15]. It was concluded that UHPC connections with short, straight rebar provided considerable resistance to separation and transferred the load effectively under seismic condition. In another study, UHPC material was used to provide the longitudinal connection between lightweight bridge decks and steel beams to form composite bridge girder [16]. The T-shape connections proved to perform better than the traditional connection due to the better control of crack width.

Attempts have also been made to introduce environmentally friendly materials in bridge deck applications. For instance, Venancio [17] tested half-scale partial-depth precast bridge deck panels developed with sustainable UHPC. Quartz powder as a whole and a fraction of fine sand content were replaced with ground granulated blast furnace slag (GGBS). UHPC panels were tested in flexure and shear, and comparison was made with reinforced concrete control panels. Panels were prepared with 51 and 76 mm depths. UHPC with conventional mild-steel reinforcement achieved the highest ultimate load. UHPC panels tested under shear configuration also failed in flexure, demonstrating excellent performance under shear force.

Another way to increase the sustainability of UHPC material is by introducing coarse aggregates. Coarse aggregates have been used in various concretes to enhance the economic and technical advantages of the cementitious products [18–20], and their long-term performance has been verified [21]. Prior research also demonstrated that the UHPC material with coarse aggregate and 2% volume fraction of steel fiber could provide as much as 5.5 times of shear resistance of the conventional RC beams with the same size and configuration [22]. The flexural failure of beams made of UHPC with coarse aggregate failed in a ductile manner, and it was found out that the maximum aggregate size of 5 or 8 mm has no significant impact on its structural behavior. At the same time, the longitudinal reinforcement ratios control the stiffness and ultimate strength of the beam [23]. Using recycled aggregate to produce sustainable ultra-high-performance concrete (sUHPC) can further enhance the sustainability of the material. Recycled aggregates have been used for paving blocks to enhance sustainability [20]. It was proved that using RCA can reduce the strength of the concrete block, and adding GGBS can help partially recover the strength loss. Introducing recycled aggregate in UHPC material can help reduce both the usage of cement and virgin aggregate.

Optimization in design is another aspect of precast deck panels, which can lead to improved service. One way to achieve this is to use prestressing. This combination of high-strength or fiber reinforced concrete with prestressing has excellent potential in bridge deck applications because not only can it decrease the self-weight of the deck system by reducing the thickness, but also offers other advantages such as less maintenance, improved durability, reduced crack width, and high tensile strength.

Limited research is available on the application of combining prestressing with the application of sustainable high strength fiber reinforced concrete in the development of bridge deck panels. With this background, the primary objective of this research was to develop and characterize different configurations of full-depth precast prestressed high strength concrete—with and without fibers—bridge deck panels, and partial-depth precast sustainable ultra-high performance concrete (sUHPC)-normal/recycled aggregate concrete bridge deck panels. The collective efforts aimed to explore different optimization methods considering reduced self-weight and increased sustainability while focusing first on their conformation to the existing design standards.

2. Experimental Work

Experimental work was carried out in two phases. The first phase consisted of testing six prestressed deck panels and the second phase consisted of testing of six partial-depth precast non-prestressed sUHPC deck panels. Table 1 presents the test matrix of the entire stock of specimens, as well as some specimens from prior research programmes.

Five types of reinforcement were listed in Table 1: Type A is high strength steel rebar with yielding strength of 690 MPa and ultimate strength 1030 MPa; Type B and Type C are #10 and #13 carbon fiber reinforced plastic rebar, respectively, with ultimate tensile strength equal 2172 and 2068 MPa. Type D is the prestressing steel strand with yielding strength of 416 MPa and ultimate strength of *1380* MPa; Type E is the normal strength steel rebar with yielding strength of 250 MPa.

Prestressed deck panels were divided into three groups, with two identical specimens in each group. The first group consisted of two high strength concrete (HSC) panels, which had no steel fibers and coarse aggregates but included conventional shear reinforcement (#10-s:75 mm c/c). The flexural reinforcement comprised of 5.5 mm dia. high-strength steel wires. The second group consisted of prestressed high strength fiber reinforced concrete (HSFRC) deck panels with no shear reinforcement and having 2% steel fibers by volume. The third group was optimized prestressed HSFRC (O-HSFRC) panels. Optimization was carried out by modifying the thickness of panels and prestressing force. Depth of panels for the first two groups was 127 mm, and for the optimized panels, the depth was reduced to 102 mm. Width for all the panels was fixed as 457 mm. The span length was 1.22 m with prestressing carried out in the direction perpendicular to traffic.

Design of HSFRC deck panels was carried out following the Precast/Prestressed Concrete Institute (PCI) recommended guidelines [24]. Prestressing was carried out using the method of pre-tensioning. Force transfer was carried out on the 28th day of casting, and then the panels were moved to the lab to perform the test, within a week after force transferring. Average concrete strength was 69 MPa with a standard deviation of 4.42 MPa. The ultimate tensile strength of prestressing wires was 1380 MPa. An initial prestressing of 965 MPa was applied. Prestress losses were assumed to be 25% at the preliminary design stage. Deck panel sections were designed to satisfy the allowable stress limits for concrete and steel, both at the mid-span and support sections. The ultimate condition was also checked for the sections. The compressive strength of HSC/HSFRC was obtained by testing three 100×200 mm cylinders for each batch. Figure 1 presents some details of specimen preparations.

Table 1. Test matrix and specimen details.

Ref./Phase	Group	Span	Width	Depth	Reinforcement			28-Days Concrete Compressive Strength
					Flexural—Top	Flexural Bottom	Shear	
		m	mm	mm	φ (mm)	φ (mm)	c/c (mm)	MPa
Saleem et al. [5]	1T1S-127#1 1T1S-127#2	1.22	300	127	No. 13M-Type A	No. 22M Type A	-	124 186
Ghasemi et al. [13]	1T1S-114#1 1T1S-114#2 1T1S-102	1.22	381	114 114 102	No. 10M-Type A	No. 16M Type A	-	166 166 186
Ghasemi et al. [6]	1T1S-102#1 1T1S-102#2 1T1S-127#1 1T1S-127#2 1T1S-102	1.22	381	102 127 102	No. 10 M-Type B	No. 13 M -Type C	-	166 186
Prestressed HSC/HSFRC deck panels	HSC-1 (*) HSC-2 (*) HSFRC-1 HSFRC-2 O-HSFRC-1 O-HSFRC-2	1.22	457	127 127 102	-	5.5 mm Type D	#10-s:75 mm - -	48 (SD:3.45) 69 (SD:4.42) 49 (SD:3.83)
Sustainable UHPC-NA/RA deck panels	A-NA A-RA B-NA B-RA C-NA D-RA	1.52	406 457 457	127(87) ** 127 (97) 127 (97)	No. 10M Type E - -	No. 10M Type E No. 10M Type E No. 10M Type E	Inclined 5 mm 5 mm N/A	sUHPC:167 (SD:11.45) NC: 25.6 (SD:1.26) RA: 26.7 (SD: 1.45)

Note: *: Except for HSC-1 and 2, all other concrete material contains 2% volume fraction of steel fibers. **: Number in parentheses is the thickness of the UHPC layer.

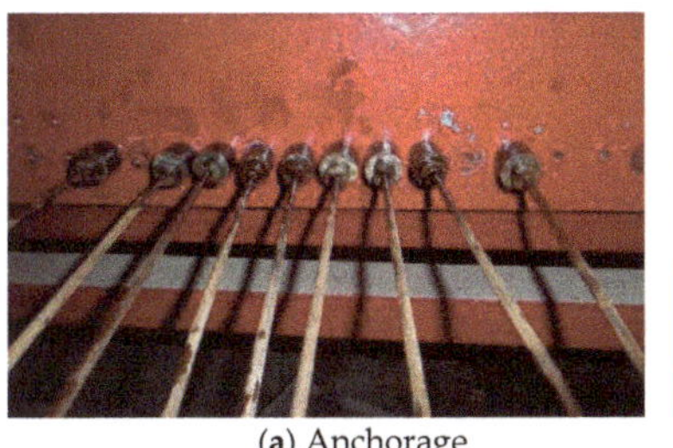

(a) Anchorage

(b) Demolding

(c) Final specimens

Figure 1. Preparation of high strength concrete (HSC)/high strength fiber-reinforced concrete (HSFRC) prestressed deck panels.

Panels tested in the second phase were also divided into three groups. In each group, one panel had normal aggregate (NA) concrete as the top layer, whereas the other panel had recycled aggregate (RA) concrete as the top layer. All of these panels had an overall depth of 127 mm and a span of 1.52 m. Widths of panels varied between 0.41 and 0.46 m, depending on the reinforcement configurations. The designate Type A-NA represents deck type A with natural aggregate concrete on top of sUHPC. Similarly, Type A-RA denotes deck type A with recycled aggregate concrete on top. Details of geometry, steel fibers volume fraction and reinforcement are provided in Table 1. The cross-sectional and longitudinal configurations of the three types of panels are shown in Figures 2–4. Design of sUHPC panels was experimental because no specific guidelines are available for such type of composite concrete deck panels. For deck type A (as shown in Figure 2), the bottom steel plate is stay-in-place welded with the bottom tension reinforcement, inclined continuous shear reinforcement, and compression reinforcement. Conventional shear reinforcement was used in deck type B (as shown in Figure 3). Deck type C as shown in Figure 4 had two 40 × 40 mm sUHPC ribs running in the longitudinal direction. Transverse reinforcement along the width at regular intervals was passed through the ribs to develop the composite connection between two discrete concrete layers. Rebars with the yield strength of 250 MPa were used in all specimens. Compressive strength was determined by testing three 150 × 300 mm cylinders for each concrete type and is reported in Table 1. Reference mix design for UHPC by Graybeal [25] was adopted for the preparation of sUHPC, except that 30% of the cement content was replaced with GGBS. Similarly, in RA concrete, 30% of natural/virgin coarse aggregates were replaced with recycled aggregate. Physical properties of materials for sUHPC are presented in Table 2. Table 3 shows the mix designs for sUHPC, NA/RA concrete.

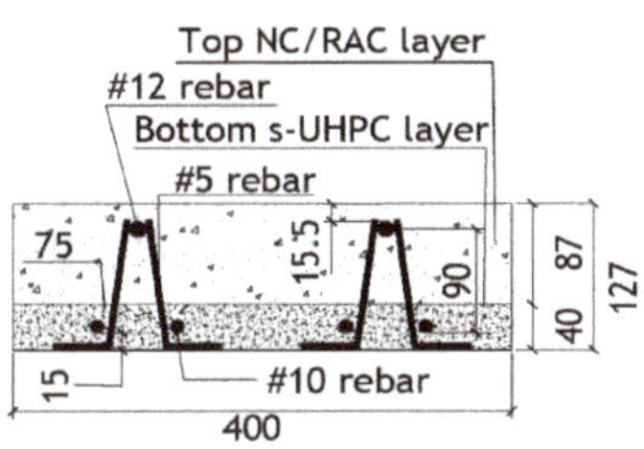

(a) Schematic cross-section

(b) Steel reinforcement layout

Figure 2. Details of deck type A (units in mm).

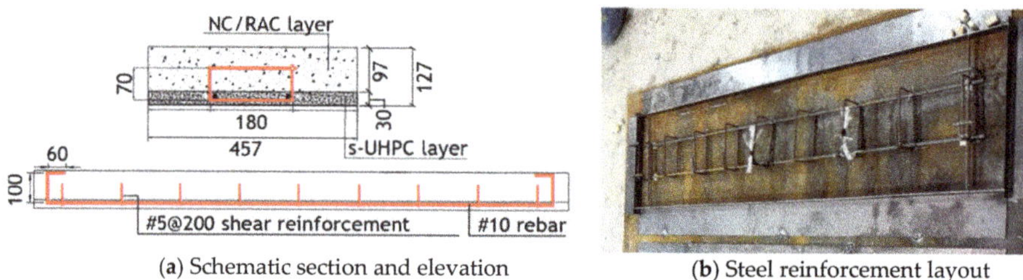

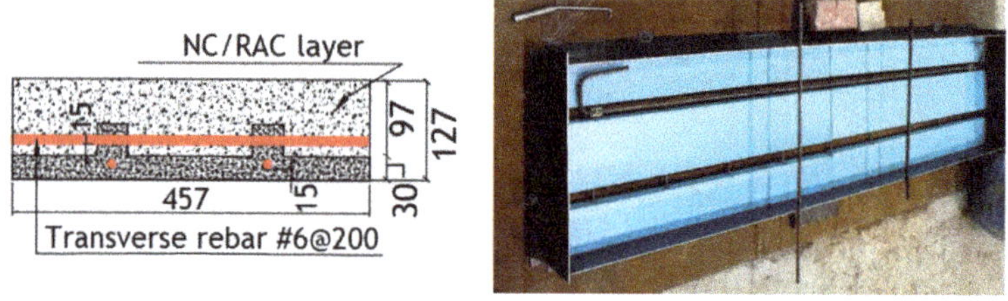

(**a**) Schematic section and elevation (**b**) Steel reinforcement layout

Figure 3. Details of deck type B (units in mm).

(**a**) Cross-sectional schematic (**b**) Steel reinforcement layout

Figure 4. Details of deck type C (units in mm).

Table 2. Constituent material properties for sustainable ultra-high performance concrete (sUHPC).

Constituents	Physical Properties
Cement	Grade 52.5 OPC
GGBS	Grade 100, 12.5 µm, activity index = 98%
Recycled CA	Crushed RC; particle size range = 5–10 mm
Sand	Fineness modulus > 2.4
Silica fume	Nano-silica, grade 900, 0.15 µm
Superplasticizer	Poly-carboxylate type
Steel fibers	Mild steel, plain, ϕ0.2 × 12.5 mm

Table 3. Mix designs for sUHPC, normal aggregate (NA), and recycled aggregate (RA) concrete.

	sUHPC		NA Concrete		RA Concrete	
Constituents	Mix Design (kg/m^3)	Design Ratios	Mix Design (kg/m^3)	Design Ratios	Mix Design (kg/m^3)	Design Ratios
Cement	498	1.00	403.3	1.00	403.3	1.00
GGBS	214	0.43	-	-	-	-
Natural Aggregate	-	-	935.2	2.32	654.6	1.62
Recycled Aggregate	-	-	-	-	280.6	0.70
Fine sand	1020	2.05	863.2	2.14	863.2	2.14
Quartz powder	211	0.42	-	-	-	-
Silica fume	231	0.46	-	-	-	-
Superplasticizer	40	0.08	4.03	0.01	4.03	0.01
Steel fibers	156	0.31	-	-	-	-
Water	109	0.22	188.4	0.47	188.4	0.47

During casting, sUHPC was introduced from one end of the formwork and was allowed to flow to the other end. Whenever more material was required, it was introduced behind the leading edge of the flowing material. The same casting method was used for all the specimens to maintain similar fiber distribution. The traditional method of moist curing was used. Specimens were sealed with plastic sheets for 28 days to prevent the loss of moisture.

All the deck panels were tested as simple-span against AASHTO HS-20 truck dual-tire wheel with a footprint measuring 510×250 mm [14], which was applied using a neoprene pad with the longer side placed perpendicular to traffic direction. A steel plate with size in 50 mm excess on all four sides was placed on top of a neoprene pad to distribute the load over the pad uniformly (Figure 5). Linear variable displacement transducers (LVDT) and strain gauges were installed at strategic locations to acquire the deflection and strain data as shown in Figure 6, in which D and S represent LVDT and strain gauge, respectively. Loading was applied in displacement control mode at a rate of 1 mm/min. All instruments were connected to a high-speed data acquisition system which recorded the data at a frequency of 1 Hz. Tests were stopped at 25% load drop unless preceded by excessive cracking or deflection, which could make the test setup unstable.

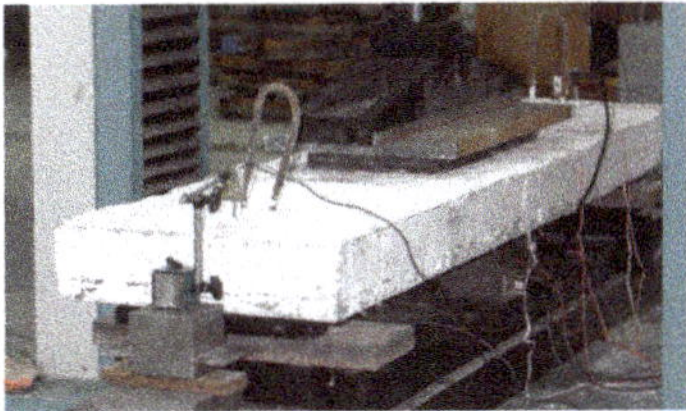

(**a**) Prestressed HSC/HSFRC deck panels   (**b**) Sustainable UHPC-NA/RA panel

Figure 5. Test setups.

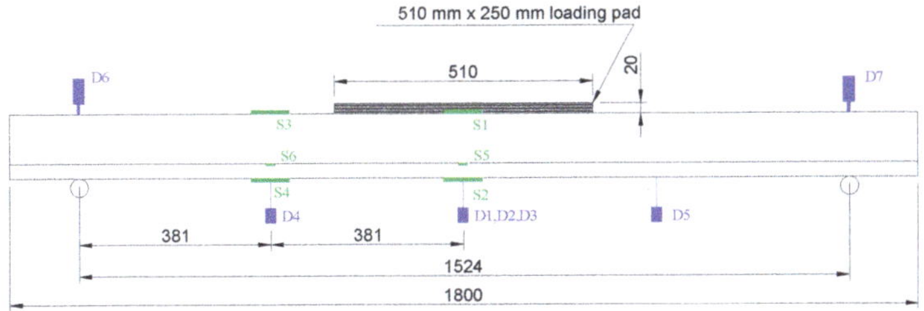

Figure 6. Instrumentation plan.

3. Sectional Analyses

Sectional analyses were performed to predict the critical moment of the composite section for sUHPC decks. The modulus of elasticity of normal strength concrete (either with or without recycled aggregates) and s-UHPC are estimated to be 20 and 45 GPa, respectively. The compressive strength of normal concrete and sUHPC are assumed to be 26 and 150 MPa, respectively. The modulus of elasticity of steel rebar is set to be 200 GPa, with yielding strength of 250 MPa while the ultimate tensile strength of sUHPC is 10 MPa. By using the material properties and geometries of three types of decks, the location of the neutral axis and the second moment of inertia can be calculated, as shown in Table 4. Based on the elastic analyses, the critical moment corresponding to curvature when the concrete start to crush, the rebar starts to yield, and the sUHPC reaching ultimate tensile strength can be predicted. Plastic section analyses lead to another set of results on the plastic moment by considering the yielding of rebar as the only source of sectional tensile force or consider the contribution from both the rebar and sUHPC. It was found out that the tensile strength of sUHPC makes a big difference regarding the plastic sectional moment, which increases from 8.1 to 22.5 kN·m for deck type A. The summary of the results is listed in Table 4.

Table 4. Summary of sectional analyses.

Deck	Neutral Axis	Second Moment of Inertia	Elastic Analyses			Plastic Analyses	
			Cracking Moment	Crushing Moment	Yielding Moment	Plastic Moment (Rebar)	Plastic Moment (Rebar + UHPC)
	(mm)	(kN·m^2)	(kN·m)	(kN·m)	(kN·m)	(kN·m)	(kN·m)
Type A	51.5	2159	34.2	37.2	85.5	8.1	22.5
Type B	51.7	2251	9.7	38.9	76.7	4.3	18.3
Type C	36.3	2639	16.1	37.8	154.6	4.3	15.8

4. Results and Discussion

4.1. Summary of the Load Demand and Capacity

Based on the approximate methods of analyses specified in the AASHTO bridge design specification [14], the width of the equivalent interior strip (W) can be calculated based on the spacing of the supporting components (S) as shown in the following Equation (1). For deck strip specimens in phases 1 and 2, the spacings are 1.22 and 1.52 m, respectively, which lead to the equivalent width of W equal to 1331 and 1496 mm. It is assumed that the deck strip with equivalent width will sustain full wheel load under "Strength I" and "Service I" limit states. Therefore, with the wheel load of 71.2 kN, the service load demand is 92.6 after considering the dynamic impact factor of 0.3 and the ultimate load demand is 162.0 kN with same impact factor and the load factor equals 1.75. By dividing the service and ultimate load by the effective width (W), the capacity demand of the unit width of the deck slab can be calculated. Table 5 summarizes the experiment results of deck strips available in the literature as well as those presented in this paper. The deflections under service and ultimate load levels are identified. Experimental ultimate load capacities of each deck strip were compared to the target load demand to calculate the capacity versus demand ratio. The unified capacity versus demand ratio based on the self-weight is also listed as the last column for an intuitive indicator of the effectiveness of the design.

$$W = 660 + 0.55S \tag{1}$$

Table 5. Summary of testing results.

Ref./Phase	Group	Deflection at Service Load	Deflection at Ultimate Load	Ultimate Load Test	Target Ultimate Load	Capacity versus Demand	Capacity/Demand per Unit Weight
		(mm)	(mm)	(kN)	(kN)		
Saleem et al. [5]	1T1S-127#1	-	25.0	178.0	36.5	4.9	3.2
	1T1S 127#2		25.0	209.0		5.7	3.7
Ghasemi et al. [13]	1T1S-114#1	-	21.0	123.0		2.7	2.6
	1T1S-114#2		22.0	110.0	45.6	2.4	2.3
	1T1S-102	-	23.0	101.0		2.2	2.3
Ghasemi et al. [6]	1T1S-102#1	-	30.2	74.6		1.6	1.8
	1T1S-102#2		27.0	76.3		1.7	1.9
	1T1S-127#1	-	26.2	95.6	45.6	2.1	2.1
	1T1S-127#2		24.6	87.0		1.9	1.9
	1T1S-102	-	26.2	83.0		1.8	2.0
Prestressed HSC/HSFRC deck panels	HSC-1	0.97	14.5	117.5		2.0	1.2
	HSC-2	1.80	19.9	120.0		2.0	1.2
	HSFRC-1	0.51	7.9	96.0	59.5	1.6	0.9
	HSFRC-2	0.97	3.9	97.0		1.6	1.0
	O-HSFRC-1	7.30	27.2	28.0		0.5	0.3
	O-HSFRC-2	-	14.4	45.0		0.8	0.6
Sustainable UHPC-NA/RA deck panels	A-NA	0.98	15.8	114.4	45.4	2.5	0.8
	A-RA	0.98	17.2	114.2		2.5	0.8
	B-NA	2.68	7.5	46.4		0.9	0.3
	B-RA	2.67	10.2	47.5		0.9	0.3
	C-NA	4.85	5.0	39.5	51.1	0.8	0.3
	C-RA	2.12	8.1	57.2		1.1	0.4

The load versus deflection curves for various prestressed deck strips tested in this research is shown in Figure 7a. Figure 7b provides a clear comparison between test and literature results. While the deck strip features a UHPC layer and high strength steel rebar (1T1S-Steel) reaches the highest ultimate load, the deck with prestressing strands and composite sUHPC deck can reach a similar level of capacity to demand ratio, which is denoted by the size of the marker in the Figure 7b. The deck strips with polymer rebars also exhibit excellent efficiency. However, generally speaking, they have more significant deflection at ultimate load under the reported configuration. The experiment observations and results of prestressed HSC/HSFRC deck and sUHPC deck will be discussed in the following sections.

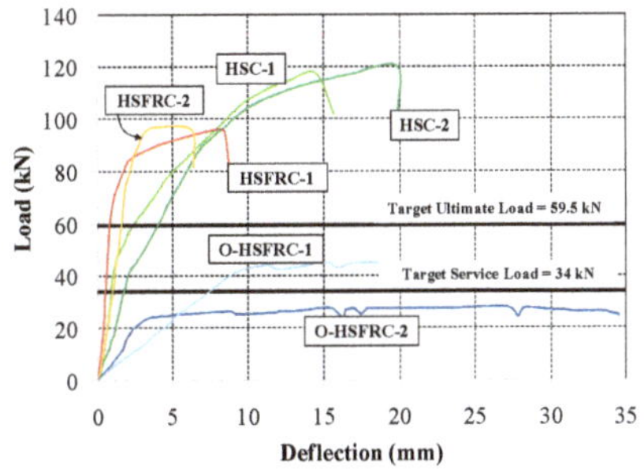
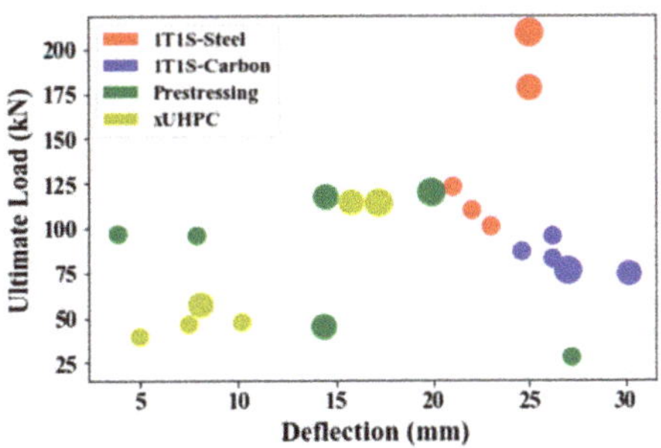

(**a**) Prestressed HSC/HSFRC deck panels   (**b**) Sustainable UHPC-NA/RA panel

Figure 7. Load deflection responses of prestressed deck panels.

4.2. Prestressed HSC/HSFRC Deck Panels

It was found out that the capacity to demand ratios of panels HSC-1 and HSC-2 are less than the single-span specimens tested by Saleem et al. [5] but are comparable to the panels tested by Ghasemi [6,13]. Both panels achieved the ultimate loads, double the target load, which shows the benefit of introducing prestressing. These two panels did not have steel fibers but only minimum conventional shear reinforcement (#10-s:75mm c/c). Deflection in panel HSC-1 at the service load was 0.97 mm, which is less than the limiting deflection of 1.53 mm (span/800) as recommended by AASHTO LRFD [14]. Deflection at the ultimate load was 14.5 mm (Table 4), which is 15 times more than the service load deflection. Load-deflection responses and failure pattern of both HSC deck panels are shown in Figures 7a and 8, respectively. The first flexural crack in the panel HSC-1 appeared at the load of 47.5 kN. This crack was at the mid-span followed by several other flexure and shear cracks under and around loading patch. The failure occurred at the peak load of 117.5 kN due to shear compression mode of failure, as shown in Figure 8a. This failure was sudden, as can be seen in the load-deflection response. However, this should not be a matter of concern because the failure load is double the target load and significantly high deflection (meaning ductility) at the ultimate load. Saleem et at. [5] and Ghasemi [6,13] also observed shear failure for similar span and load patch. However, their failure was more ductile because of the presence of steel fibers. First flexural cracking panel HSC-2 was noted at 42.5 kN, followed by several other flexure cracks. The failure occurred at a peak load of 120 kN due to rupture of prestressing wire with the complete opening of the flexural crack along the width of the panel. At 103 kN, the second prestressing wire ruptured with a loud noise and load dropped rapidly to 90 kN. No crushing of concrete or large shear cracks were observed in this case. Deflection of panel HSC-2 at the service load was 1.8 mm, which is little more than limiting value of deflection. This panel also had the capacity to demand ratio of 2.0 and a very high deflection of 19.9 mm at the ultimate load.

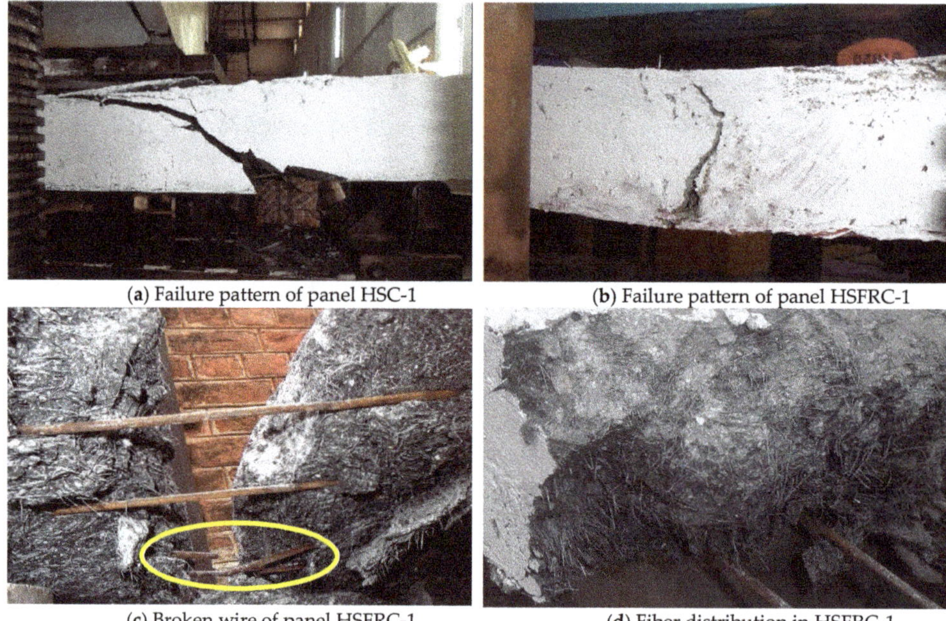

(**a**) Failure pattern of panel HSC-1   (**b**) Failure pattern of panel HSFRC-1

(**c**) Broken wire of panel HSFRC-1   (**d**) Fiber distribution in HSFRC-1

Figure 8. Failure pattern and forensics of prestressed panels.

The panels HSFRC-1 and 2 had no conventional shear stirrups but 2% steel fiber by volume of concrete. Both the identical panels exhibited a capacity to demand ratio of 1.6, which is less than the HSC panels and comparable to UHPC-CFRP deck panels tested by Ghasemi et al. [6]. Load-deflection responses are shown in Figure 7a. The first flexural crack in the panel HSFRC-1 was observed at 70 kN after which the load-deflection curve started to flatten and then dropped suddenly after achieving a peak load of 96 kN. The reason behind the sudden drop was the rupture of the prestressing wire, as shown in Figure 8c. No other flexural crack followed the first crack until the final failure. This flexural crack opened to a width of 4 mm throughout the width of the panel, as shown in Figure 8b. No crushing in compression or shear cracks were observed in this panel. Deflection at the service load was 0.51 mm, which is well below the limiting deflection. For deck panel HSFRC-2, the first flexural crack appeared at a load of 75 kN near the mid-span and continued to grow in an upward direction until the panel reached a peak load of 97 kN. The panel continued to yield at the peak load for 3 mm until the loud bursting of prestressing wire dropped the load to 82 kN. The behavior of panels HSFRC-1 and HSFRC-2 was similar except that this panel had a higher service load deflection of 0.97 mm.

In comparison to HSC panels, HSFRC panels demonstrated higher stiffness throughout the loading. However, their capacity to demand ratio was smaller, but still acceptable. The panel HSFRC-1 was broken down after the test to reveal the distribution of fibers across the depth (Figure 8d). Based on visual observations, it is evident that fibers, under the effect of gravity, concentrated more on the lower part of the panel. However, this should not be a matter of concern because these panels are proposed to be simply supported, and concrete on the top will be in compression and fibers do not significantly contribute to compressive strength.

The optimized high strength fiber reinforced concrete (O-HSFRC) panels had reduced depth of 102 mm, which was intended to decrease the self-weight. Figure 7a presents the load-deflection responses of O-HSFRC panels. Panel O-HSFRC-1 reached the peak load of 28 kN with 24 kN as first crack load. The panel failed in flexure with the crack appearing at the location of the joint between two concrete batches. Three prestressing wires also ruptured before the test was stopped at a deflection of

35 mm. For panel O-HSFRC-2, the first flexural crack was observed at a load level of 39 kN. The peak load of 45 kN was reached, followed by flexural failure. Rupturing of two prestressing wires was also observed before the test was stopped at a deflection of around 20 mm. Both panels could not achieve the target load. However, this should not lead to the conclusion that they do not have the potential to meet the target load. The main reason behind lower failure load could be a lower quality of concrete due to high environmental temperature, error in following proper mixing regime, and joint formation due to delay in successive batches. With better quality control and casting the whole specimen with one batch will surely increase the failure load to meet the target.

4.3. Sustainable UHPC-NA/RA Deck Panels

Load-deflection curves for all the deck panels tested in phase 2 are presented in Figure 9a while the moment versus curvature plot for deck types A, B, and C are shown in Figure 9b–d, respectively. The moments were calculated based on the recorded load value and an assumed uniform distribution in the load patch area. The curvature was calculated based on the strain recording S1 and S5, which are the strain of top concrete and strain of the rebar. Moment predictions from the sectional analyses are also presented for comparisons.

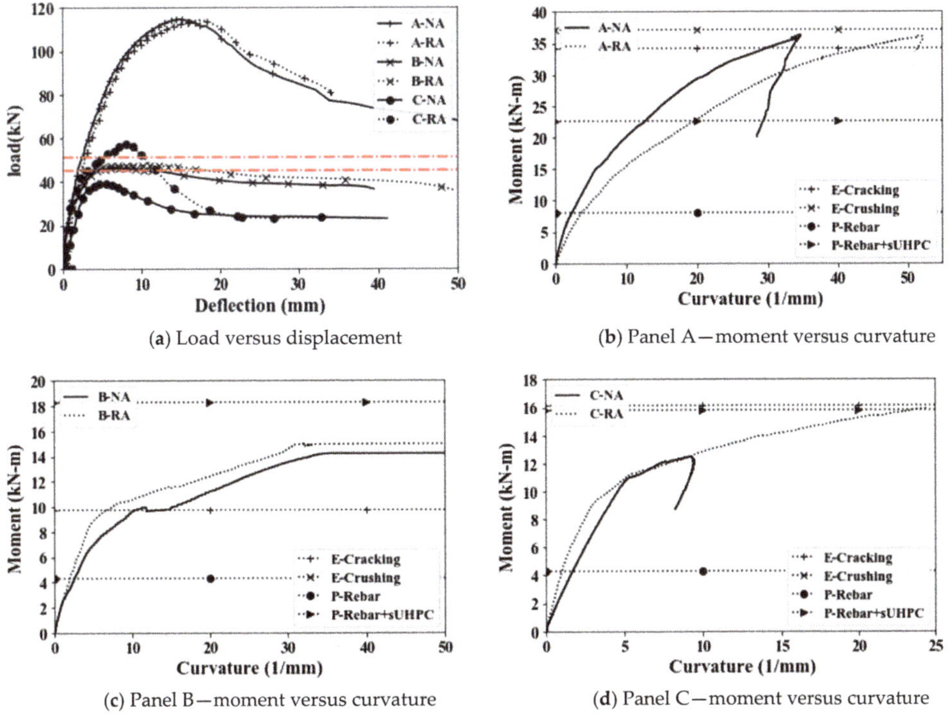

(a) Load versus displacement

(b) Panel A—moment versus curvature

(c) Panel B—moment versus curvature

(d) Panel C—moment versus curvature

Figure 9. Load deflection responses of composite deck panels.

The failure pattern of composite deck Type A and Type B are shown in Figure 10. For panel A-NA having the top layer prepared with regular concrete, the first flexural crack was seen on the side surface at 53 kN, followed by delamination of bottom steel plate. The peak load of 114.4 kN was recorded. Crushing of concrete was not observed. However, the main steel bar at the bottom yielded, as shown in Figure 11a at the 83 kN, making it a ductile failure, as shown in Figure 10a. Mid-span tensile strain at the peak load in the bottom UHPC layer was recorded as 0.0062. At the failure stage, the flexural crack reached to the top, and the bottom steel plate got detached significantly under the loading area, as shown in Figure 10a. Deflection at the peak load was 15 mm beyond which the panel deflected another 45 mm and the test was stopped due to safety concerns. The ductile behavior can be attributed to dowel action of the longitudinal bars and the fiber pullout mechanism of the bottom UHPC layer. Deflection at the service load was 0.98 mm, which is well below the AASHTO LRFD [14] limit. Capacity to demand ratio was 2.5, indicating the provision of optimization through a reduction in thickness.

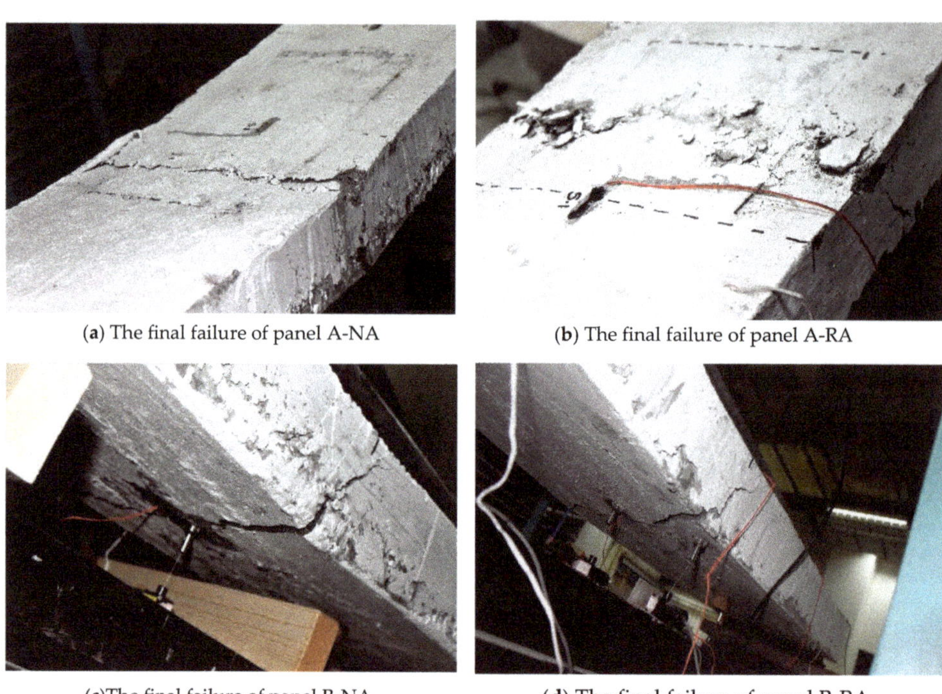

<table>
<tr><td>(a) The final failure of panel A-NA</td><td>(b) The final failure of panel A-RA</td></tr>
<tr><td>(c)The final failure of panel B-NA</td><td>(d) The final failure of panel B-RA</td></tr>
</table>

Figure 10. Failure patterns of composite deck panels.

For the deck type A-RA, the first flexural cracking load was noted as 47.5 kN. After the appearance of two flexural cracks, one near the edge of the neoprene pad and other close to the mid-span, failure occurred at a load-level of 114.18 kN. At the failure, the flexural crack had reached the compression face, as shown in Figure 10b. Compressive strain in concrete at the failure load was recorded as 0.0026, which is close to the crushing strain of 0.003. Some chipping-off at the top was observed in concrete directly under the load patch. Strain in steel at the mid-span at peak load was recorded as 0.003, and the yield strain was reached at around 80 kN, as shown in Figure 11b. Service load-deflection was 0.98 mm, and the capacity to demand ratio was 2.5.

Panels A-NA and A-RA behaved very similarly, both in the pre-peak and post-peak regions. Type A deck panels not only exceeded the target load (45.4 kN) by far but also exhibited higher stiffness compared to the rest of the panels tested in this phase. The sectional moment passed the predicted plastic moment, indicating good composite action that leads to a more evenly distributed stress along with the section height. The ultimate sectional moment was close to the predicted values for elastic failure due to concrete cracking and crushing, as shown in Figure 11b, which leads to more efficient use of materials. The yielding of the steel rebar provides sufficient ductility; however, based on the analyses, the contribution of sUHPC regarding the tensile forces are significant. The welded connection between the bottom, top, inclined shear reinforcement, and bottom steel plate is effective in controlling the cracking development. This developed outstanding composite behavior leading to higher load capacity and stiffness.

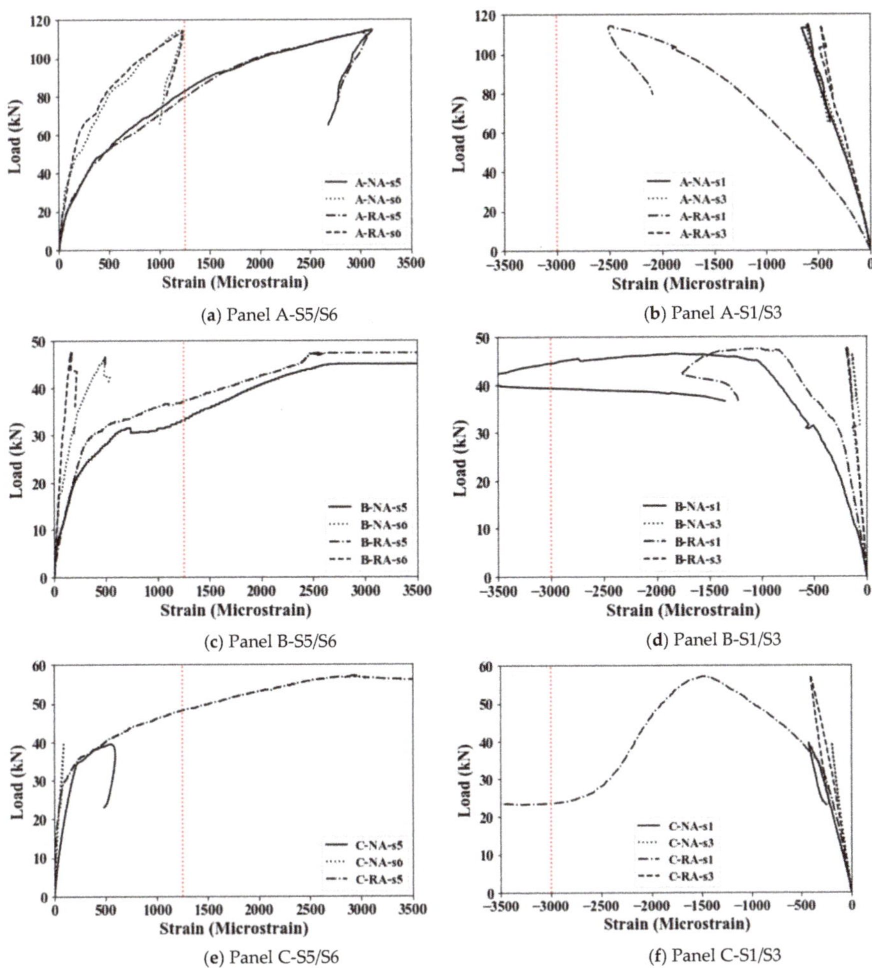

Figure 11. Load–strain responses for the composite deck panels.

For the deck panel, B-NA failure load was observed as 46.4 kN, and failure mode was flexural, caused by the rupture of steel reinforcement. From Figure 11c,d, it is clear that the bottom rebar yielded at a load level around 32 kN while concrete was crushed when approaching the ultimate load. The stress distribution of the section at the ultimate load level led to the ultimate moment close to the fully plastic sectional moment, as shown in Figure 9c. The ultimate moment falls short of the prediction due to the cracking opening at the middle span and reduction of the contribution from sUHPC. Deflection at the service load was recorded as 2.68 mm, which is more than the AASHTO LRFD limit. Demand to capacity ratio was 0.9. This panel is 10% short of the target load and failed by breaking for bottom rebar (No. 10). This problem can be solved by using one size bigger rebar (No. 13) as the primary reinforcement or increasing the grade of steel. This will also increase the stiffness leading to smaller service load deflection. The provided shear stirrups are effective as no shear cracks were observed. The stirrups also act as shear connectors between the layers of UHPC and regular concrete, leading to better composite behavior and enhanced flexural capacity. However, the widening of the flexural cracks, especially those at the middle span is not constrained. The post-peak response of this panel is more ductile compared to both panels of type A, and load drop was much more gradual, as can be seen in Figure 9a, because the failure is mainly due to the plastic deformation of steel rebar rather than the crushing of the top concrete. The test was stopped at a deflection of 43 mm due to safety concerns.

For the panel, B-RA first flexural crack was observed at a load of 32 kN while failure occurred at a load of 47.5 kN. The test was stopped at a deflection of 50 mm. Strain in steel reinforcement at peak load was recorded as 0.0025, which is twice the yield strain (0.00125) for 250-grade steel. Yielding of steel bars was reached at a load level of 37 kN while the concrete strain on compression face at that stage is around 500 microstrain, which is way less than the crushing strain of concrete. Compressive strain at peak load was recorded as 0.0012, which is half the value of crushing strain for concrete. Deflection at the service load was 2.67 mm, and the capacity to demand ratio was 0.9. The usage of concrete containing recycled aggregates did not cause a drastic change in either the pre-peak or post-peak behaviors.

Deck panel C-NA got damaged during transportation. Therefore, it failed at a much smaller load of 39.5 kN. Failure was flexural with significant post-peak deflection (43 mm) till the stoppage of the test. For the panel, C-RA first flexural crack appeared near the edge of loading pad at a load of 40 kN followed by a series of other flexural cracks near the mid-span. Ultimate failure load was recorded at 57.20 kN with the flexural mode of failure at a mid-span deflection of 8 mm. The deck panel offered sufficient ductility as the panel continued to take load till a deflection of 37 mm when the test was stopped. As can be seen from Figure 9d, specimen C-NA has a premature failure due to the pre-damage while specimen C-RA presents the typical performance for this type and is discussed. Steel reinforcement was recorded to yield at a load level of 48 kN when the compressive strain at the top surface was only 0.0008, as shown in Figure 11e,f. Concrete strain at the ultimate load was recorded as 0.0015 while it approached the crushing strain of 0.003 as the crack entered the top surface of recycled aggregate concrete. Deflection at the service load was 2.12 mm, which is more than the limiting deflection. Capacity to demand ratio was 1.1. The drop of load for this panel was much quicker than the type B panels (Figure 9a). The post-peak capacity of this panel is almost half of the type B panels. This can be attributed to reduced composite action between the two layers of concrete, leading to the conclusion that UHPC ribs and horizontal transverse rebars are not as effective as shear stirrups in developing composite action, as observed in type B panels.

5. Conclusions and Recommendations

This research work focused on developing low-profile precast bridge deck panels for accelerated bridge construction using a combination of prestressing, high-strength, and ultra-high strength concrete, and recycled aggregate concrete. Use of prestressing and high strength materials enables a design with an overall depth of 102 and 127 mm, which are significantly less than the conventional deck thicknesses of 200–250 mm. In total, 12 deck panels were tested: six prestressed panels prepared with high strength/fiber reinforced concrete and six composite deck panels made of sustainable UHPC overlaid with normal or recycled aggregate concrete. In the light of experimental work, the following conclusions may be drawn:

The prestressed concrete deck panels with/without steel fibers exceeded the loading conditions expected from the HS20 truck and met the serviceability requirements as well. The capacity to demand ratios of HSC and HSFRC panels are comparable to those tested by Ghasemi [6,13]. By steel fiber, cracking load of HSFRC panels increased by 60%, on average, in comparison with HSC panels. This leads to the conclusion that replacing shear stirrups with steel fibers significantly improved the stiffness of the system. However, the optimized panels (O-HSFRC) could not meet the requirements but demonstrated potential. With a slight improvement in design and preparation, these panels should be able to meet the strength and stiffness demand. The governing mode of failure in the prestressed HSC panels was a sudden shear failure, as can be seen in the load-deflection responses. The HSFRC and O-HSFRC panels, however, failed in flexure having a plateau before the rupture of wire. Especially, the O-HSFRC panels exhibited significant deflection beyond the peak load, indicating substantial ductility and warning before failure.

Deck panels A-NA and A-RA behaved identically. Both exceeded the demand by 2.5 times. Service load deflections and first crack loads were also similar and well within limits. Significantly high post-peak deflections were observed. Replacement of conventional constituents of concrete with sustainable materials displayed satisfactory performance with no detrimental effect on the mechanical behavior of the deck panels. Welded connection of steel (flexure and shear) and bottom plate proved very effective in achieving the high capacity to demand ratio. This design can be optimized further to reduce the self-weight. Panels B-NA and B-RA fell slightly short of meeting the strength and serviceability requirements. However, this design is relatively simple and has exhibited a long post-peak plateau pointing towards ductile behavior. With a slight improvement in design, these panels should able to meet the desired load and serviceability needs. Panel C-NA met the load demand but not the deflection limit. The first flexural crack, however, appeared after service load. The design needs modification to improve composite action between the two layers of concrete, which will enhance stiffness and will thus help in meeting the deflection limit.

The sUHPC A-NA/RA deck panels offered the highest capacity to demand ratio (2.5) among the entire stock of 12 specimens and their deflection was the lowest of the tested specimens. When the test was stopped at 60 mm, panel A-NA was still holding more load than the load target, and a similar case was also observed for panel A-RA. Hence, the sustainable UHPC NA/RA composite deck panels may be the most suitable alternative for replacing the conventional bridge deck systems, especially for the situations when self-weight is a concern, for instance, moveable bridges.

Although the proposed deck designs show great promise, further studies are still needed before any field implementation: (1) more precise measurement of the prestressing losses that help lead to more accurate design; (2) optimization of design by decreasing the self-weight; (3) design of connections with girders and adjacent deck panels; (4) testing of full-scale panels (multi-unit, multi-span) with the connection under static and dynamic effects of moving wheel load to assess the long-term behavior; (5) implementation of prestressing in the sUHPC-NA/RA deck panels.

Author Contributions: All authors have read and agree to the published version of the manuscript. Conceptualization, M.A.S. and J.X.; methodology, M.A.S. and J.X.; formal analysis, M.N.Z.; investigation, M.N.Z. and M.M.S.; writing—original draft preparation, M.N.Z.; writing—review and editing, M.A.S. and J.X.; visualization, M.M.S.; supervision, J.X.; project administration, M.A.S.; funding acquisition, M.A.S. and J.X.

Funding: This research study was jointly funded by Bridge Engineering Lab. at the Department of Civil Engineering, UET Lahore, Pakistan and Department of Civil Engineering, Xi'an Jiaotong-Liverpool University (XJTLU), Suzhou, China (funding # RDS10120190072).

Acknowledgments: Raw material sponsorship from Imporient Chemicals® Ltd., SIKA Pakistan® and Lahore Ceramics Pvt. Ltd. is highly acknowledged.

Conflicts of Interest: The authors declare no conflicts of interest.

References

1. Keller, T.; Rothe, J.; Castro, J.; de Osei-Antwi, M. GFRP-Balsa Sandwich Bridge Deck: Concept, Design, and Experimental Validation. *J. Compos. Constr.* **2014**, *18*, 04013043. [CrossRef]
2. Manalo, A.; Aravinthan, T.; Fam, A.; Benmokrane, B. State-of-the-Art Review on FRP Sandwich Systems for Lightweight Civil Infrastructure. *J. Compos. Constr.* **2017**, *21*, 04016068. [CrossRef]
3. Al-Ramahee, M.; Chan, T.; Mackie, K.R.; Ghasemi, S.; Mirmiran, A. Lightweight UHPC-FRP Composite Deck System. *J. Bridg. Eng.* **2017**, *22*, 04017022. [CrossRef]

4. Federal Highway Administration, U.S.; Department of Transportation. FHWA: Tables of Frequently Requested NBI Data [Internet]. 2017. Available online: http://www.fhwa.dot.gov/bridge/britab.htm (accessed on 5 February 2020).

5. Saleem, M.A.; Mirmiran, A.; Xia, J.; Mackie, K.R. Ultra-High-Performance Concrete Bridge Deck Reinforced with High-Strength Steel. *ACI Struct. J.* **2011**, *108*. [CrossRef]

6. Ghasemi, S.; Mirmiran, A.; Xiao, Y.; Mackie, K.R. Novel UHPC-CFRP Waffle Deck Panel System for Accelerated Bridge Construction. *J. Compos. Constr.* **2016**, *20*, 04015042. [CrossRef]

7. Keierleber, B.; Bierwagen, D.; Wipf, T.; Abu-Hawash, A. Design Of Buchanan County, Iowa, Bridge, Using Ultra High-Performance Concrete And Pi Beam Cross Section. In Proceedings of the Mid-Continent Transportation Research Symposium, Ames, IA, USA, 15–16 August 2007.

8. Habel, K.; Viviani, M.; Denarié, E.; Brühwiler, E. Development of the mechanical properties of an Ultra-High Performance Fiber Reinforced Concrete (UHPFRC). *Cem. Concr. Res.* **2006**, *36*, 1362–1370. [CrossRef]

9. Graybeal, B.A. Compressive Behavior of Ultra-High-Performance Fiber-Reinforced Concrete. *ACI Mater. J.* **2007**, *104*, 146–152.

10. Graybeal, B.A. *Structural Behavior of Ultra-High Performance Concrete Prestressed I-Girders Final Report*; Federal Highway Administration: McLean, VA, USA, 2006.

11. Graybeal, B.A. *Characterization of the Behavior of Ultrahigh Performance*; University of Maryland: College Park, MD, USA, 2005.

12. Ahlborn, T.M.; Peuse, E.J.; Misson, D.L. *Concrete For Michigan Bridges Material Performance—Phase I Final Report*; Michigan Department of Transportation: Lansing, MI, USA, 2008.

13. Ghasemi, S.; Zohrevand, P.; Mirmiran, A.; Xiao, Y.; Mackie, K.R. A super lightweight UHPC–HSS deck panel for movable bridges. *Eng. Struct.* **2016**, *113*, 186–193. [CrossRef]

14. Watts, R.; Mills, R.W.; Fish, R. The Highway Coalition Revisited: Using the Advocacy Coalition Framework to Explore the Content of the American Association of State Highway and Transportation Officials' Daily Transportation Update. *Public Voices* **2016**, *13*, 79. [CrossRef]

15. Huang, C.; Song, J.; Zhang, N.; Lee, G.C. Seismic Performance of Precast Prestressed Concrete Bridge Girders Using Field-Cast Ultrahigh-Performance Concrete Connections. *J. Bridg. Eng.* **2019**, *24*, 04019046. [CrossRef]

16. Deng, S.; Shao, X.; Yan, B.; Wang, Y.; Li, H. On Flexural Performance of Girder-To-Girder Wet Joint for Lightweight Steel-UHPC Composite Bridge. *Appl. Sci.* **2020**, *10*, 1335. [CrossRef]

17. Venancio, V.G. *Behavior of Ultra-High Performance Concrete Bridge Deck Panels Compared to Conventional Stay-In-Place Deck Panels in Partial Fulfillment of the Requirements for the Degree*; Missouri University of Science and Technology: Rolla, MO, USA, 2016.

18. Containing, A.; Sludge, M. Properties of Concrete with Recycled Concrete Aggregate Containing Metallurgical Sludge Waste. *Materials* **2020**, *13*, 1448.

19. Duarte, G.; Gomes, R.C.; de Brito, J.; Bravo, M.; Nobre, J. Economic and Technical Viability of Using Shotcrete with Coarse Recycled Concrete Aggregates in Deep Tunnels. *Appl. Sci.* **2020**, *10*, 2697. [CrossRef]

20. Wang, X.; Chin, C.S.; Xia, J. Material Characterization for Sustainable Concrete Paving Blocks. *Appl. Sci.* **2019**, *9*, 1197. [CrossRef]

21. Tam, V.W.; Kotrayothar, D.; Xiao, J. Long-term deformation behaviour of recycled aggregate concrete. *Constr. Build. Mater.* **2015**, *100*, 262–272. [CrossRef]

22. Qi, J.; Wang, J.; Feng, Y. Shear performance of an innovative UHPFRC deck of composite bridge with coarse aggregate Shear performance of an innovative UHPFRC deck of composite bridge with coarse aggregate. *Adv. Concr. Constr.* **2019**, *7*, 219–229.

23. Feng, Y.; Qi, J.; Wang, J.; Liu, J.; Liu, J. Flexural Behavior of the Innovative CA-UHPC Slabs with High and Low Reinforcement Ratios. *Adv. Mater. Sci. Eng.* **2019**, *2019*, 1–14. [CrossRef]

24. PCI Industry Handbook Committee. *PCI Design Handbook*, 7th ed.; Precast/Prestressed Concrete Institute PCI: Chicago, IL, USA, 2010; 226p.

25. Graybeal, B.A. *Material Property Characterization of Ultra-High Performance Concrete No. FHWA-HRT-06-103*; Federal Highway Administration: Washington, DC, USA, 2006.

Article

Economic and Technical Viability of Using Shotcrete with Coarse Recycled Concrete Aggregates in Deep Tunnels

Gonçalo Duarte [1], Rui Carrilho Gomes [1], Jorge de Brito [1,*], Miguel Bravo [2] and José Nobre [1]

[1] Instituto Superior Técnico, Universidade de Lisboa, 1649-004 Lisbon, Portugal;
    goncalo.f.m.duarte@tecnico.ulisboa.pt (G.D.); rui.carrilho.gomes@tecnico.ulisboa.pt (R.C.G.);
    jose.nunes.nobre@tecnico.ulisboa.pt (J.N.)
[2] CERIS, Setúbal's Polytechnic Institute, 2910-761 Setúbal, Portugal; miguelnbravo@gmail.com
*   Correspondence: jb@civil.ist.utl.pt

Received: 10 February 2020; Accepted: 3 April 2020; Published: 14 April 2020

Abstract: This work analyzes the technical and economic viability of using coarse recycled aggregates from crushed concrete in shotcrete, as a primary lining support in tunnels. Four incorporation ratios of coarse natural aggregate (CNA) with coarse recycled concrete aggregates from concrete (CRCA) were studied: 0%, 20%, 50% and 100%. The mechanical properties of the dry-mix shotcrete were obtained in an independent experimental campaign. Initially, the technical viability of CRCA shotcrete was validated for deep rock tunnels, based on the convergence-confinement method. Two cases were studied to determine the equivalent thickness for each combination of replacement ratio using CRCA shotcrete: (i) similar stiffness and (ii) similar yield stress. Subsequently, an economic assessment was performed. The stiffness criterion increased the thickness below 10% in both the 20% and 50% replacement ratios, which shows their technical viability with very marginal cost increase (<5%). On the other hand, the maximum pressure criteria required higher increments, close to 30% in the 50% replacement ratio. A full replacement was proven to be impracticable in both analyses.

Keywords: shotcrete; deep tunnels; convergence-confinement method; coarse recycled concrete aggregate; dry-mix process

---

1. Introduction

The construction sector is considered one of the main players in the generation of environmental impact. Besides cement production, which is responsible for more than 6% of all global CO_2 emissions, the total amount of construction and demolition waste (CDW) represents a third of all waste generated in the European Union. However, there are barriers to its reuse, one of the main of which lies in the lack of trust in the quality of CDW [1].

The incorporation of recycled aggregates (RA) in concrete, by replacing a percentage of natural aggregates (NA), usually leads to poor mechanical and durability-related performances [2,3].

It is therefore important to understand to what extent the behavior of the RA differs from that of the NA. According to Bravo et al. [2], the main differences between these aggregates are:

-   Higher water absorption capacity of the RA, leading to concrete mixes with lower workability and effective water/cement ratio;
-   Lower compactness and thus lower resistance to crushing, lower modulus of elasticity and higher short- and long-term deformability.

Chan [4] and Bravo et al. [2] evaluated the performance of recycled aggregate concrete (RAC), finding that the nature of RA strongly influences their binding capacity to the new cement paste, thus affecting the mechanical properties of the resulting concrete. Nevertheless, these authors also concluded that the use of these aggregates in concrete production is feasible. However, considering that their use affects the mechanical and durability-related properties of RAC, imposition of limitations and cautions on their inclusion is required.

Shotcrete is widely used in the underground mining and tunneling industry, slope stability, rehabilitation works, and in many other areas.

The tunneling industry is growing because the available space in dense urban environment tends to be used for nobler purposes than transit subways, parking lots, electric lines, water supply and sewer lines [5].

Nowadays, the widespread use of the New Austrian Tunneling Method (NATM) has increased the use of shotcrete for linings [6]. In recent years, several studies on the use of fibres to replace the traditional electro-welded meshes have been published [7–10].

To the authors' knowledge, the incorporation of recycled aggregates from CDW in deep tunnels has not yet been studied in detail. This work studies the viability of using CDW in urban tunnels, because CDW is usually generated in urban environments. A technical and economic assessment of different dry-mix shotcrete compositions, incorporating coarse recycled concrete aggregates (CRCA) in substitution of coarse natural aggregates (CNA), intends to contribute to the sustainability of the construction sector.

For generalization purposes, the analysis is made for a reference case of a deep tunnel where analytical equations proposed by Carranza-Torres and Fairhurst [11] are valid (see Appendix A), based on the convergence-confinement method, which is broadly used to describe the ground-support interaction in design practice [12,13].

It is demonstrated that limited incorporation ratios of CRCA can be a valid alternative from a technical and economic perspective.

2. Experimental Program

2.1. Tests

Firstly in this experimental campaign, a detailed characterization of all aggregates used in the production of concrete was made.

The physical characterization of the aggregates involved the following tests: particle density and water absorption [14], bulk density and volume of voids [15], shape index [16], Los Angeles wear [17] and humidity content [18].

Table 1 shows the tests performed to characterize the mechanical properties of the shotcrete mixes with RA.

Table 1. Hardened properties of shotcrete.

	Age [Days]	Number of Specimens	Standard
Bond Strength (by Pull-off)	40	3–5	EN 1542 [19]
Compression Strength	7	3	EN 12,390-3 [20]
	28	5	
	56	3	
Splitting Tensile Strength	28	3	EN 12,390-6 [21]
Modulus of Elasticity	28	3	LNEC E-397 [22]
Abrasion Resistance	91	3	DIN 52,108 [23]
Ultra-Sound Pulse Velocity	28	5	EN 12,504-4 [24]

It should be referred that the mechanical tests were always performed on core specimens (extracted from slabs produced with shotcrete) with a diameter of 98 mm. The ultrasound pulse velocity determination was carried out in core specimens which were then tested for compressive strength at 28 days.

2.2. Materials

In this research, limestone gravel and alluvial rolled sand were used as NA, respectively, as coarse and fine aggregates. RA from concrete were obtained by crushing a C30/37 strength class concrete with a maximum aggregate size of 14 mm. All the aggregates were added to the mixes completely dry. Type I 42.5 R cement was also used.

Table 2 shows the physical properties of the aggregates used. When compared to NA, it is confirmed (as stated in the introduction) that RA have lower density and higher water absorption capacity. This derives from the nature and porosity of the RA from concrete. For the same reason and because they possess more elongated shapes, RA present a lower bulk density than NA.

Table 2. Physical properties of the aggregates.

	Oven-Dry Particles Density (kg/m^3)	Water Absorption (%)	Bulk Density (kg/m^3)	Shape Index (%)	Los Angeles Wear (%)
Gravel 1	2624	1.65	1406	27	25
Granule	2634	1.02	1409	22	25
Coarse Sand	2574	0.71	1544	-	-
Fine Sand	2548	0.59	1557	-	-
RA	2370	4.55	1287	45	36

Table 2 also shows that the RA display a higher shape index than NA. This can lead to a lower workability of the concrete mix with RA compared to that of the reference concrete (RC), which has to be balanced by a higher w/c ratio.

It was also found that, due to their composition, RA show a lower resistance to fragmentation than NA.

2.3. Mixes Design

The composition of the reference concrete was determined by the Faury method for a target strength belonging to class C30/37 (Figure 1). The maximum aggregate size used was 12.5 mm. Concrete compositions with RA were based on that of reference concrete. However, in order to keep the concrete workability constant (125 ± 25 mm slump), the w/c ratios of the concrete mixes with RA were increased. Table 3 shows the final composition of the concrete mixes studied.

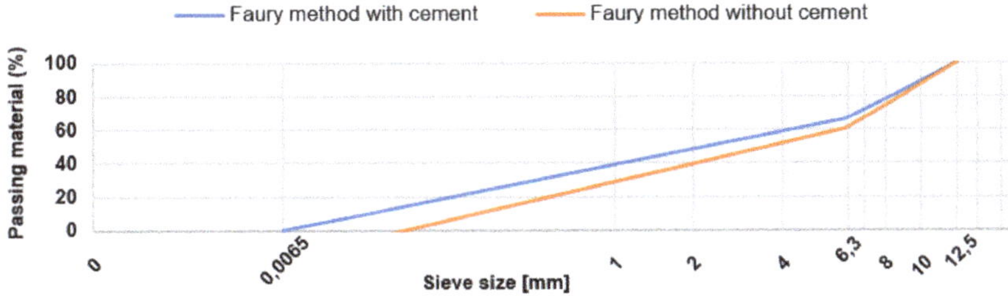

Figure 1. Faury method for reference concrete.

Table 3. Composition of the concrete mixes studied (m^3/m^3) and estimated w/c ratios.

	RC	C20	C50	C100
Coarse Natural Aggregates	0.331	0.264	0.164	0.000
Coarse Recycled Aggregates	0.000	0.066	0.164	0.327
Fine (Natural) Aggregates	0.354	0.352	0.350	0.348
Total Aggregates	0.685	0.681	0.678	0.674
Type I 42.5 R Cements	0.115	0.115	0.115	0.115
Water	0.175	0.179	0.182	0.186
Voids	0.025	0.025	0.025	0.025
Total	**1.000**	**1.000**	**1.000**	**1.000**
w/c Ratios	**0.46**	**0.47**	**0.50**	**0.50**

3. Mechanical Properties

Regarding the mechanical properties, as stated, the bond strength (by pull-off), compressive and splitting tensile strengths, secant modulus of elasticity, abrasion resistance and ultrasound pulse velocity were determined. Table 4 provides a summary of the hardened state mechanical properties.

Table 4. Mechanical properties of shotcrete mixes (adapted from Duarte et al. [25]).

		RC	C20	C50	C100
Pull-off bond [MPa]		1.11	0.8	1.11	1.14
			−2.91%	−0.15%	+2.45
Compressive strength [MPa]	**7-day**	18.27	17.72	15.99	14.09
			−3.01%	−12.48%	−22.88%
	28-day	37.18	33.83	30.35	27.25
			−9.01%	−18.37%	−26.71%
	56-day	43.24	36.58	35.77	31.91
			−15.40%	−17.28%	−26.20%
Splitting tensile strength [MPa]		3.33	3.1	3.01	2.84
			−6.84%	−9.61%	−14.77%
Ultrasound pulse velocity [km/s]		4.82	4.79	4.54	4.48
			−0.45%	−5.66%	−6.97%
Modulus of elasticity [GPa]		26.14	25.41	24.34	18.08
			−2.79%	−6.89%	−30.83%
Abrasion resistance [%]		8.13	7.49	7.73	7.06
			−7.87%	−4.92%	−13.16%

Legend: Green—Better result in comparison with reference concrete (RC); Yellow—Worse result than RC, with reductions lower than 10%; Red—Worse result than RC, with reductions higher than 10%.

Regarding bond strength, no clear variation of this property with the incorporation of CRCA in concrete was found. As for compressive strength, it decreased as the percentage of the replacement of CNA with CRCA increased. Maximum reductions between 20% and 30% were obtained. This result is in agreement with those of several authors who analyzed cast RAC and obtained maximum reductions within this range. Even so, at seven days, the strength values of (RAC) shotcrete compared to cast RAC were lower, which supports the hypothesis that hardening is slower in the former than in the latter.

With respect to tensile strength, the incorporation of CRCA led to non-significant influence, as also reported by other authors, with maximum decreases of 15%, which are clearly lower than those in compressive strength. The reason for this is that the bond between the cement paste and CRCA is improved (compared to CNA) due to their shape and higher water absorption capacity.

The ultrasound pulse velocity decreased as the CRCA incorporation increased, which is consistent with the higher porosity of these aggregates that leads to a concrete mix with lower quality.

The modulus of elasticity was one of the properties mostly affected by the replacement of CNA with CRCA, with reductions of up to 31%. This stems from the higher deformability of the CRCA and lower strength of the bonded cement paste, the latter promoting further degradation of stress transmission capacity between the various concrete constituents. With respect to cast concrete, the reductions obtained (by the incorporation of CRCA) were more significant. The maximum aggregate size adopted also played a role, since the smaller the size the higher the porosity, and thus the higher the deformability and the lower the modulus of elasticity.

Finally, the abrasion resistance increased with the incorporation of CRCA as a result of the higher roughness and porosity of these aggregates, which yielded a better bonding to the cement paste and thus a better adhesion to concrete. Hence, the prevailing tendency was the improvement of this property with the incorporation of this type of aggregate. This result is in agreement with those of other researches, in which reductions in depth loss between 10% and 15% were obtained, showing that this property indeed improves when CRCA are incorporated in concrete.

4. Technical Viability

4.1. Case Study

Nowadays, the analysis and design of tunnels in urban environments require the use of numerical simulations to simulate with accuracy complex geometries of the tunnel and ground layering. For simplicity's sake, the case study selected is a deep tunnel in a rock massif (hydrostatic stress field of 5 MPa), with a 3 m radius circular cross-section, where the analytical method proposed by Carranza-Torres and Fairhurst [11] is valid and is briefly described in Annex 1. In this example, stress relief factor λ is taken equal to 85% and lining thickness, t_c, is equal to 200 mm for CNA shotcrete.

For the lining of the tunnel, shotcrete with different CRCA incorporation ratios (IR) was considered, whose mechanical properties were determined experimentally by Duarte et al. [25]. Table 5 shows the values of the design compressive strength and modulus of elasticity obtained, including the percentage of reduction of the mechanical properties in relation to the shotcrete mix without incorporation of CRCA (IR0).

Table 5. Design compressive strength and modulus of elasticity of each composition.

	IR0	IR20	IR50	IR100
f_{cd} [MPa]	22.1	19.9	17.6	15.5
($\Delta f_{cd,IR0}$ [%])	(-)	(−10%)	(−21%)	(−30%)
E_c [GPa]	26.1	25.4	24.3	18.1
($\Delta E_{c,IR0}$ [%])	(-)	(−3%)	(−7%)	(−31%)

Table 5 shows that incorporating CRCA in the mix results in lower mechanical properties, ranging from about −3% for IR20 to −31% for IR100. For tunnels, the reduction of strength and stiffness in the shotcrete with recycled aggregates can be balanced by an increase of the lining's thickness. The lining's equivalent thickness for different CRCA incorporation ratios was computed for two criteria: (i) lining with similar stiffness and (ii) lining with similar yield stress.

4.2. Elastic Radial Stiffness Criterion, K_s

In this section, it is assumed that each shotcrete mix considered has a similar stiffness, namely the elastic radial stiffness, K_s, is equal to 636 MPa/m (t_c = 200 mm for CNA shotcrete).

Table 6 quantifies the differences in yield pressure and equivalent thickness. For IR20 and IR50, the thickness increase is small (<15 mm, 3% and 7%, respectively). The thickness increase is significant (42%, 83 mm) for IR100 only.

Table 6. Variation of yield stress and equivalent thickness when a similar stiffness criterion is adopted.

	IR0	IR20	IR50	IR100
Yield Stress: $p_s^{m\acute{a}x}$ [MPa]	1.43	1.32	1.21	1.39
($\Delta p_s^{m\acute{a}x}$ [%])	(-)	(−8%)	(−15%)	(−2%)
Equivalent Thickness: e [mm]	200	206	214	283
(Δe [%])	(-)	(3%)	(7%)	(42%)

Table 6 supports the conclusion that both IR20 and IR50 are competitive solutions from a technical point of view, with small increases of the lining's thickness (<7%).

4.3. Maximum Pressure Criterion, $p_s^{m\acute{a}x}$

In this section, it is assumed that each shotcrete mix considered has a similar yield stress: 1.43 MPa (t_c = 200 mm for CNA shotcrete). Table 7 quantifies the differences in stiffness and equivalent thickness. For IR20, the equivalent thickness increases 12% (24 mm), which can be considered acceptable, while IR50 and IR100 require thickness increments over 28% (more than 50 mm).

Table 7. Results for the maximum pressure criterion.

	IR0	IR20	IR50	IR100
K_s [MPa/m]	636	697	766	653
(ΔK_s [%])	(-)	(10%)	(20%)	(3%)
Equivalent Thickness: e [mm]	200	224	255	290
(Δe [%])	(-)	(12%)	(28%)	(45%)

For cases where yield stress criterion is relevant, only the IR20 mix can be considered equivalent to IR0, as the other options may be compromised by an increase of the lining thickness over 50 mm.

5. Economic Viability

Adopting the equivalent thicknesses determined in Section 3, the economic viability analysis of CRCA shotcrete is presented in this section. Based on data collected from Portuguese tunneling companies, the estimation of the current costs range to produce dry-mix shotcrete is shown in Table 8.

Figure 2 shows the unit costs, in €/m^3, of the different shotcrete scenarios, assuming three possible costs for CRCA to assess its impact:

- Scenario 1 (50% CNA)—the recycled aggregates' cost is taken as equal to 50% of the natural aggregates (7.5 €/m^3);
- Scenario 2 (NULL) —the recycled aggregates' cost is zero (taking into account that CDW producers are ready to pay to get rid of it);
- Scenario 3 (SUB) —there is a subsidy equal to 7.5 €/m^3 to promote recycled aggregates' use. Because the cost of CRCA is smaller than that of CNA, the unit cost of shotcrete with incorporation of CRCA is smaller.

Table 8. Material costs per m^3 of dry-mix shotcrete produced.

	Cost Range [€/m^3]	Cost Adopted	
		[€/m^3]	[%]
Equipment	16 to 20	18	13
Cement and Fly Ash	45 to 50	47.5	34
Labour	30 to 40	35	25
Additives	17 to 23	20	14
Plasticizers	4 to 5	4,5	3
Natural Aggregates	13 to 17	15	11
Total		140	100

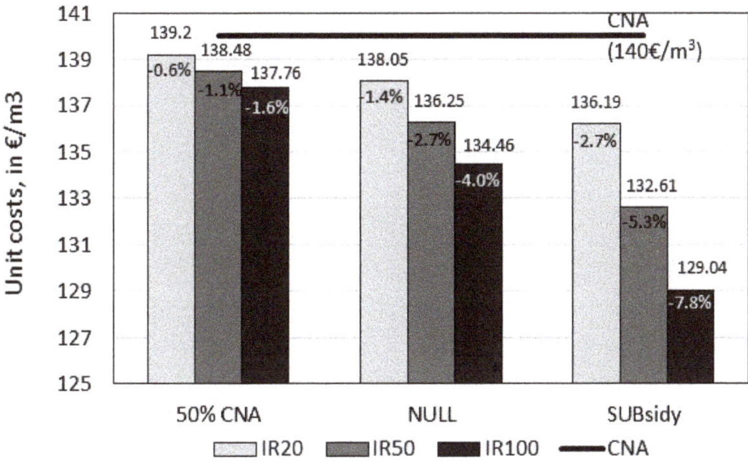

Figure 2. Cost of shotcrete for each composition, assuming different cost scenarios of the recycled aggregates.

Figure 2 shows that incorporating recycled aggregates in shotcrete generates small variations is cost per cubic meter (<8%), for the three scenarios analyzed.

Table 9 shows the cost of the lining per unit length of the tunnel for the two cases for which equivalent thickness was determined: similar stiffness and similar yield stress.

In all cases, for simplicity's sake, rebound losses equal to 25% of the shotcrete's theoretical volume were assumed.

For the case in which CRCA cost is 50% of the cost of CNA, the lining's cost per unit length of the tunnel with CRCA is slightly higher than without CRCA for IR20 (2.3%) and IR50 (5.3%) for Case 1. For the other IR's and cases, the increase in cost is higher than 10%.

Because the recycled aggregates' cost represents less than 15% of the total production costs, an increase in thickness due to the lower mechanical properties leads to higher consumption of the other components, and thus costs increase.

Table 9. Lining cost per unit length of the tunnel for Scenario 1 (50% Coarse Natural Aggregates).

CRCA Incorporation Ratio [%]	Thickness [mm]	Theoretical Volume [m^3/m]	Estimated Volume with Rebound Losses [m^3/m]	Shotcrete Cost Per Unit Volume [€/m^3]	Lining's Cost Per Unit Length [€/m]	Variation [€/m]
Case 1—Similar stiffness criterion						
0	200	3.64	4.86	140.00	680.26	-
20	206	3.75	5.00	139.20	695.94	+15.68 (2.3%)
50	214	3.89	5.19	138.05	716.00	+35.74 (5.3%)
100	283	5.08	6.78	136.19	922.98	+242.7 (35.7%)
Case 2—Similar yield stress criterion						
0	200	3.64	4.86	140.00	680.26	-
20	224	4.06	5.42	139.20	754.40	+74.14 (10.9%)
50	255	4.60	6.14	138.05	847.13	+166.87 (24.5%)
100	290	5.20	6.94	136.19	944.66	+264.40 (38.9%)

Figure 3 shows the variation of the lining's cost per unit length of tunnel for the three hypothetical costs of CRCA.

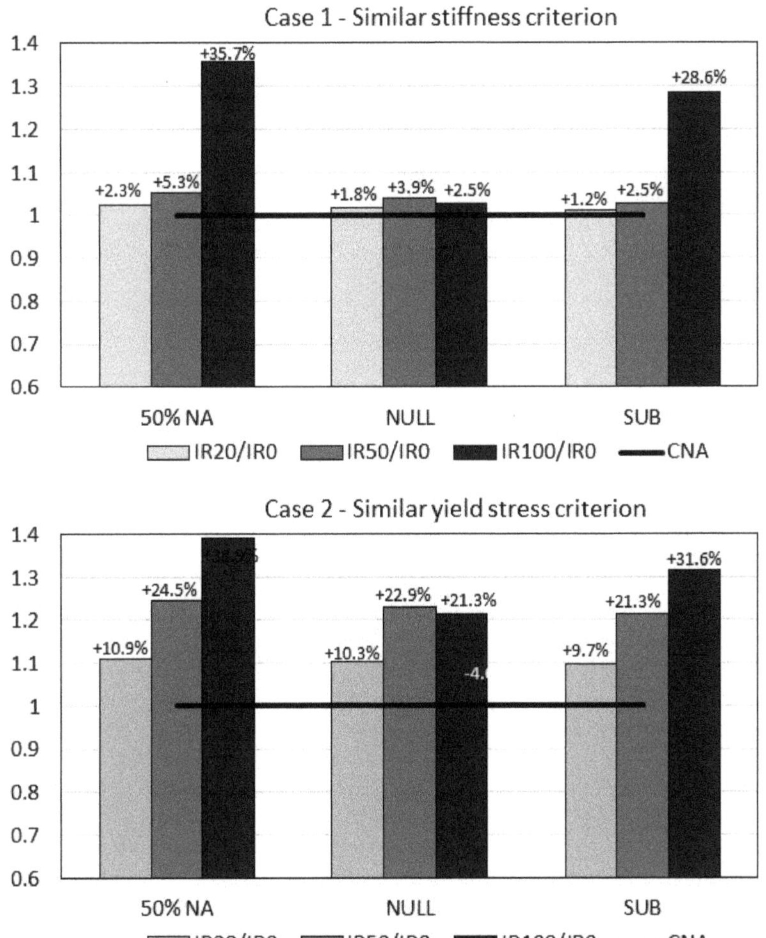

Figure 3. Variation of lining's cost per unit length of tunnel for three different costs of Coarse Recycled Concrete Aggregates.

Figure 3 shows that the lining's cost per unit length of the tunnel is fairly insensitive to the cost of CRCA. In particular for Case 1, similar stiffness criterions-mixes IR20 and IR50 lead to marginal increases in the lining's cost (between 1.2% and 5.3%), while the other cases studied lead to variations higher than 10%.

So, IR20 and IR50 are interesting alternatives for CRCA incorporation in dry-mix shotcrete solutions. The remaining cases require a larger increase in lining's thickness, which is economically unattractive.

6. Conclusions

The technical economic analysis of dry-mix shotcrete incorporating CRCA was performed considering three different replacement ratios of CNA with CRCA, for two cases: similar stiffness criterion and similar yield stress criterion.

For deep tunnels, a small increase in the lining's thickness is required to have linings of similar stiffness for the IR20 and IR50 mixes, which shows their technical viability. An economic assessment proved that both mixes can be competitive against a conventional shotcrete, with very small increases in the lining's cost per unit length of the tunnel.

A sensitivity study to assess the impact of CRCA cost on lining's cost per unit length shows that CRCA cost is very small, and therefore its influence on the overall costs is also small.

Based on this study, it can be concluded that using coarse recycled concrete aggregates in tunnels can be an interesting solution for the IR20 and IR50 mixes, with marginal impact on thickness and the lining's cost per unit length of tunnel.

The viability of using CRCA in tunnel construction practice should be analyzed considering the specific data from each project, such as costs to produce dry-mix shotcrete and the strength and deformability study.

Author Contributions: Conceptualization, G.D., R.C.G., J.d.B., M.B., J.N. methodology, G.D., R.C.G, J.d.B., M.B., J.N. formal analysis, G.D., R.C.G., J.d.B. validation, J.d.B. investigation, G.D., R.C.G., J.N. data curation, G.D., R.C.G., J.N. writing—original draft, G.D., R.C.G. writing—review and editing, G.D., R.C.G., J.d.B., M.B. All authors have read and agreed to the published version of the manuscript.

Funding: This research received no external funding.

Acknowledgments: The authors wish to thank CERIS (Civil Engineering Research and Innovation for Sustainability) research centre and FCT (Foundation for Science and Technology).

Conflicts of Interest: The authors declare no conflict of interest.

Appendix A The Convergence-Confinement Method

The stress redistribution and three-dimensional deformation induced by tunnel excavation and lining installation during construction is a very complex three-dimensional problem. To simulate tunnel construction, the convergence-confinement method in plane strain conditions is widely used. The method relies on the ground reaction curve, which describes the response of the ground around the tunnel [26,27], and the support reaction curve, which describes the lining response.

The two-dimensional tunneling problem is illustrated in Figure A1a. The tunneling process is represented by increasing the λ value from zero to one. Figure A1b shows the main characteristics of the convergence-confinement method. The convergence curve corresponds to the internal pressure versus the tunnel radial displacement. The radial displacement of the tunnel increases as the internal pressure decreases. The tunnel can be self-stabilized without a liner (Curve [a]) in Figure A1b) or the surrounding ground can fail which leads to an increase in the ground load acting on the tunnel lining (Curve [b] in Figure A1). In the latter case, a liner must be installed to keep the stability of the tunnel.

Based on the convergence-confinement method, Carranza-Torres and Fairhurst [11] proposed a set of analytical equations valid for deep circular tunnels in rock. These equations are used to ensure the technical viability of the shotcrete solutions proposed in a simplified manner, bypassing complex numerical simulations necessary for shallow tunnels and non-circular shapes, but without loss of generalization capacity of the conclusions.

The problem is studied as a two-dimensional plane strain case, in which the pressure, p_i, and the radial displacement, u_r, are constant in the tunnel walls.

The ground reaction curve is obtained from the elastoplastic solution of a circular opening subjected to a hydrostatic stress field, σ_0. Equations (A1) and (A2) present the elastic and plastic portions of the curve, respectively.

$$u_r^{el} = \left(\frac{S_0 - p_i}{2G_{rm}}\right)R \tag{A1}$$

where u_r^{el} is the elastic radial displacement of the massif, S_o the uniform tension around the tunnel, p_i the pressure of the walls at the tunnel, and G_{rm} the shear modulus of the rock massif.

$$u_r^{pl} = u_r^{el}\left\{\frac{K_\psi-1}{K_\psi+1} + \frac{2}{K_\psi+1}\left(\frac{R_{pl}}{R}\right)^{K_\psi+1} + \frac{1-2\nu}{4\left(S_o-P_i^{cr}\right)}\left[ln\left(\frac{R_{pl}}{R}\right)\right]^2 - \left[\frac{1-2\nu}{K_\psi+1}\frac{\sqrt{P_i^{cr}}}{S_o-P_i^{cr}} + \frac{1-\nu}{2}\frac{K_\psi-1}{\left(K_\psi+1\right)^2}\frac{1}{S_o-P_i^{cr}}\right]\left[(K_\psi+1)ln\left(\frac{R_{pl}}{R}\right) - \left(\frac{R_{pl}}{R}\right)^{K_\psi+1} + 1\right]\right\} \tag{A2}$$

where K_ψ is the dilatancy coefficient, P_i^{cr} the critical pressure, related with the transition of an elastic to plastic behavior of the tunnel walls, and R_{pl} the radius of the failed region, developed when $p_i < P_i^{cr}$.

The shotcrete support is modeled with perfect elastoplastic behavior, with an elastic radial stiffness k_s, and maximum pressure, $p_{s,max}$. Equations (A3)–(A5) describe the support behavior [28].

$$p_s = K_s u_r \tag{A3}$$

$$p_s^{máx} = \frac{f_{cd}}{2}\left[1 - \frac{(R-t_c)^2}{R^2}\right] \tag{A4}$$

$$K_s = \frac{E_c}{(1+\nu_c)R}\frac{R^2-(R-t_c)^2}{(1-2\nu_c)R^2+(R-t_c)^2} \tag{A5}$$

where p_s is the pressure in the support, f_{cd} the design compressive strength of the shotcrete, u_r the radial displacement of the support, ν_c the Poisson's coefficient of the shotcrete, E_c the modulus of elasticity of the shotcrete, R the radius of the shotcrete ring and t_c representing its thickness.

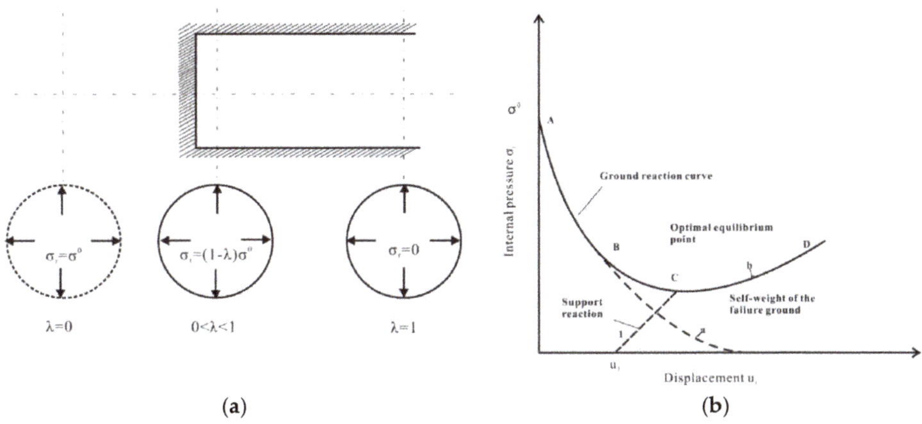

Figure A1. Convergence confinement method: (**a**) evolution of tunnel convergence with face advance; (**b**) ground reaction curve and support reaction curve (adapted from Orestes [13]).

References

1. European Commission. *EU Construction & Demolition Waste Management Protocol.* 2017. Available online: https://www.google.com.hk/url?sa=t&rct=j&q=&esrc=s&source=web&cd=3&ved=2ahUKEwjNvaWNjODoAhUiGaYKHfWHBWUQFjACegQIAxAB&url=https%3A%2F%2Fec.europa.eu%2Fdocsroom%2Fdocuments%2F26368%2Fattachments%2F3%2Ftranslations%2Fen%2Frenditions%2Fnative&usg=AOvVaw3cvos29Vbsj1923H9gahcM (accessed on 10 October 2019).
2. Bravo, M.; de Brito, J.; Pontes, J.; Evangelista, L. Mechanical performance of concrete made with aggregates from construction and demolition waste recycling plants. *J. Clean. Prod.* **2015**, *99*, 59–74. [CrossRef]

3. Bravo, M.; de Brito, J.; Pontes, J.; Evangelista, L. Durability performance of concrete with recycled aggregates from construction and demolition waste plants. *Constr. Build. Mater.* **2015**, *77*, 357–369. [CrossRef]
4. Chan, C. *Use of Recycled Aggregate in Shotcrete and Concrete*; The University of British Columbia: Vancouver, Canada, 1998.
5. Broere, W. Urban underground space: Solving the problems of today's cities. *Tunn. Undergr. Space Technol.* **2016**, *55*, 245–248. [CrossRef]
6. Kolymbas, D. *Tunnelling and Tunnel Mechanics: A Rational Approach to Tunnelling*; Springer: Berlin/Heidelberg, Germany, 2005.
7. Bentur, A.; Mindess, S. *Fibre Reinforced Cementitious Composites*, , 2nd ed.; Taylor & Francis: Abingdon, UK, 2006.
8. Jovicic, V.; Šušteršic, J.; Vukelic, Z. The application of fibre reinforced shotcrete as primary support for a tunnel in flysch. *Tunn. Undergr. Space Technol.* **2009**, *24*, 723–730. [CrossRef]
9. Massone, L.; Nazar, F. Analytical and experimental evaluation of the use of fibers as partial reinforcement in shotcrete for tunnels in Chile. *Tunn. Undergr. Space Technol.* **2018**, *77*, 13–25. [CrossRef]
10. Yang, J.M.; Kim, J.; Yoo, D. Performance of shotcrete containing amorphous fibers for tunnel applications. *Tunn. Undergr. Space Technol.* **2017**, *64*, 85–94. [CrossRef]
11. Carranza-Torres, C.; Fairhurst, C. Application of the convergence-confinement method design to rock masses that satisfy the Hoek-Brown failure criterion. *J. Tunn. Undergr. Space Technol.* **2000**, *15*, 187–213. [CrossRef]
12. Panet, M.; Guenot, A. Analysis of convergence behind the face of a tunnel: Tunnelling 82, proceedings of the 3rd international symposium, Brighton, 7–11 June 1982, P197–204. Publ London: IMM, 1982. *Int. J. Rock Mech. Min. Sci. Geomech. Abstr.* **1983**, *20*, A16.
13. Oreste, P. Analysis of structural interaction in tunnels using the convergence-confinement approach. *J. Tunn. Undergr. Space Technol.* **2003**, *18*, 347–363. [CrossRef]
14. NP EN 1097-6. *Tests for Mechanical and Physical Properties of Aggregates—Part 6: Determination of Particle Density and Water Absorption*; IPQ: Lisbon, Portugal, 2003.
15. NP EN 1097-3. *Tests for Mechanical and Physical Properties of Aggregates—Part 3: De-termination of Loose Bulk Density and Voids*; IPQ: Lisbon, Portugal, 2003.
16. NP EN 933-4. *Tests for Geometrical Properties of Aggregates—Part 4: Determination of Particle Shape e Shape Index*; IPQ: Lisbon, Portugal, 2002.
17. NP EN 1097-2. *Tests for Mechanical and Physical Properties of Aggregates—Part 2: Methods for the Determination of Resistance to Fragmentation*; IPQ: Lisbon, Portugal, 2002.
18. NP EN 1097-5. *Tests for Mechanical and Physical Properties of Aggregates—Part 5: Determination of the Water Content by Drying in a Ventilated Oven*; IPQ: Lisbon, Portugal, 2002.
19. EN 1542. *Products and Systems for the Protection and Repair of Concrete Structures. Test Methods*; Measurement of Bond Strength by Pull-Off: Brussels, Belgium, 1999.
20. EN 12390-3. *Testing Hardened Concrete. Compressive Strength of Test Specimens*; Committee European for Normalization: Brussels, Belgium, 2003.
21. EN 12390-6. *Testing Hardened Concrete. Tensile Splitting Strength of Test Specimens*; Committee European for Normalization: Brussels, Belgium, 2003.
22. LNEC E-397. *Determination of the Modulus of Elasticity in Compression (in Portuguese)*; National Laboratory of Civil Engineering (LNEC): Lisbon, Portugal, 1993.
23. DIN 52108. Testing of Inorganic Non-metallic Materials—Wear Test Using the Grinding Wheel According to Boehme. In *Deutsches Institut Fur Normung E.V*; German National Standard: Berlin, Germany, 2010.
24. EN 12504-4. *Testing Concrete. Determination of Ultrasonic Pulse Velocity*; Committee European for Normalization: Brussels, Belgium, 2004.
25. Duarte, G.; Bravo, M.; de Brito, J.; Nobre, J. Mechanical performance of shotcrete produced with recycled coarse aggregates from concrete. *Constr. Build. Mater.* **2019**, *210*, 696–708. [CrossRef]
26. Amberg, W.; Lombardi, G. An elasto-plastic analysis of the stress-strain state around an underground opening, Part II. In Proceedings of the 3rd Congress of ISRM, Denver, Colorado, USA, 1974; Volume 1, pp. 1049–1060.

27. Panet, M. Understanding deformations in tunnels. In *Comprehensive Rock Engineering*; Pergamon Press: Oxford, UK, 1993; Volume 1, pp. 663–690.
28. Hoek, E.; Brown, E.T. *Underground Excavations in Rock*; CRC Press: Boca Raton, FL, USA, 1980; 532p, ISBN 9780419160304.

Article

High Performance Self-Compacting Concrete with Electric Arc Furnace Slag Aggregate and Cupola Slag Powder

Israel Sosa, Carlos Thomas *, Juan Antonio Polanco, Jesus Setién and Pablo Tamayo

LADICIM (Laboratory of Materials Science and Engineering Division), University of Cantabria, 39005 Santander, Spain; sosai@unican.es (I.S.); polancoa@unican.es (J.A.P.); setienj@unican.es (J.S.); pablo.tamayo@unican.es (P.T.)
* Correspondence: thomasc@unican.es

Received: 17 December 2019; Accepted: 19 January 2020; Published: 22 January 2020

Abstract: The development of self-compacting concretes with electric arc furnace slags is a novelty in the field of materials and the production of high-performance concretes with these characteristics is a further achievement. To obtain these high-strength, low-permeability concretes, steel slag aggregates and cupola slag powder are used. To prove the effectiveness of these concretes, they are compared with control concretes that use diabase aggregates, fly ash, and limestone supplementary cementitious materials (SCMs, also called fillers) and intermediate mix proportions. The high density SCMs give the fresh concrete self-compacting thixotropy using high-density aggregates with no segregation. Moreover, the temporal evolution of the mechanical properties of mortars and concretes shows pozzolanic reactions for the cupola slag. The fulfillment of the demands in terms of stability, flowability, and mechanical properties required for this type of concrete, and the savings of natural resources derived from the valorization of waste, make these sustainable concretes a viable option for countless applications in civil engineering.

Keywords: self-compacting concrete; high-performance concrete; EAFS; cupola slag; electric arc furnace slag; mechanical properties

---

1. Introduction

The production of crude steel in Europe in the year 2017 was 168.3 Mt, almost 4% higher than in 2016, while the production by electric arc furnaces stands at 40.3% of total production (67.8 Mt) [1]. This steel production determines the amount of electric arc furnace slag (EAFS) generated during the fusion processes of scrap, totalling 18 Mt for the year 2016 according to the Euroslag. In Europe, 12 Mt of cast iron was generated in 2017. These values of production reveal the volume of slag resulting from the steel and cast iron industries that ends in landfills.

Several authors have incorporated different types of industrial waste as SCM's to concrete to modify its properties [2–4]. The pozzolanic properties of the slag generated in steel processes depend on the cooling process. With rapid cooling (as is the case of cupola slag) using water, the vitrification of the slag occurs, leaving the silica in an amorphous form and, therefore, susceptible to reaction. On the contrary, slow cooling (case of EAFS) promotes the complete crystallization of the phases and the inertization of the final product, thus not compromising its dimensional stability. The pozzolanicity of EAFS has been studied, but its reactivity has been reported to be rather weak [5] although it can be improved by remelting treatments [6]. The content of periclase (MgO) in the slag causes a risk of potential expansion because the process of transformation into brucite [Mg(OH)$_2$] by hydration is slow or even delayed, putting the dimensional stability at risk. Given the possible expansive reactions, it is important to verify the efficiency of the stabilization treatments [5]. Another problem present in these aggregates is the significant deficit of particle sizes that pass through the smaller sieves in the sands. Therefore, the manufacture of mortars and concretes with EAFS sand entails mixing by combining them, either with natural sand or with inert filler [7,8].

Another interesting practice in recovery/recycling strategies (in addition to incorporating EAFS aggregates) is to incorporate recycled aggregates from Construction and Demolition Wastes (C&DW) to self-compacting concrete [2,9,10]. It has been demonstrated that their mechanical properties and durability [2,11–13] are suitable for structural concrete and they can be recycled several times [14].

SCMs are important for the self-compacting concrete (SCC), since they reduce the intergranular interaction [14], increase the cohesion and the flowability of the mixture, improve the hydration of the paste [15,16], and strengthen the resistance to segregation. In hardened concrete, SCMs typically reduce capillarity and permeability but also mechanical properties can be reduced [14,17]. The cements most used for the manufacture of high strength SCC are Portland type I, however, blends containing one or more SCMs [18] and 350 kg/m^3 of CEM can be used readily [19]. It is recommended not to exceed 500 kg/m^3 (to avoid shrinkage problems) and to use SCMs to improve the workability of the fresh concrete. Several authors [20,21] have established that drying shrinkage increases with drying speed and is proportional to the volume of cement paste, while the opposite occurs by increasing the lime filler content.

The use of cupola slag as a concrete SCM is not a common application despite of its sustainable benefits. Nevertheless, the use of granulated cupola furnace slag (GCFS) as fine and coarse aggregate (0–16 mm) in concrete does not seem a viable option [22]. However, some authors have demonstrated by the manufacture of mortars with various replacements [23] that is suitable to be used as SCM with the right activation process. Nevertheless, great reactivity has been reported when acting as substitution of ordinary Portland cement (OPC), showing 30% compressive strength gains for 15% replacements at 28 days [22].

The use of SCMs increases the concrete strength, and in greater proportion, by using SCMs with pozzolanic properties. According to Domone [23], the type and proportion of SCM has greater influence on the compressive strength than the water/filler ratio (cement + SCMs). Rozière et al. [20], reported the rise in compressive strength by increasing the limestone filler content, keeping the effective water/cement ratio and the amount of cement.

The use of SCC brings with it a series of advantages with respect to the conventional one such as better adhesion between the paste and the aggregates or the uniform distribution of the stresses during load applications [19], even having the same w/c ratio [24,25]. Some studies claim that indirect tensile strength is higher in SCC [25] due to its packed structure, other studies argue that there are no differences because this property does not depend on the paste content [23,24] and other studies argue that this greater paste content affects negatively [20]. Tensile splitting strength of SCC and conventional concrete with the same amount of cement and water/cement ratio is directly influenced by the type of aggregate used [19] and elastic modulus in SCC is lower because it has a lower proportion of coarse aggregate [21], and for the same compressive strength the SCC presents a bigger strain [19]. The stiffness of SCC can be 40% lower but in high strength concrete (the concrete considered in this study) the difference is reduced to 5%.

The mechanical properties of concretes with EAFS are superior to those obtained with conventional aggregate concretes [26,27], mainly due to an improvement in the bond with the cement paste owing to the quality of the paste-aggregate interfacial transition zone (ITZ), which can be observed by means of a scanning microscope [23,28]. Experimentally, it has been found that the coarse fraction of EAFS contributes to the increase in compressive strength, tensile splitting strength, and elastic modulus. Similarly, the total substitution of fine aggregate leads to a reduction in compressive strength [29]. In terms of durability, concrete with EAFS is more vulnerable to frost and freeze–thaw cycles [30].

The use of EAFS in SCC is a challenge and very few studies have been done to date, mainly due to a decrease in flowability, also due to intergranular friction and a slight increase in density, although it has been possible to obtain stronger concretes than conventional ones [31], always using a significant amount of superplasticizer additive. Recently, Santamaría et al. [32,33] have demonstrated the feasibility of manufacturing SCC using EAFS as both coarse and fine, obtaining consistency classes of S4 and SF2 and reasonably good mechanical properties. Likewise, Qasrawi [34] advises not to use replacements greater than 50% of EAFS so as not to negatively affect the properties in the fresh state, mainly density, air content, and stability. With regard to these studies and broadly speaking, in this paper new concrete mixes are developed through the use of two different wastes, to obtain high-performance concrete. Analyzing in detail the specific differences with the works found in the literature, this paper presents these main novelties:

- The EAFS aggregate has not been separated by screening operations for mixing (only 2 fraction ranges have been used: 0/6 and 6/12).
- Two wastes from two different industrial processes are used in this research, both of the oxidation or reduction stages.
- The developed concrete mixes seek to obtain very high strength concrete without damaging the properties in the fresh state.
- A total replacement of the natural coarse and a partial replacement of the natural fines have been made.
- A comparison with a control concrete with high quality aggregates (diabase) has been made.
- It has been possible to value a new waste that would otherwise end up in landfills (cupola slag).

The aim of this study is to demonstrate that it is possible to obtain a high-performance concrete (HPC), considering this to have a strength between 70 and 150 MPa, which is self-compacting and also uses steel slag aggregate in all fractions (coarse, fine, and SCMs), obtaining a concrete with countless potential applications. For this purpose, the work consisted of two phases; in the first, mortar mixes were produced with different cement replacements for cupola slag, thus demonstrating the pozzolanicity of this material. In the second phase, self-compacting concretes have been compared with different types of coarse, fine, and locally available SCMs, demonstrating the improvement of mechanical performance with the use of EAFS aggregates and cupola slag (as SCM) with respect to conventional materials such as diabase (high quality aggregate) or siliceous sand.

2. Materials and Methods

2.1. Materials

Standardized siliceous sand, CEM I 52.5 R and cupola slag have been used for the manufacture of conventional mortars. For the manufacture of the different concrete proportions, coarse 6/12 (DC) diabase and coarse 6/12 (SC) slag have been used, limiting their quantities to avoid problems of blockage and segregation. Diabase sand 0/6 (DS), electric arc furnace slag 0/6 (SLS), and silica sand 0/2 (SIS) have been used as fine aggregates. The SCM used include limestone filler (LF), fly ash (FA), and cupola slag filler (CS). The granulometric distribution of the aggregates used (coarse and fine) according to EN 933-1 is shown in Figure 1.

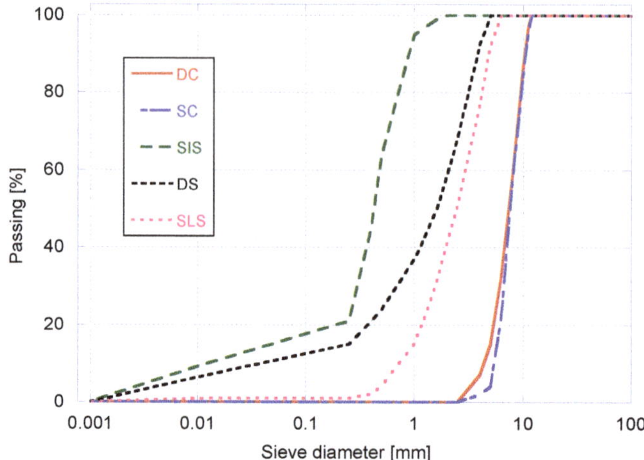

Figure 1. Aggregates granulometric grading.

The physical-mechanical properties of the aggregates have been determined by characterizing the bulk specific gravity and porosity according to EN 1097-3, the apparent specific gravity and water absorption according to EN 1097-6, the resistance to fragmentation determining Los Angeles coefficient according to EN 1097-2 and the aggregates crushing value according to UNE 83112. The fine aggregates (cement and SCM) have been characterized by determining the actual density according to UNE 80103 and by the specific Blaine surface according to EN 196-6. The results obtained are shown in Tables 1 and 2. Both types of aggregates have excellent mechanical properties and a 35% higher density in the case of EAFS.

Table 1. Main properties of the coarse aggregates used in the self-compacting concrete (SCC).

Material	Bulk Specific Gravity [g/cm^3]	Apparent Specific Gravity [g/cm^3]	Absorption [vol.%]	Open Porosity [wt.%]	Los Angeles [%]	Aggregates Crushing Value [%]
EAFS coarse	3.65	3.85	1.43	5.22	15	18
Diabase coarse	2.72	2.81	1.11	3.02	15	15

Table 2. Main properties of the sands, supplementary cementitious materials (SCMs) and cement used in the SCC.

Material	Actual Density [g/cm^3]	Blaine Surface [m^2/kg]
EAFS sand	3.77	-
Diabase sand	2.89	-
Silica sand	2.61	-
Cupola slag SCM	2.89	429.4
Limestone filler	2.65	274.1
Fly ash SCM	2.13	400.3
CEM I 52.5 R	3.11	495.7

The chemical characterization of the slags used was carried out by means of X-ray fluorescence (XRF), in order to determine the semi-quantitative concentration of compounds in oxides represented by percentages (Table 3). For this characterization an ARL-ADVANT-XP Thermo-spectrometer was used. It was observed that the main components of slags were iron, calcium, and silicon oxides, although there were traces of chromium and titanium oxides, probably generated during the injection of oxygen.

On the other hand, the cupola slag shows high concentrations of silicon, calcium, and aluminum oxides, the first being an indicator of the possible reactivity of the material, since it is in an amorphous state. The compositions of the rest of the materials were obtained by Energy-dispersive X-ray spectroscopy (EDX) using a Zeiss EVO MA15 scanning electron microscope (SEM) equipped with an Oxford Instruments X-ray detector, selecting different representative areas of particles chosen randomly.

Table 3. Chemical composition of the materials used.

Compound [wt.%]	Fe_2O_3	CaO	SiO_2	Al_2O_3	MgO	MnO	Cr_2O_3	TiO_2	Na_2O	SO_3	K_2O	Others
EAFS	37.90	30.26	12.00	7.4	4.93	4.53	1.15	0.53	-	-	-	<0.5
Cupola slag	6.34	29.97	43.56	13.64	2.1	2.80	-	0.51	-	-	-	<0.5
Fly ash	7	6.1	55	20.4	2.6	-	-	0.9	1.2	4	2	-
CEM I 52.5 R	3.38	66.6	17.81	4.79	1.3	-	-	0.2	-	4.49	0.78	-

The expansiveness of the EAFS is one of its main problems, and can cause cracking of concrete in the medium/long term. To avoid this phenomenon, EAFS aggregates have been submerged in pools for 24 h and have remained wet in storage stacks for 3 months, in order to hydrate free lime and magnesia. The expansiveness of the aggregates was determined according to EN 1744-1, obtaining values of 0.16 vol.% at 24 h and 0.17% vol. at 168 h, the first value being those required when the MgO content is less than 5%.

The mortar proportions (Table 4) were carried out with CEM I 52.5 R, CEN standardized sand and in accordance with the amounts proposed by EN 196-1 (except the amount of sand, slightly higher). The mortars were cured submerged in water at a temperature of 20 ± 1 °C. The manufacture of these mortars (M) was carried out using replacements of 0, 10, 20, 30 vol.% of cement by cupola slag filler, in order to establish the pozzolanity of the cupola slag compared to cement.

Table 4. Mix proportions of the mortars [g].

Material [g]	M-0%	M-10%	M-20%	M-30%
CEM I 52.5 R	450	405	360	315
Silica sand	1450	1450	1450	1450
Cupola slag filler	-	42	84	126
Water	225	225	225	225
w/c ratio	0.5	0.6	0.7	0.8

For the mix proportions of the self-compacting control concrete, the methodology proposed by Dinakar et al. [35] was used, based on compressive strength. The design goal was to obtain a high-performance concrete with 100 MPa strength at 90 days, for which an amount of cement of 450 kg/m^3, and the use of 2% (of the cement weight) of a superplasticizer additive (enabling a more viscous paste to be obtained) was selected. A limestone filler quantity of 100 kg/m^3 was used, thus using a total amount of cement, SCMs, and filler of 550 kg/m^3, less than the maximum 600 kg/m^3 recommended by EHE-08 [36] and EFNARC [37]. An attempt was made to maximize the coarse content (50 vol.%) to obtain greater use of the by-product without affecting segregation or blocking. Likewise, the amount of water used was optimized so as not to adversely affect the strength without affecting the flowability or segregation.

Four concrete mixes were made: three control mixes that use diabase coarse with three different filler materials (limestone, fly ash and cupola slag) and a fourth that uses EAFS coarse with cupola slag filler. The first three dosages enable the comparison of the SCM used, while the fourth enables the comparison of high strength natural aggregates with siderurgical aggregates. The three control dosages are analogous, while the fourth had to be modified because it presented notable deficiencies in the fresh state. The sand content was increased because SLS has less fine aggregates than DS and the latter has a much more cavernous and angular geometry. In addition, the reduction of SC enabled slump to be improved and prevented concrete blockage. A slightly lower w/c ratio was used in this case due to the limitation due to the segregation of the EAFS aggregate.

The preparation of these self-compacting concretes was similar to that of conventional concretes, with the exception of kneading time (12 min), considerably increased to ensure the complete distribution of the superplasticizer additive. The mixtures were made in a 120 L rotating drum mixer, with 30 L batches. The samples were been demolded at 24 h and were cured in a moisture chamber, at a constant temperature of 20 ± 2 °C and constant humidity of 95 ± 5%. The final proportions, in kg/m^3, of the mixtures appear in Table 5.

Table 5. Concrete mix proportions (kg/m^3).

Concrete	SCC-DC-LF	SCC-DC-FA	SCC-DC-CS	SCC-SC-CS
Diabase coarse (DC)	896	896	896	-
Slag coarse (SC)	-	-	-	1101
Diabase sand (DS)	411	411	411	-
Silica sand (SIS)	386	386	386	605
Slag sand (SLS)	-	-	-	444
Limestone filler (LF)	100	-	-	-
Fly ash (FA)	-	80	-	-
>Cupola slag filler (CS)	-	-	109	109
CEM I 52.5 R	450	450	450	450
Water	180	180	180	174
Superplasticizer additive	9	9	9	9
w/c ratio	0.40	0.40	0.40	0.39

2.2. Properties of Fresh Concrete

The self-compactability of a concrete depends essentially on its filling ability, its passing ability, and its static and dynamic stability or segregation resistance. The characterization of the properties in the fresh state of self-compacting concrete is different from conventional concretes, using in this study the slump flow, the L-box and the V-funnel tests. The slump flow test has been carried out according to EN 12350-8, to obtain the flow capacity of the concrete, as well as its stability. To carry out this test, the Abrams cone (EN 12350-2) and a 900 × 900 mm metal steel plate have been used, on which three concentric circles are marked, measuring the conjugated diameters of the drained concrete and determining the average diameter (SF). This test also determines the time it takes for the concrete to cover the 500 mm circle (t_{500}), which allows the relative viscosity of the concrete to be evaluated, as well as the flow rate. According to EN 206-9 the permissible range for this test is between 550 and 850 mm for the SF and the t_{500} must be less than or equal to 8 s.

The L-box test was carried out according to EN 12350-10 using 3 bars. It measures the ability to pass through the reinforcements, as well as stability. The concrete was poured through the upper opening, measuring the height of the concrete in the vertical section (H1) as well as the height at the end of the horizontal section (H2). The "passing ability" of concrete for the PL test is defined as the H2/H1 ratio. According to EN 206-9, there are two kinds of passing capacity delimited by a PL of 0.8. The V-funnel test has been carried out according to EN 12350-9 to assess the viscosity, the capacity to pass through confined spaces and the mold filling capacity. The test measures the time it takes for the concrete to flow through the V-mold, from the opening of the lower gate until the mold is emptied. According to EN 206-9, the permissible t_v range is 0–25 s, with 9 s being the limit between the two existing categories.

2.3. Physical Properties of the Concrete Mixes

The apparent, bulk and saturated-surface-dry (SSD) specific gravities were determined following EN 12390-7 and, additionally, both accessible porosity (vol.%) and water absorption (wt.%) were obtained according to UNE 83980 applying air vacuum. The physical properties were determined on nine thirds of standard cylindrical specimens of 150 × 300 mm at 28 days per mix (a total of 36 specimens), specimens in which the upper and lower ends were cut in order to avoid edge effects.

2.4. Mechanical Properties of the Mortars

The mortars underwent the mechanical tests proposed by and according to EN 196-1. These tests consisted first of all in the determination of flexural strength and then in the determination of compressive strength on each of the halves obtained in the first test. Three specimens were manufactured per age and per mix (64 specimens in total), with tests being performed at 7, 28, 60, and 90 days. The tests were carried out on a servohydraulic machine with a capacity of 250 kN at rates of 0.05 mm/s for flexural tests and 0.1 mm/s for compression tests.

2.5. Mechanical Properties of the Concrete

The uniaxial compressive strength was determined according to EN 12390-3 on four cubic specimens of 100 mm at 7, 28, 90, 180 and 365 days per mix (a total of 80 specimens). The use of these specimens is compatible with the maximum aggregate size and correction coefficients were applied if necessary. This test was carried out on a servohydraulic machine with a capacity of 1500 kN at a rate of 0.5 MPa/s, after removing the specimens from the moisture chamber and waiting for their surface drying.

The compressive elastic modulus was determined following EN 12390-13 on 1 standardized cylindrical specimen at 7, 28, 90, 180, and 365 days of age per mix (a total of 20 specimens). Method A was applied, which includes 3 preload cycles to check the correct positioning of the specimens and subsequently 3 loading/unloading cycles were applied. The initial and final stress and strain values enable the initial elastic modulus (first cycle) and the stabilized elastic modulus (third cycle) to be obtained. The specimens were capped with sulfur on their upper face and fitted with strain gauges 120 mm in length. The test was carried out on a servohydraulic machine with a capacity of 2000 kN at a load/unload rate of 0.7 MPa/s.

The splitting tensile strength test was carried out according to EN 12390-6, on 9 thirds of standard cylindrical specimens of 150 × 300 mm at 90 days per mix (a total of 36 specimens). A servohydraulic testing machine of 1500 kN capacity and a load rate of 0.05 MPa/s was used for this purpose.

3. Results and Discussion

3.1. Properties of Fresh Concrete

The properties in the fresh state of the different mixes of SCC appear in Table 6. The mixes that incorporate diabase coarse and limestone filler or slag filler have an average slump flow of 750–850 mm corresponding to the upper category or SF3 according to EN 206-9, equivalent to category AC-E3 according to EFNARC. In the case of using slag coarse or diabase aggregates with fly ash, the average slump flow is reduced to 650–750 mm, corresponding to category SF2 according to EN 206-9 or AC-E2 according to EFNARC, ideal workability for conventional applications as pillars and walls. The use of slags as coarse aggregate penalizes slump flow by 10%, which is compatible with the results obtained in conventional concretes, where the use of slags leads to a general decrease in workability [29,30]. The decrease in the flowability of concrete with slags is due to their increased friction, as well as their angular geometry and their high density. The t_{500} for the SCC-DC-FA and SCC-DC-CS mixes are classified as VS2 according to EN 206-9 and AC-V1 according to the EFNARC, with values at the limit of the recommendations. On the other hand, the mixes with limestone filler and slag coarse fall outside the categories proposed by the recommendations, the latter due to the viscosity of the paste itself, since a greater amount of fines is used. The slump test of the SCC-SC-CS mix can be seen in the following link: https://www.youtube.com/watch?v=rZUtHjiP4dw.

Table 6. Rheological properties of the self-compacting concrete in fresh state.

Concrete	V Funnel Test [s]	L-Box Test [%]	Slump Flow Test		
			D_1 [mm]	D_2 [mm]	t_{500} [s]
SCC-DC-LF	80	0.89	740	770	9
SCC-DC-FA	46	0.84	740	730	7
SCC-DC-CS	89	0.91	760	780	8
SCC-SC-CS	136	0.80	690	680	10

The flowability is at odds with viscosity, incorporating high-density aggregates requires a viscous paste (high fine content and low w/c ratio) that prevents the segregation of aggregates. Increasing viscosity also means obtaining a higher volume of occluded air or a poorer surface finish, which is why it is important to establish a compromise between both variables. In Figure 2, the correct distribution of aggregates in all mixes can be observed. In the case of SCC-DC-LF, areas with higher aggregate concentration and a mortar band of about 10 mm thick and slight bleeding are observed on the perimeter of the disk. The behavior of the SCC-DC-FA is more stable, favored by the shape of the fly ash particles that provide greater homogeneity in slump. Although it also has a mortar band about 10 mm thick, no bleeding is reported. In the case of SCC-DC-CS there is a homogeneous distribution of aggregates but a decrease in the viscosity of the paste and consequently greater bleeding and exudation due to incorporating the vitreous powder. The disk of the SCC-SC-CS shows a very homogeneous, symmetrical, and stable distribution. There is no segregation or concentration of coarse aggregates despite the high density of slags although there is a mortar band of about 10 mm around the disk.

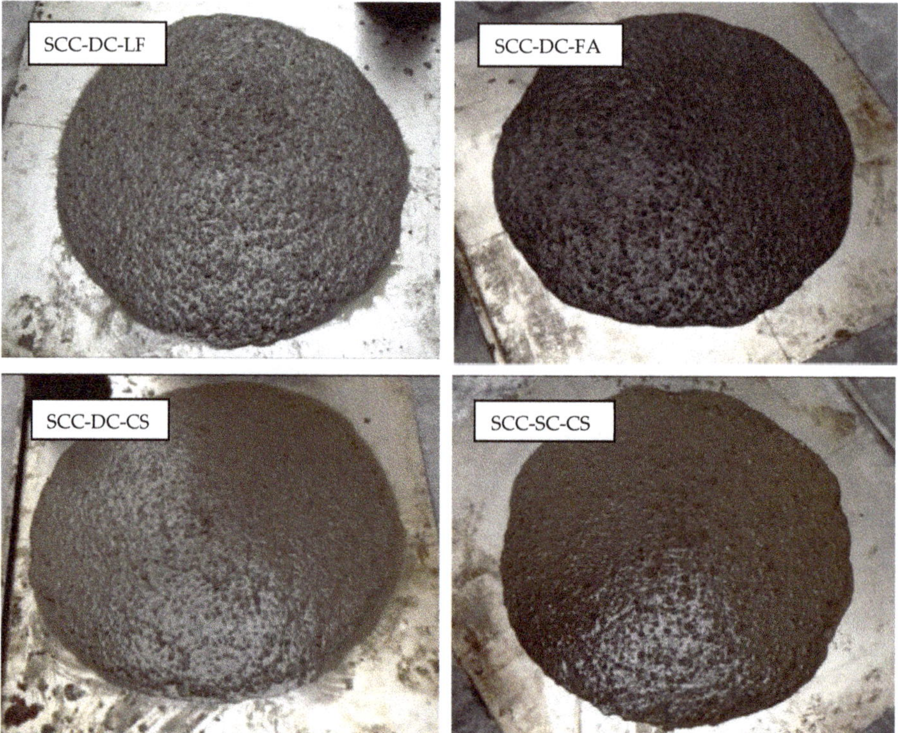

Figure 2. Slump disks detail of all mixes.

The values obtained from PL are consistent with the slump measured. The results of the L-box test (Table 6) show a passing capacity (PL) greater than 0.8 in all cases, which allows the mixes to be classified as PL1 according to EN 206-9 and as AC-RB2 according to EFNARC. In none of the cases has there been blockage in the bars and it should be noted that the SCC-SC-CS mix has shown a flow rate through confined spaces well below the other mixes, taking twice as long to stabilize the final height H2. This low-passing speed is associated with the high friction in the sliding between layers of paste with slag sand and cupola filler.

The results for the V funnel test also appear in Table 6. As was already apparent in the L-box test, the greater times correspond to the mixes that use cupola slag filler and slag sand. The difference between using limestone filler and cupola slag filler is almost negligible, although the use of fly ash improves the mobility of the mixture due to its spherical shape. On the other hand, the values obtained with the SCC-SC-CS are 70% higher than with SCC-DC-LF and SCC-DC-SC, a value that rises to 300% higher in comparison with SCC-SC-FA. The SCC-SC-CS is the most affected in this parameter due to the high viscosity of the paste, this viscosity also generates thixotropy gelation, observed after keeping the fresh mixture at rest. In all mixes the values are higher than the 25 s established by EN 206-9 for conventional SCC due to the high coarse content, which makes it difficult to pass through the funnel. Therefore, the designed mixtures are not recommended for highly confined areas, obliging modification of the mixture.

3.2. Physical Properties of the Concrete Mixes

The physical properties of the concrete mixes are shown in Table 7. The high density of EAFS gives the concrete with slags and the mixture of cupola slag around 15% more density than the reference concrete with diabase aggregates, except the mix that contains fly ash, where the ratio goes up to 20%. This is partly due to the low actual density of the fly ash (2.13 g/cm^3).

Table 7. Physical properties of the concrete mixes.

Concrete	Apparent Specific Gravity [g/cm^3]	Bulk Specific Gravity [g/cm^3]	Bulk SSD Specific Gravity [g/cm^3]	Open Porosity [vol.%]	Water Absorption [wt.%]
SCC-DC-LF	2.47	2.63	2.53	5.98	2.41
SCC-DC-FA	2.29	2.58	2.4	11.59	5.08
SCC-DC-CS	2.48	2.66	2.55	6.9	2.78
SCC-SC-CS	2.88	3.03	2.93	5.21	1.82

Due to the difference in densities between the mixes, it is ideal to perform a comparison between the open porosity (vol.%) and not the water absorption (wt.%). The mix with the lowest open porosity is the one that uses EAFS and cupola slag due, in part, to having a slightly lower w/c ratio and due to the higher density of aggregate particles, which facilitate the expulsion of trapped air (greater self-compaction). Conversely, the mix with greater open porosity is the one that uses diabase coarse and fly ash (220% greater than the previous one), this is due to a reaction between the plasticizer additive and the fly ash, which generates bubbles of around 1 mm.

In the comparison between SCC-DC-LF and SCC-DC-CS, it can be concluded that there are no significant differences in any of the properties shown between the limestone filler and the cupola slag SCM, it being necessary to analyze the mechanical properties of both mixes to establish the benefits of each filler. The visual aspect of each mix in the hardened state is shown in Figure 3, where a better orientation of the aggregates in the filling can be seen in the SCC-SC-CS mix.

Figure 3. Appearance of the concrete mixes.

3.3. Mechanical Properties of the Mortars

Figure 4 shows the results of the mechanical properties of mortars that incorporate several replacements of Portland cement. Compressive strength increases with the age for all replacements following a logarithmic trend ($R^2 \approx 0.85$), while intermediate replacements shows intermediate behavior between maximum replacement and reference mortars. After 7 days, mortars with a 30% replacement show a 27% loss in compressive strength with respect to the reference mortar, showing a slow rate of hydration reactions. This loss is reduced to 12% at 28 days and 5% at 60 days. At 90 days, the strength shown for all replacements tends to converge because the start of hydration of the alite and the peak of hydration occur much later when using cupola slag SCM than when using Portland cement (the speed of the reactions is slower when using cupola slag) [38].

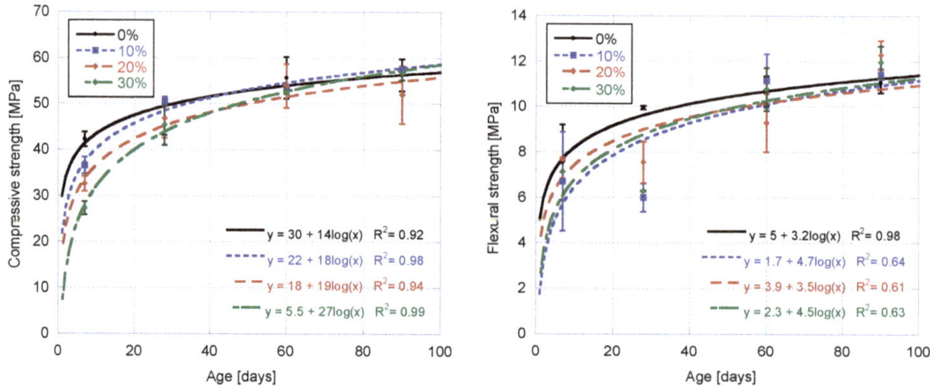

Figure 4. Mechanical properties of mortars with several cement replacements with cupola slag.

Flexural strength (Figure 4) also increases with age for all replacements and analogously to compressive strength. After 7 days, the strength shown by 30% replacements is 25% lower than that obtained with the reference mortars. For 28 and 60 days the losses reach 10% and 6% respectively. Again, there is a development of strength with age, more pronounced when cupola slag SCM is incorporated into the mortar. For ages over 90 days the values tend to converge again, showing the slowness of the reactions and demonstrating the strong pozzolanic character of the cupola slag.

3.4. Mechanical Properties of the Concrete Mixes

The evolution of the compressive strength over time for the different mixes is shown in Figure 5. Regarding the SCM's, comparing the 3 mixes that use natural aggregates (identical proportions), it is observed that the cupola slag (SCC-SC-CS and SCC-DC-CS mixes) provides more strength than the limestone filler and the fly ash (SCC-DC-LF and –SCC-DC-FA), both at short and long ages, despite the instability of the mix. This demonstrates the pozzolanic character of the cupola slag, showing a 13% greater compressive strength at 28 days and 11% greater at 360 days than both mixes and showing once again the importance of the promotion of the pozzolanic reaction by the SCM's in high-performance self-compacting concrete [39]. Of these last two (SCC-DC-LF and –SCC-DC-FA), it is surprising that with the fly ash similar strength is achieved as with the limestone filler, despite the greater porosity of the first, due to the high alkalinity of concrete, which favors the reaction of the SCM. In the comparison of the SCC-SC-CS with the rest of the mixes, great improvements in compressive strength are observed at all ages due mainly to three factors: better compressive strength of EAFS, higher coefficient of friction between particles due to the EAFS roughness, and a better bond between paste and aggregates. Comparing SCC-SC-CS with SCC-DC-CS, there is an increase in resistance of 5% at 28 days and 13% at 360 days due to the incorporation of the EAFS aggregate and a slight reduction in the w/c ratio. Comparing SCC-SC-CS with SCC-DC-LF and SCC-DC-FA, the increase is 20% at 28 days and 25% at 360 days. Figure 6 (obtained by scanning electron microscopy) shows the appearance of the cupola slag particles at 5000× after reacting, generating compounds in the form of parallel hexagonal plates, perfectly integrated in the paste and in combination with the cement hydration products. The composition of these plates, determined by EDX, corresponds to hydrated calcium aluminosilicates.

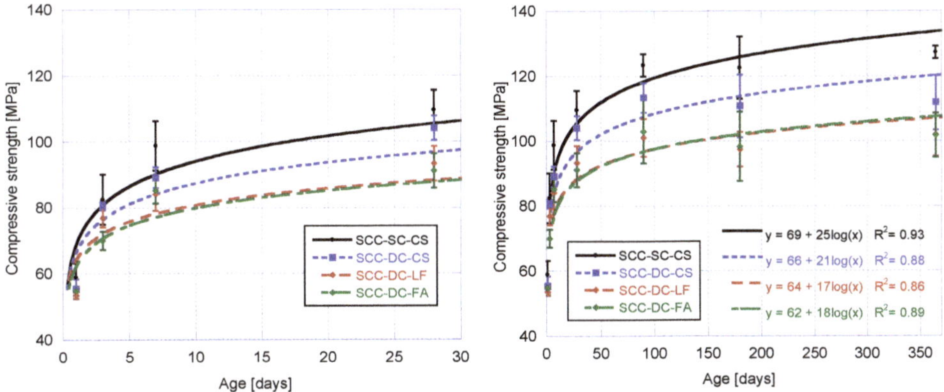

Figure 5. Evolution of the compressive strength for the different mixes: the first at 28 days (left) and the second at 360 days (right).

Figure 6. Appearance of the cupola slag after reacting on the cement paste (5000×).

In absolute terms, the mixes that incorporate cupola slag exceed 100 MPa at 28 days while the rest of the mixes are over 90 MPa, all of which can be considered high-performance concretes. From 28 to 360 days, the most evolved mix is SCC-SC-CS (16.5%) followed by SCC-DC-FA (12%) and SCC-DC-LF (7.5%) and SCC-DC-CS (7.5%). Also, at 360 days, the mix with cupola SCM and EAFS reaches 130 MPa. Comparing Figures 4 and 5, a remarkable increase in strength can be observed when cupola slag is used as a filler, rather than a cement substitute, which indicates cupola slag's potential as a reactive filler.

In Figure 7, the appearance of some of the cracking of the cubic specimens tested at 365 days is presented. As can be seen, due to the high stress applied to these concretes and the excellent paste–aggregate interface, the crack planes have propagated through the paste and aggregates, releasing all their energy through explosive cracks.

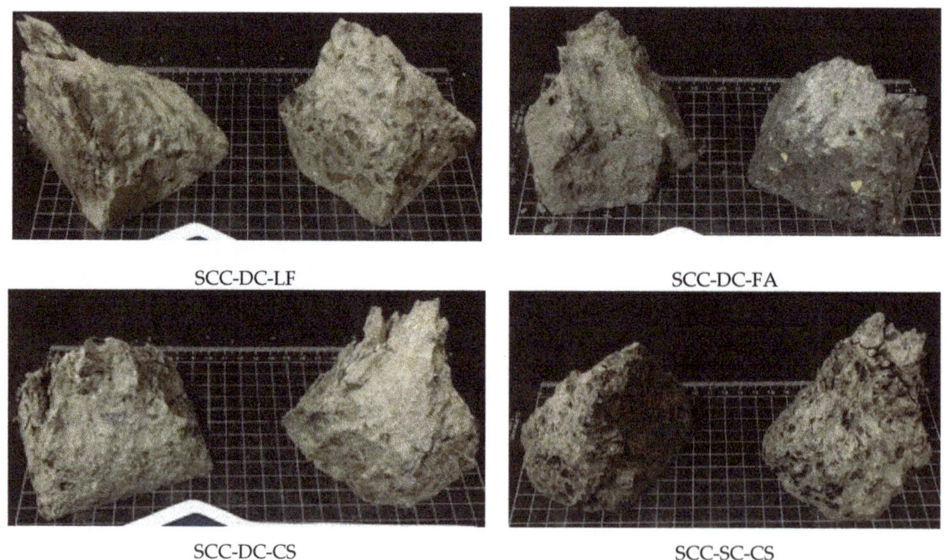

Figure 7. Detail of the compressive strength cracking (365 days).

The analysis of the cracking behavior of these materials is important, and in addition to the visual analysis, there are methodologies to determine the cracking load from deflection–load curves in flexural tests [40].

The evolution of the elastic modulus over time is shown in Figure 8. Among the mixes that use diabase aggregate, it is observed that SCC-DC-FA has the smallest elastic modulus (37 and 40 GPa at 28 and 360 days), which is due to the great porosity of the mix, allowing greater deformations. The SCC-DC-CS mix shows a modulus 10% higher than SCC-DC-LF, having a slightly lower porosity, due to the pozzolanic character and the greater Blaine surface of the cupola slag SCM. On the other hand, the SCC-SC-CS mix shows an elastic modulus far superior to the rest of the mixes (56 and 59 GPa at 28 and 360 days), being 24% and 18% higher than for SCC-DC-CS at 28 and 360 days respectively. This is due to the high EAFS modulus (and its iron nature), a slightly lower w/c ratio and a 25% lower porosity (Table 1). Compared with compressive strength, the elastic modulus has increased less with age, obtaining at 7 days approximately 90% of the final elastic modulus in all mixes.

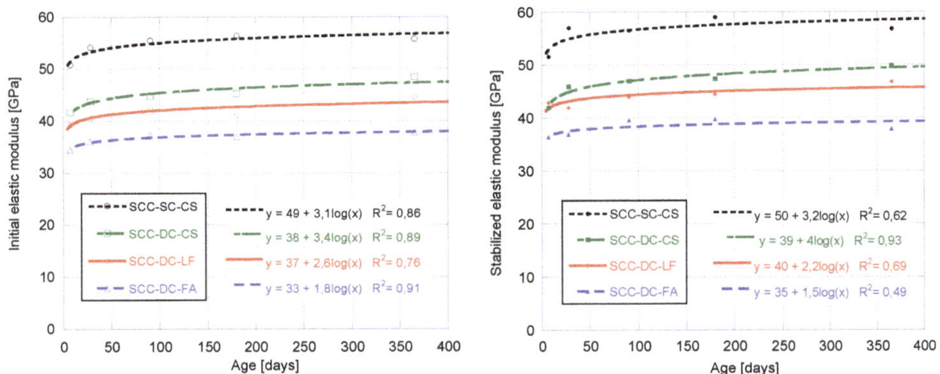

Figure 8. Evolution of the initial elastic modulus (**left**) and the stabilized elastic modulus (**right**) for the different concrete mixes.

On the other hand, the stabilized elastic modulus is slightly higher in all cases, since the material undergoes small permanent deformations for loads below the elastic limit in all cycles

The results of tensile splitting (Brazilian) strength for all mixes at 90 days are presented in Figure 9. Among all the mixes that use natural aggregate, it is observed that SCC-DC-FA has the lowest tensile splitting strength (4.5 MPa), due to a high porosity and a lower bonding in the interface transition zone favored by the smooth surface of the fly ash particles. The SCC-DC-LF mix is the next one with the highest tensile splitting strength (5.4 MPa), slightly lower than SCC-DC-CS (5.7 MPa), where once again the reactive character of the cupola slag SCM shows an improvement of the mechanical properties. For this property, there are no differences between using natural aggregate or siderurgical aggregate, as shown by the comparison between SCC-DC-CS and SCC-SC-CS (5.6 MPa), showing that the bonding strength in the interfacial transition zone (ITZ) is similar (good quality) for both aggregates.

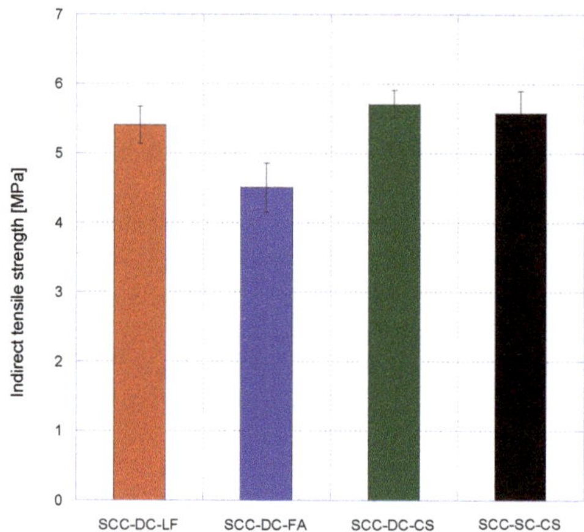

Figure 9. Tensile splitting strength at 90 days for all mixes.

The main mechanical properties of these mixes have been established, however, as future work it is proposed to check the functionality of conventional mechanical models for these materials and the creation of new models if necessary. These models will allow predicting the post-crack behavior of the material and the damage mechanisms for complex structures, using finite-element numerical approaches, as they are widely addressed in the literature [41].

4. Conclusions

This research deals with the recycling of two by-products, cupola slag and electric arc furnace slag, to obtain high-performance self-compacting concrete and the comparison of these with similar concrete mixes using high-quality natural aggregates (diabase) and the most common SCMs. After analyzing the rheological and physical-mechanical results obtained, the following conclusions can be drawn:

- It is possible to obtain a viscous paste that is capable of maintaining high-density EAFS particles in suspension by incorporating fillers. A very homogeneous, symmetrical, and stable distribution of the mixture has been achieved, without segregation or concentration of coarse aggregates, despite the high density of slags. In turn, EAFS reduces concrete slump and passing ability by 10%, while the flow rates are reduced more noticeably compared to reference mixes, due to their high viscosity.
- The incorporation of EAFS coarse and fine aggregates increases the density by 15% with respect to the reference mixes, obtaining high density concretes (>2800 kg/m^3). The open porosity of all mixes is very similar (5–6%) except for the mix that uses fly ash, due to a gasifying reaction with the superplasticizer additive. Likewise, the high content of rapid cement and the low w/c ratio have resulted in all mixes thixotropy gelation of the paste, more pronounced effect on concrete with EAFS.
- For the different cement replacements by cupola slag in mortars, it is worth mentioning the low speed of the reactions of the cupola slag, showing compressive and flexural strengths similar to mortars without replacement for ages of 90 days.
- The incorporation of cupola slag leads to an increase in compressive strength after 28 days of about 10% with respect to other SCM's. The use of EAFS aggregate provides a 5% improvement in the compressive strength with respect to diabase using the same SCM. The mixes that incorporate cupola slag SCM have evolved twice as much as the rest of the mixes from 28 to 360 days. The elastic modulus of mixes with EAFS is 24% higher than the mixes with diabase and the elastic

modulus of mixes that use slag cupola powder is 10% higher than the mixes with traditional fillers. In the case of the elastic modulus, practically 90% of the value at 360 days is obtained after 7 days. No significant differences were found in the tensile splitting strength at 90 days for all the mixes.

Author Contributions: Conceptualization, J.A.P. and C.T.; methodology, C.T. and I.S.; validation, J.A.P. and J.S.; formal analysis, I.S. and P.T.; investigation, I.S.; data curation, I.S.; writing—original draft preparation, I.S. and P.T.; writing—review and editing, C.T. and P.T.; visualization, I.S.; supervision, J.A.P. and C.T.; project administration, C.T. and J.A.P. All authors have read and agreed to the published version of the manuscript.

Funding: This research received no external funding.

Acknowledgments: The authors of this research would like to thank GLOBAL STEEL WIRE for the EAF slag supply and Saint Gobain Pam España for the Cupola Furnace Slag as well as ROCACERO for providing the cement, natural aggregates and superplasticizer additive.

Conflicts of Interest: The authors declare no conflict of interest.

References

1. World Steel Association (Worldtsteel), World Steel in Figures. 2018. Available online: https://www.worldsteel.org/en/dam/jcr:f9359dff-9546-4d6b-bed0-996201185b12/World+Steel+in+Figures+2018.pdf (accessed on 17 December 2019).

2. Fiol, F.; Thomas, C.; Muñoz, C.; Ortega-López, V.; Manso, J.M. The influence of recycled aggregates from precast elements on the mechanical properties of structural self-compacting concrete. *Constr. Build. Mater.* **2018**, *182*, 309–323. [CrossRef]

3. Martuscelli, C.C.; Santos, J.C.d.; Oliveira, P.R.; Panzera, T.H.; Aguilar, M.T.P.; Garcia, C.T. Polymer-cementitious composites containing recycled rubber particles. *Constr. Build. Mater.* **2018**, *170*, 446–454. [CrossRef]

4. Norambuena-Contreras, J.; Thomas, C.; Borinaga-Treviño, R.; Lombillo, I. Influence of recycled carbon powder waste addition on the physical and mechanical properties of cement pastes. *Mater. Struct. Constr.* **2016**, *49*, 5147–5159. [CrossRef]

5. Rojas, M.F.; de Rojas, M.I.S. Chemical assessment of the electric arc furnace slag as construction material: Expansive compounds. *Cem. Concr. Res.* **2004**, *34*, 1881–1888. [CrossRef]

6. Muhmood, L.; Vitta, S.; Venkateswaran, D. Cementitious and pozzolanic behavior of electric arc furnace steel slags. *Cem. Concr. Res.* **2009**, *39*, 102–109. [CrossRef]

7. Manso, J.M.; Gonzalez, J.J.; Polanco, J.A. Electric arc furnace slag in concrete. *J. Mater. Civ. Eng.* **2004**, *16*, 639–645. [CrossRef]

8. Manso, J.M. Manufacture of Hydraulic Concrete with Electric Arc Furnace Slags. 2001. Available online: https://dialnet.unirioja.es/servlet/tesis?codigo=246266 (accessed on 17 December 2019).

9. Sainz-Aja, J.; Carrascal, I.; Polanco, J.A.; Thomas, C.; Sosa, I.; Casado, J.; Diego, S. Self-compacting recycled aggregate concrete using out-of-service railway superstructure wastes. *J. Clean. Prod.* **2019**, *230*, 945–955. [CrossRef]

10. Sainz-Aja, J.; Carrascal, I.; Polanco, J.A.; Thomas, C. Fatigue failure micromechanisms in recycled aggregate mortar by μCT analysis. *J. Build. Eng.* **2020**, *28*, 101027. [CrossRef]

11. Thomas, C.; Setién, J.; Polanco, J.A.; de Brito, J.; Fiol, F. Micro- and macro-porosity of dry- and saturated-state recycled aggregate concrete. *J. Clean. Prod.* **2019**, *211*, 932–940. [CrossRef]

12. Thomas, C.; Setién, J.; Polanco, J.A.; Cimentada, A.I.; Medina, C. Influence of curing conditions on recycled aggregate concrete. *Constr. Build. Mater.* **2018**, *172*, 618–625. [CrossRef]

13. Thomas, C.; Setién, J.; Polanco, J.A.; Alaejos, P.; de Juan, M.S. Durability of recycled aggregate concrete. *Constr. Build. Mater.* **2013**, *40*, 1054–1065. [CrossRef]

14. Uysal, M.; Yilmaz, K. Effect of mineral admixtures on properties of self-compacting concrete. *Cem. Concr. Compos.* **2011**, *33*, 771–776. [CrossRef]

15. Puertas, F.; Santos, H.; Palacios, M.; Martínez-Ramírez, S. Polycarboxylate superplasticiser admixtures: Effect on hydration, microstructure and rheological behaviour in cement pastes. *Adv. Cem. Res.* **2005**, *17*, 77–89. [CrossRef]

16. Ye, G.; Liu, X.; de Schutter, G.; Poppe, A.-M.; Taerwe, L. Influence of limestone powder used as filler in SCC on hydration and microstructure of cement pastes. *Cem. Concr. Compos.* **2007**, *29*, 94–102. [CrossRef]

17. Ramezanianpour, A.A.; Ghiasvand, E.; Nickseresht, I.; Mahdikhani, M.; Moodi, F. Influence of various amounts of limestone powder on performance of Portland limestone cement concretes. *Cem. Concr. Compos.* **2009**, *31*, 715–720. [CrossRef]

18. Domone, P.L. Self-compacting concrete: An analysis of 11 years of case studies. *Cem. Concr. Compos.* **2006**, *28*, 197–208. [CrossRef]

19. Khayat, K.; de Schutter, G. *Mechanical Properties of Self-Compacting Concrete*; State of the Art Report RILEM TC 228-MPS; Springer: Berlin/Heidelberg, Germany, 2014.

20. Rozière, E.; Granger, S.; Turcry, P.; Loukili, A. Influence of paste volume on shrinkage cracking and fracture properties of self-compacting concrete. *Cem. Concr. Compos.* **2007**, *29*, 626–636. [CrossRef]

21. Loser, R.; Leemann, A. Shrinkage and restrained shrinkage cracking of self-compacting concrete compared to conventionally vibrated concrete. *Mater. Struct.* **2009**, *42*, 71–82. [CrossRef]

22. Arum, C.; Mark, G.O. Partial Replacement of Portland Cement by Granulated Cupola Slag—Sustainable Option for Concrete of Low Permeability. *Civ. Environ. Res.* **2014**, *6*, 17–26.

23. Domone, P.L. A review of the hardened mechanical properties of self-compacting concrete. *Cem. Concr. Compos.* **2006**, *29*, 1–12. [CrossRef]

24. EFNARC. *Specification and Guidelines for Self-Compacting Concrete*; European Federation for Specialist Construction Chemicals and Concrete Systems: Norfolk, UK, 2005.

25. Holschemacher, K. Hardened material properties of self-compacting concrete. *J. Civ. Eng. Manag.* **2004**, *10*, 261–266. [CrossRef]

26. Papayianni, I.; Anastasiou, E. Production of high-strength concrete using high volume of industrial by-products. *Constr. Build. Mater.* **2010**, *24*, 1412–1417. [CrossRef]

27. Arribas, I.; Santamaría, A.; Ruiz, E.; Ortega-López, V.; Manso, J.M. Electric arc furnace slag and its use in hydraulic concrete. *Constr. Build. Mater.* **2015**, *90*, 68–79. [CrossRef]

28. Sedran, T.; de Larrard, F.; Hourst, F.; Contamines, C. Mix Design of Self-Compacting Concrete. In Proceedings of the RILEM International Conference on Production Methods and Workability of Concrete, Glasgow, Scotland, 3–5 June 1996; pp. 439–450.

29. Pellegrino, C.; Cavagnis, P.; Faleschini, F.; Brunelli, K. Properties of concretes with Black/Oxidizing Electric Arc Furnace slag aggregate. *Cem. Concr. Compos.* **2013**, *37*, 232–240. [CrossRef]

30. Manso, J.M.; Polanco, J.A.; Losanez, M.; Gonzalez, J.J. Durability of concrete made with EAF slag as aggregate. *Cem. Concr. Compos.* **2006**, *28*, 528–534. [CrossRef]

31. Sheen, Y.-N.; Le, D.-H.; Sun, T.-H. Innovative usages of stainless steel slags in developing self-compacting concrete. *Constr. Build. Mater.* **2015**, *101*, 268–276. [CrossRef]

32. Santamaría, A.; Orbe, A.; Losañez, M.M.; Skaf, M.; Ortega-Lopez, V.; González, J.J. Self-compacting concrete incorporating electric arc-furnace steelmaking slag as aggregate. *Mater. Des.* **2017**, *115*, 179–193. [CrossRef]

33. Santamaría, A.; Ortega-López, V.; Skaf, M.; Chica, J.A.; Manso, J.M. The study of properties and behavior of self compacting concrete containing Electric Arc Furnace Slag (EAFS) as aggregate. *Ain Shams Eng. J.* **2019**. [CrossRef]

34. Qasrawi, H. Towards sustainable self-compacting concrete: Effect of recycled slag coarse aggregate on the fresh properties of SCC. *Adv. Civ. Eng.* **2018**, *2018*, 1–9. [CrossRef]

35. Dinakar, P.; Sethy, K.P.; Sahoo, U.C. Design of self-compacting concrete with ground granulated blast furnace slag. *Mater. Des.* **2013**, *43*, 161–169. [CrossRef]

36. Ministerio de Fomento de España (Spain), Spanish Structural Concrete Standard (EHE-08)—In Spanish. 2008. Available online: https://www.fomento.gob.es/organos-colegiados/mas-organos-colegiados/comision-permanente-del-hormigon/cph/instrucciones/ehe-08-version-en-castellano (accessed on 17 December 2019).

37. BIMB; CEMBUREAU; ERMCO; EFCA; EFNARC. The European Guidelines for Self-Compacting Concrete. Specification, Production and Use. 2005. Available online: https://www.theconcreteinitiative.eu/images/ECP_Documents/EuropeanGuidelinesSelfCompactingConcrete.pdf (accessed on 17 December 2019).

38. Stroup, W.W.; Stroup, R.D.; Fallin, J.H. Cupola Slag Cement Mixture and Methods of Making and Using the Same. U.S. Patent 6,521,039, 18 February 2003.

39. Kang, S.-H.; Hong, S.-G.; Moon, J. The use of rice husk ash as reactive filler in ultra-high performance concrete. *Cem. Concr. Res.* **2019**, *115*, 389–400. [CrossRef]

40. Stochino, F.; Pani, L.; Francesconi, L.; Mistretta, F. Cracking of Reinforced Recycled Concrete Slabs. *Int. J. Struct. Glass Adv. Mater. Res.* **2017**, *1*, 3–9. [CrossRef]
41. Amadio, C.; Bedon, C.; Fasan, M.; Pecce, M.R. Refined numerical modelling for the structural assessment of steel-concrete composite beam-to-column joints under seismic loads. *Eng. Struct.* **2017**, *138*, 394–409. [CrossRef]

Article

Mechanical and Durability Properties of Concrete with Coarse Recycled Aggregate Produced with Electric Arc Furnace Slag Concrete

Pablo Tamayo [1], Joao Pacheco [2], Carlos Thomas [1,*], Jorge de Brito [2] and Jokin Rico [3]

[1] LADICIM (Laboratory of Materials Science and Engineering), University of Cantabria. E.T.S. de Ingenieros de Caminos, Canales y Puertos, Av./Los Castros 44, 39005 Santander, Spain; pablo.tamayo@unican.es

[2] CERIS, DECivil, Instituto Superior Técnico, Universidade de Lisboa, Av. Rovisco Pais, 1049-001 Lisbon, Portugal; ceris@tecnico.ulisboa.pt (J.P.); jb@civil.ist.utl.pt (J.d.B.)

[3] INGECID S.L. (Ingeniería de la Construcción, Investigación y Desarrollo de Proyectos), E.T.S. de Ingenieros de Caminos, Canales y Puertos, Av./Los Castros 44, 39005 Santander, Spain; jokinrico@ingecid.es

* Correspondence: carlos.thomas@unican.es

Received: 19 November 2019; Accepted: 23 December 2019; Published: 27 December 2019

Abstract: The search for more sustainable construction materials, capable of complying with quality standards and current innovation policies, aimed at saving natural resources and reducing global pollution, is one of the greatest present societal challenges. In this study, an innovative recycled aggregate concrete (RAC) is designed and produced based on the use of a coarse recycled aggregate (CRA) crushing concrete with electric arc furnace slags as aggregate. These slags are a by-product of the steelmaking industry and their use, which avoids the use of natural aggregates, is a new trend in concrete and pavement technology. This paper has investigated the effects of incorporating this type of CRA in concrete at several replacement levels (0%, 20%, 50% and 100% by volume), by means of the physical, mechanical and durability characterization of the mixes. The analysis of the results has allowed the benefits and disadvantages of these new CRAs to be established, by comparing them with those of a natural aggregate concrete (NAC) mix (with 0% CRA incorporation) and with the data available in the literature for concrete made with more common CRA based on construction and demolition waste (CDW). Compared to NAC, similar compressive strength and tensile strength values for all replacement ratios have been obtained. The modulus of elasticity, the resistance to chloride penetration and the resistance to carbonation are less affected by these CRA than when CRA from CDW waste is used. Slight increases in bulk density over 7% were observed for total replacement. Overall, functionally good mechanical and durability properties have been obtained.

Keywords: recycled aggregate concrete; electric arc furnace slags; mechanical properties; durability

1. Introduction

Approximately 90% of construction and demolition wastes (CDW) are currently going to landfills even though they are potentially recyclable [1]. The use of this waste should be a priority to achieve the sustainable development objectives set by the European Commission, although this action is hindered due to lack of facilities and standards, lack of support from governments or lack of users' confidence [1,2]. The use of CDW as aggregates in concrete production, mostly coarse recycled aggregates (CRA), not only means a saving of natural resources derived from the extraction of aggregate, but also economic savings. Analogously, concrete with electric arc furnace slags (EAFS) as aggregate is based on the use of waste (from the steel industry) that would otherwise be deposited in landfills. In this case, the reduction of CO_2 emissions in the processes without taking into account the transport and manufacturing of the materials can be as high as 35% [3]. On the other hand, CO_2 emissions induced by concrete crushing are not very different from those generated in the production of natural aggregates [4].

To improve the understanding of the paper, the acronyms used and their meaning are shown in Table 1.

Table 1. Acronyms used in the article.

Acronym	Meaning
RA	Recycled aggregate
CRA	Coarse recycled aggregate
RCA	Recycled concrete aggregate
RAC	Recycled aggregate concrete
NA	Natural aggregate
NAC	Natural aggregate concrete
CDW	Construction and demolition waste
EAFS	Electric arc furnace slags

Most CRA are produced by crushing concrete that has ended its service life, i.e., they are composed mainly of natural stone and attached mortar. Typically, CRA are materials with lower density and porosity than natural aggregates (NA) because the attached mortar is less dense and more porous than the natural aggregate that it covers. The average density can be 8% lower and the average water absorption 5–6 times those of the natural aggregates [5,6]. According to the current Spanish concrete standards, the aggregates' water absorption must be less than 5% to be used in structural concrete [7,8]. According to Etxeberría et al. [9], the shape index of NA is 25% and 28% for CRA produced in quarries, although its value depends on the crushing process. Typically, laboratory-produced CRA are made with a single crushing stage (usually a jaw crusher), whilst NA are produced with multiple crushing (primary, secondary and sometimes tertiary). De Brito et al. [10] found that when CRA go through the same crushing process as NA their shape index is expected to be lower than that of NA. The water absorption and the shape index are vital for the calculation of the compensation water to determine the total water/cement (w/c) ratio [11].

There are many studies on the use of recycled aggregate (RA) in the production of structural concrete. Most of them only consider replacement of the coarse fraction of the aggregates, because the fine fraction has a great cohesion and water absorption that make it difficult to control the quality of the aggregates [12] and reduce the workability in the fresh state [13]. The risk of contamination of the finer fraction is also higher [14].

The use of CRA is more common also because it has less porosity and adhered paste. Typically, CRA concrete is around 4%–8% less dense [15–17], although this effect can be the opposite for high density CRA (e.g., CRA based on EAFS). The water absorption of these CRA concretes can be 500% higher than that of NA concrete (NAC) [18,19], although it tends to decrease due to the crystallization of hydration products, depending on the crushing age of the source concrete and curing conditions [20–22]. The fresh workability with CRA is lower than that of NAC for equal w/c effective ratios, so it is important to correct the water content to achieve similar slump without the help of admixtures [16,23–28].

One of the main characteristics of this concrete is the presence of three interfacial transition zone (ITZ). One is between the original aggregate of CRA and the cement paste in the source concrete and it is formed by dense hydrates, and in the case of EAFS concrete is of higher quality than with NA [29]. Another is between the old cement paste and the new paste, and the third ITZ is between the NA of the recycled aggregate and the new cement paste. The two ITZs between recycled aggregate (stone and mortar) and the new paste are where the chemical reactions between both generate loose and pore interfaces [30,31]. These ITZs are thus weaker and limit the mechanical properties of CRA [30]. Concrete with 100% CRA shows a loss in compressive strength with respect to coarse NA at 28 days, for the same effective w/c ratio and amount of cement, from 10% to 37% [9,17,27,32–34], approximately proportional to the replacement level [35,36] and depending on the relative strength of the new paste and CRA [37]. This decrease in compressive strength may make it necessary to use about 5% more cement to achieve the same strength as with NA, thus compromising the cost-effectiveness and sustainability of CRA [9,12]. From 28 to 91 days, the relative increase in compressive strength with recycled aggregate concrete (RAC) is sometimes greater than with NA due to the hydration of unhydrated cement grains [37]. The splitting tensile strength is typically lower but often not by as much as compressive strength, but there are studies where the tensile strength can even be higher with CRA than with NA [9,38], while the modulus of elasticity can decrease by 15% for 30% replacement and 45% for 100% replacement [35,39]. This reduction is because CRA has a lower modulus of elasticity than NA and due to the increase in the effective w/c ratio to maintain workability constant. Reductions in the modulus of elasticity result in increases in the peak strain of concrete under monotonic compression—[32] reports an increase of 20% for aggregate replacement of 100%. It is also known that the use of CRA but also the new cement paste affect the fatigue behaviour of concrete, reducing the fatigue limit and fatigue life [40,41]. Other authors have shown that the multi-recycling of the concrete that contain CRA is limited and that after three recycling cycles, CRA are mostly composed of mortar and new NA are needed in the mix design [42].

The drying shrinkage of concrete with CRA can be 50% higher [43] than with coarse NA, while chloride ions' penetration can reach up to 150% increases [24,44] due to the high permeability of this concrete. Resistance to carbonation is linked to the porosity of concrete [15] and, therefore, to the porosity of CRA, although it is also strongly linked to the chemical composition of concrete [24]. The high porosity of CRA makes carbonation depths increase between 22%–187% for 100% replacement [24,33,45] in comparison to NAC. Some authors propose the use of more crushing stages to eliminate the attached mortar and thus obtain rounder and less porous aggregates [45,46], the use of acids or heat to disaggregate the mortar of RCA [47,48], thermo-mechanical processes [49,50] or the use of several mechanical systems for on-site processing [51]. Another solution presented by the literature is the use of crushed bricks or steel slag [13,52] to compensate the loss of strength and durability, due to their pozzolanicity. However, these beneficiation techniques add new steps to the aggregate production process and increase production costs and the environmental impact of CRA production.

Currently, there is an increasing trend towards the use of steel slag in concrete [23]. The use of EAFS as CRA in concrete is a novelty (to the best of the authors' knowledge, it has never been done before) and its use is justified by the potential benefits of this aggregate and by the boom of its use in recent years mostly in road pavements or hydraulic structures [53]. The characteristics of the source concrete determine the behaviour of the recycled aggregates concrete. Concrete with EAFS offers an improvement in compressive strength by 50% compared to concrete with NA, a slightly higher modulus [54] and a generalized improvement in durability (low water absorption and permeability) [55,56], whereas the roughness of the aggregates allows improving the quality of the new ITZ of CRA. The quality of the CRA from EAFS concrete will compensate the loss of mechanical and durability properties that more common CRA provides, saving natural resources. This concrete has potential applications in foundations, plain concrete walls, or structures where a high self-weight is important (e.g., radiation-proof structures). The results of the physical-mechanical and durability tests, for concrete with coarse replacements of 0%, 20%, 50% and 100% by CRA, will be analysed and discussed, establishing the suitability of their use.

2. Materials and Methods

2.1. Materials

In this research, the NA used were: limestone gravel (2/6, 6/12 and 12/20 mm), and silica sand (0/2 and 0/4 mm) to produce the reference concrete. For the manufacture of RAC, CRA obtained from the crushing of concrete with EAFS (using a jaw crusher) 2 months old has been used. The resulting crushing material has a range of grading of 0/25 mm. This EAFS concrete has been manufactured with cement (CEM) I 52.5 R and a w/c ratio of 0.47. This source material presents at 28 days a compressive strength of 88 MPa, a modulus of elasticity of 52 GPa and an oxygen permeability of 6.48×10^{-18} m^2. The physical properties of both the NA and RA are shown in Table 2, after performing a measurement. The specific gravity and the water absorption have been determined according to EN 1097-6, the shape index was obtained following the EN 933-4 and Los Angeles wear has been determined according to EN 1097-2. The Portland cement used in RAC is CEM I 42.5 R (European standard), whose density is 3.15 g/cm^3 according to UNE 80103, and the mix has been made with tap water.

Table 2. Characterization of the aggregates.

Material	Bulk Density [g/cm^3]	Apparent Bulk Density [g/cm^3]	Water Absorption [%wt.]	Shape Index [%]	Los Angeles Wear [%]
12/20	2.66	1.36	1.3	14.5	26
6/12	2.66	1.38	1.5	19.7	28
2/6	2.66	1.41	1.0	16.4	-
0/4	2.67	1.54	0.3	-	-
0/2	2.67	1.57	0.2	-	-
CRA 4/20	2.96	1.52	3.5	13.4	26.9

The obtained water absorption of the CRA meets the requirements of the Spanish standard EHE-08 for structural concrete [7] and is more than twice that of coarse NA. However, this value is very small compared to more common CRA that normally exceeds 5% [5,33,57], although it depends on the size range. The shape index of the CRA is slightly lower than that of the NA and approximately one and a half times the values obtained by Etxeberría et al. [9] for conventional CRA. In terms of workability, these aggregates show a low shape index, but EAFS exposed surface is very cavernous and cause mesh between aggregates, demanding an extra volume of cement paste or mortar to fill the holes in their surface. However, the solid fraction of EAFS is generally much less absorbent than that of NA. Due to its properties, good mechanical and durability properties are expected [39].

2.2. Mix Design

The design of the mix has been made using the Faury method, to obtain maximum compactness. The maximum aggregate size has been set at 20 mm, in accordance with EHE-08. The coarse aggregates (>4 mm, EN 13139) of the reference concrete (NAC) have been replaced at several ratios (20%, 50%, and 100% vol.) with CRA. The content of cement has been set at 350 kg/m^3 and the effective w/c ratio of the reference concrete (NAC) at 0.5. The total w/c ratio has been determined by adding compensation water equal to that estimated for the mixing time (10 min) from the water absorption over time test, according to the method proposed by Rodrigues et al. [58]. The strength class of concrete has been defined as C30/37 in accordance with EN 1992-1-1. The slump has been defined as 70 ± 10 mm (S2) according to EN 12350-2 and without plasticizers (in order not to introduce more variables) for all replacement ratios. To maintain the same slump in all the mixes, the effective w/c has been slightly modified. RAC has been manufactured from the theoretical curve of NAC (Figure 1), maintaining between mixes the same volume of aggregates of each sieve fraction, so the mix grading for NAC and for RAC is the same. The mix proportions used are shown in Table 3.

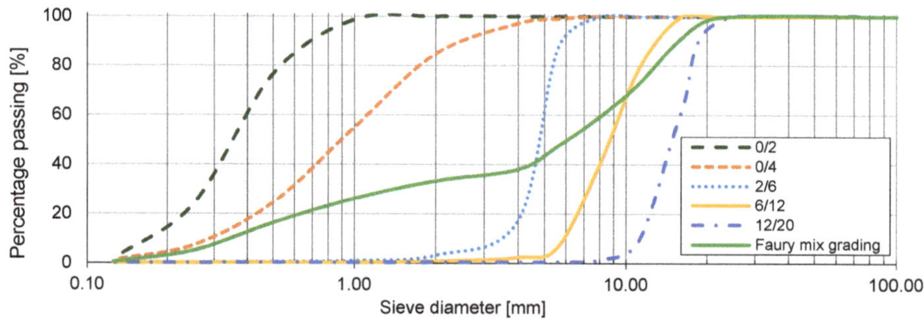

Figure 1. Theoretical curve and grading of the different aggregates according to EN 933-1.

Table 3. Concrete mix proportions.

Material/Property	Mix Proportions [kg/m³]			
CRA replacement	0%	20%	50%	100%
CEM I 42.5 R	350	350	350	350
Effective water	175	174.3	173.3	171.5
Compensation water	18.0	29.1	34.6	43.9
NA gravel (12/20)	434.4	347.5	217.19	-
NA gravel (6/12)	566.6	453.26	283.29	-
NA gravel (2/6)	207.8	173.67	122.1	36.2
NA sand (0/4)	417.1	417.1	417.1	417.1
NA sand (0/2)	265.4	265.4	265.4	265.4
CRA > 22.4 mm	-	1.5	3.9	7.7
CRA 16–22.4 mm	-	35.6	88.9	177.9
CRA 11.2–16 mm	-	76.7	191.8	383.6
CRA 8–11.2 mm	-	56.8	142,0	283.9
CRA 5.6–8 mm	-	49.8	124.6	249.2
CRA 4–5.6 mm	-	39.5	98.7	197.4
Effective w/c ratio	0.500	0.498	0.495	0.490
Slump [mm]	70	65	68	72
Fresh state density [kg/m³]	2424	2470	2563	2603

The aggregates were dried at 100 ± 2 °C until constant weight before mixing and the mixing process consisted of a sequence of 4 min with the coarse aggregates and 2/3 of the water, 2 more minutes after adding the fine aggregates and a further 4 min after adding the cement and 1/3 of the water. The concrete was demoulded after 24 h of manufacture and it has been cured in a humidity chamber at 20 ± 2 °C and 95 ± 2% humidity (except for drying shrinkage testing specimens).

A scheme illustrating the successive steps of RAC's manufacturing process is shown in Figure 2. The process describes the possible multi-recycling of RAC.

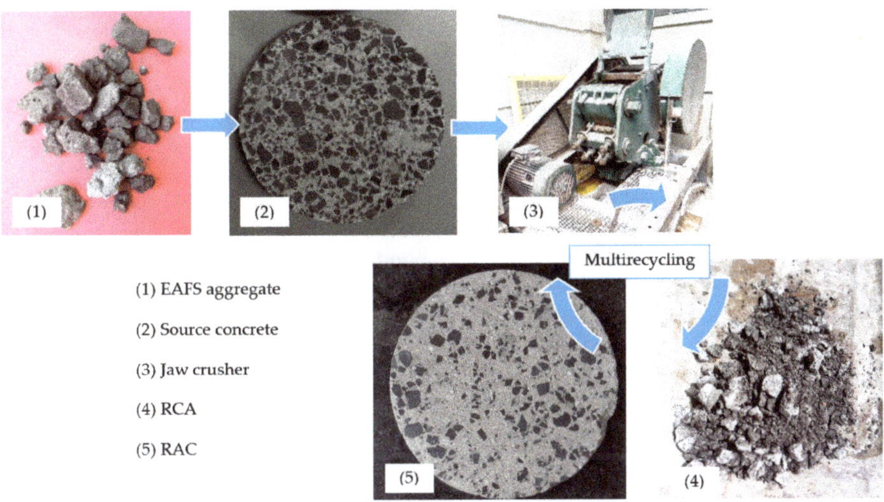

Figure 2. Recycled aggregate concrete (RAC) manufacturing process.

2.3. Physical Properties Tests

The concrete's absorption by capillarity has been determined on four cylindrical specimens with 150 mm diameter and 100 mm in length per mix proportion, after 28 days of curing in a humidity chamber and 14 days in an oven at 60 ± 5 °C. The test consists of measuring the mass evolution after 3,

6, 24 and 72 h of immersion, according to LNEC (National Laboratory for Civil Engineering) standards following the LNEC E-393. The absorption by immersion has been determined on four 100 mm cubic samples after 28 days of curing in a humidity chamber according to LNEC E-394. In addition, the apparent bulk, bulk, and saturated surface dry (SSD) density have been determined according to EN 12390-7 and the open porosity according to UNE 83980, for all the mixes produced.

2.4. Mechanical Properties Tests

The compressive strength (f_c) has been obtained on 150 mm cubic samples (EN 12390-1) per mix at ages of 7, 28 and 91 days. The test specimens have been tested using a load application rate of 0.6 MPa/s in a servo-hydraulic press of 3000 kN capacity and in accordance with EN 12390-3. The ultrasonic pulse velocity test has been performed on the specimens intended for the compressive strength test, prior to testing the compressive strength and just after their surface is dry. The measurement has been carried out according to EN 12504-4 with the transducers in direct transmission placed in collinear directions between two parallel faces with Vaseline on the contact surface between transducers and the concrete surface. The pulse velocity is calculated as the $\frac{length\ of\ specimen}{pulse\ travel\ time}$ ratio. The compressive modulus of elasticity (E) has been determined in a servo-hydraulic press of 250 kN capacity, on three cylindrical specimens with 150 mm and 300 mm in length per mix, after 28 days of curing in a humidity chamber. The upper and lower faces of the test specimens have been levelled before the test and a compressometer/extensometer equipped with high precision displacement transducers is used to measure the micro-deformation. Four loading/unloading cycles have been used, applying an initial stress of 1 MPa (17.6 kN) and a load application speed of 0.5 MPa/s (8.8 kN), using a maximum load of $f_c/3$ according to LNEC E-397. The splitting tensile strength has been determined on the three specimens used in the modulus of elasticity test, for all mixes. A servo-hydraulic press of 3000 kN capacity and a load rate of 0.05 MPa/s (3.5 kN/s) was used, according to EN 12390-6.

2.5. Durability Tests

The resistance to chloride-ion penetration was determined by calculating the diffusion coefficient by means of the depth of chlorides penetration into concrete, according to LNEC E-463. Three cylindrical specimens with 100 mm diameter and 50 mm in length per mix have been used for each of the ages (28 and 91 days) and mixes. The specimens were cured in a wet chamber and moved to a dry chamber (20 ± 2 °C and 60 ± 5% relative humidity) in the last 14 days before testing. Carbonation resistance of the concrete was determined on three cylindrical specimens with 100 diameter and 50 mm in length per mix and per exposure time, stored 14 days in a humidity chamber followed by 14 days in a dry chamber (20 ± 2 °C and 60 ± 5% humidity) before being placed for 7, 28, or 91 days in the carbonation chamber. The conditions of the carbonation chamber and the test methodology are those proposed by LNEC E-391 (temperature of 23 ± 3 °C, relative humidity of 60 ± 5%, and CO_2 concentration of 5.0 ± 0.1%). The determination of the carbonation depth was carried out with the help of a pH indicator (1% phenolphthalein solution in ethanol), cutting the specimen in quarters, spraying the solution and measuring the depth of carbonation penetration (the average depth measured in the eight contact surfaces of the broken specimen). Drying shrinkage was measured on two 100 × 100 × 500 mm prismatic specimens per mix and according to LNEC E-398, from 24 h to 91 days of age. The specimens were placed in a chamber at 20 ± 2 °C and 55 ± 5% relative humidity after demoulding and during the 91 days of testing.

3. Results and Discussion

3.1. Physical Properties

Table 4 shows the average physical properties of all the mixes produced and their standard deviation. All densities increase as the replacement does, since CRA has a bulk density 0.3 g/cm^3 higher than that of NA (Table 2). The increase in bulk density is close to 7% for 100% replacement and opposite to that obtained by other authors using more common RCA [20–22]. This increase in density allows a saving of volume with respect to NAC in applications where self-weight is important (i.e., bridge counterweights or seawalls). To compare the amount of voids in mixes with different density, the property to be analyzed is open porosity, since the water absorption is the $\frac{dry\ oven\ mass}{aparent\ volume}$ ratio and hence depends on the bulk density of the material. Open porosity increases by 18% for 100% replacement, which is approximately 70% of the relative increase obtained by Thomas et al. [15] for more common RCA mixes with a w/c ratio of 0.5. The increase in open porosity with respect to the NAC is due to RA being on average 160% more porous than coarse NA (Table 2) and coarse aggregate represents 44% of RAC's volume. This large increase is due to the porous nature of the source concrete mortar. Analogously, the increase in porosity obtained is relatively low because the source concrete has a low water/cement ratio (0.47).

Table 4. Physical properties of hardened concrete mixes.

Replacement	Bulk Density [g/cm³]	Apparent Density [g/cm³]	SSD Density [g/cm³]	Fresh State Density [g/cm³]	Open Porosity [% vol.]	Water Absorption [% wt.]
0%	2.280 ± 0.01	2.620 ± 0.008	2.410 ± 0.01	2.42	12.70 ± 0.28	5.56 ± 0.14
20%	2.300 ± 0.008	2.680 ± 0.018	2.440 ± 0.01	2.47	14.31 ± 0.31	6.23 ± 0.12
50%	2.390 ± 0.009	2.740 ± 0.01	2.520 ± 0.015	2.56	12.73 ± 0.41	5.33 ± 0.21
100%	2.460 ± 0.005	2.920 ± 0.018	2.620 ± 0.005	2.60	15.62 ± 0.53	6.34 ± 0.22

Figure 3 shows the capillarity absorption over time for the different replacement ratios used.

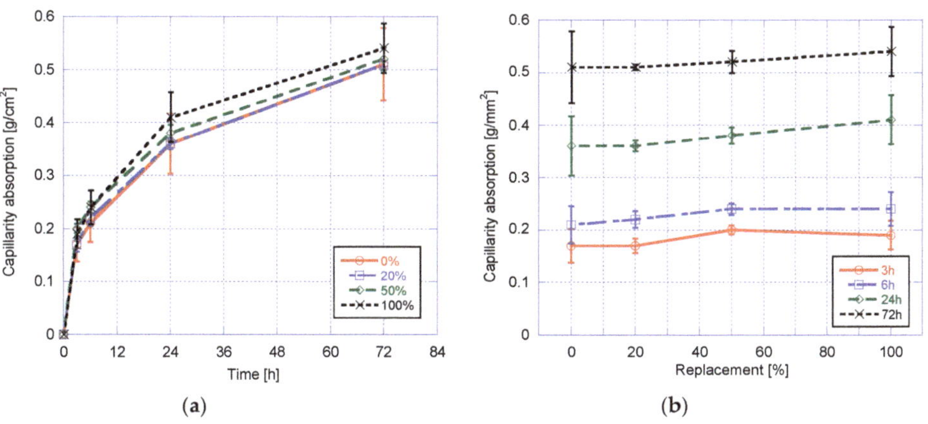

Figure 3. Capillarity absorption vs. time (**a**) and capillarity absorption vs. replacement ratio (**b**).

Capillarity absorption increases with immersion time (Figure 3a) and the level of replacement (Figure 3b). For 20% replacement, the variation in absorption with respect to NAC is negligible for any immersion time, while for 100% replacement the values may increase by 9% at 72 h of immersion. For replacement of 50%, intermediate capillary absorption increases of around 4% are obtained. For total replacement and 72 h immersion, other authors obtained increases of around 30% for both current RAC [19] and high-quality precast concrete RAC [10]. The higher absorption shown by RAC is due to the high porosity of RCA, which contains more and longer capillaries than NAC's. Figure 3a shows that the capillarity absorption difference is maximum at 24 h between 0% and 100% replacement and the difference is reduced at 72 h. This happens because the mix with 100% replacement has larger pores size (larger capillary pores saturate the first). The low water absorption of the study concrete is compatible with the minimum capillarity absorption obtained, which shows a loss very similar to that of ultrasonic pulse velocity, strongly related to the mechanical properties of the hardened concrete.

3.2. Mechanical Properties

There are several factors that positively and negatively affect the compressive strength of the study RAC: porosity of the cement paste, the new ITZ's quality and characteristics of the RCA. Porosity negatively affects strength. The greater porosity of RAC is due to the porosity of RCA and to that generated in the ITZ between RCA and the new paste [59]. This porosity largely depends on the w/c ratio of both the source concrete and RAC. This strength loss is not high for this RAC, since the porosity of mixes with a 100% replacement is only 18% higher than of NAC, because the source concrete has a low w/c ratio and RAC mixes have a lower effective w/c ratio than the source concrete. A lower quality ITZ has been reported by various authors [30,31] as the main cause of the strength loss of RAC, justifying that this element is the weakest link of the chain. The shape of the EAFS aggregates in RCA and the high compressive strength and stiffness of the source concrete minimize this effect. On the other hand, compared with NA, RCA's shape index is 23% lower and, even having a greater roughness, these aggregates allow using slightly lower effective w/c ratios to obtain the same workability, generating lower pores in the mortar and reducing the strength loss.

The result of these effects on the compressive strength is shown in Figure 4 where the evolution over time of the compressive strength for all mixes is shown.

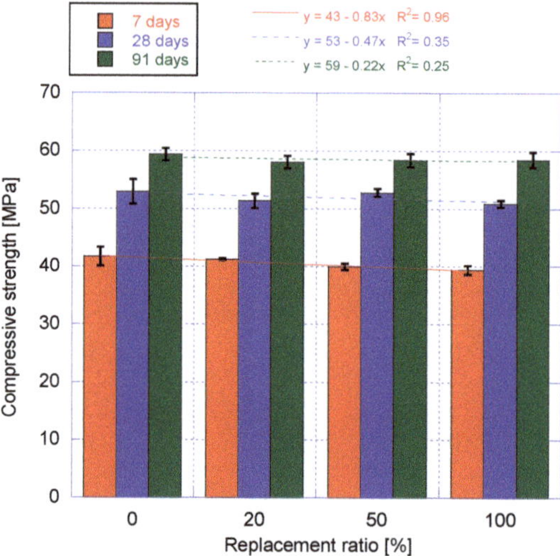

Figure 4. Compressive strength of all mixes at different ages.

At 7 days, the compressive strength decreases slightly and linearly as replacement increases. For 100% replacement, the strength loss with respect to NAC is 5.5%. At 28 days, the concrete class for all replacement is C35/45, exceeding the C30/37 class proposed at the design phase. At 28 days, a strength loss or gain depending on the replacement cannot be clearly defined, but a relative 7-day strength increase of 21% for NAC and 23% for RAC with 100% of RA is observed. This slightly higher relative increase for high replacement could be due to a hydration of the unhydrated cement grains of the old mortar, thus improving the bond in the ITZ with the new mortar. So the curing age is an important factor to check the effectiveness of these concrete mixes. At 91 days, the replacement level does not affect compressive strength. The slope of the linear adjustment is practically 0 and the standard deviation close to 1 for all replacements, proving an increase in compressive strength with age and replacement ratio, because the old mortar is still hydrating. In any case, the compressive strength obtained is much lower than that of the source concrete, although relatively higher than that obtained in other studies for current CDW [9,17,27,32–34] which on average lose around 10% compressive strength at 28 days with respect to NAC. This is due to the high quality of the RCA used in the present study, namely the high strength of the attached mortar of these CRA (due to the strength of the source concrete) and its low shape index.

The results of the splitting tensile strength tests are shown in Figure 5. The high standard deviation, which is typical of splitting tensile strength tests [33], and the similar mean values obtained show that the behaviour of RAC is very similar to NAC's for any replacement ratio. This happens because there is a good bond strength between RCA and the new paste, due to the higher roughness of RCA. This greater roughness seems to compensate for the weaker mortar surface generated in the crushing stage and exposed in the new ITZ. It also compensates for RAC not benefitting from the bond produced by the chemical interaction between calcium hydroxide and the calcareous aggregate [59] in NAC. As seen in Figure 6, fracture surfaces propagate through the aggregates (good bond strength in the ITZ), both for NAC and RAC. In the case of NAC, the crack surface is flatter, favored by the crystalline structure of calcite, while the fracture surface produced in CRA is irregular or lumpy with a smooth transition to the new mortar paste. Most of the macro-cracks present in RCA are produced in the mortar phase of the source concrete, even though sometimes they are also produced through the EAFS aggregate.

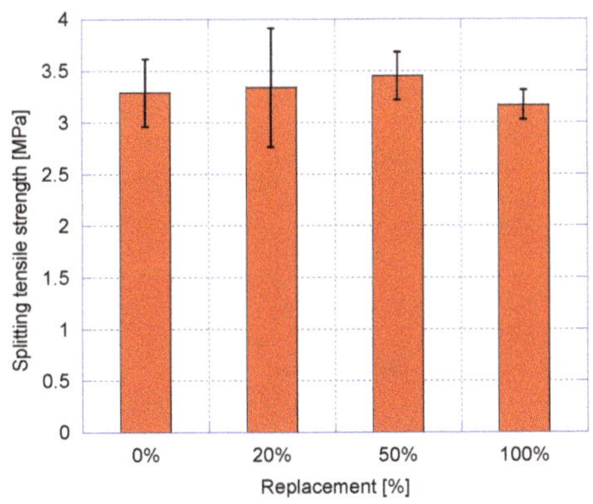

Figure 5. Tensile strength of all mixes at 28 days.

The splitting tensile strength values obtained by other authors for RAC produced with CDW are variable depending on the quality of RA. According to the classification proposed by Silva et al. [14], RA are classified in four classes according to their intrinsic properties (bulk density, water absorption and Los Angeles wear). For the most demanding category (Class A), the loss of tensile strength should be around 10% with respect to NAC [60]. The RA used in the present study, even though not complying with the water absorption requirements described for Class A, shows a tensile strength similar to NAC's.

Figure 7 shows the compressive modulus of elasticity for all mixes produced. Using a linear trend line, an inverse relationship between modulus of elasticity and replacement level can be established. While NAC shows a 40 GPa modulus of elasticity, the 100% replacement mix shows values around 33 GPa, 17% lower and with standard deviations not exceeding 0.5 MPa. This reduction is lower than that commonly reported and is related to the properties of the source concrete. The source concrete used to produce RCA has low w/c ratio, a modulus of elasticity greater than 50 GPa and a low porosity close to 10%. For similar w/c ratios and cement contents, other authors reported reductions from 15% to more than 40% [15,35,61] using RCA produced from common concrete waste.

Traditionally, density is a factor directly related to the modulus of elasticity because it is linked to the porosity of concrete [59]. The aggregate used is 7% denser than NA, but this is due more to the iron nature of RA than to the very slight increase of w/c ratio. The porosity of RCA influences its stiffness and hence the stiffness of concrete, since its increase allows greater deformations per unit of applied load.

Ultrasonic pulse velocity is intrinsically related to the modulus of elasticity and density. The ultrasonic pulse velocity obtained for all the mixes produced is shown in Figure 8. Following the trend line, it is observed that the pulse velocity decreases as the replacement ratio increases. There is a maximum loss of 6% for 100% replacement, slightly less than 9% determined by Khatib [52] for current RAC with the same w/c ratio. It should be remarked that the source concrete showed an ultrasonic pulse velocity of 5.22 km/s at the same age. On the other hand, the maximum loss obtained is similar to that obtained by de Brito et al. [10] for RAC produced from concrete precast rejects, which demonstrates the homogeneity of RAC and the high quality of the RCA used. The RCA used in this study has a density 10% and 20% higher than NA and of current RCA respectively, which a priori favours an increase in the ultrasonic pulse velocity. On the other hand, the porosity of the source concrete mortar allows RAC to strain more under sustained loads, obtaining smaller modulus of elasticity and lower ultrasonic pulse velocity than NAC, due to the discontinuities generated by the pores.

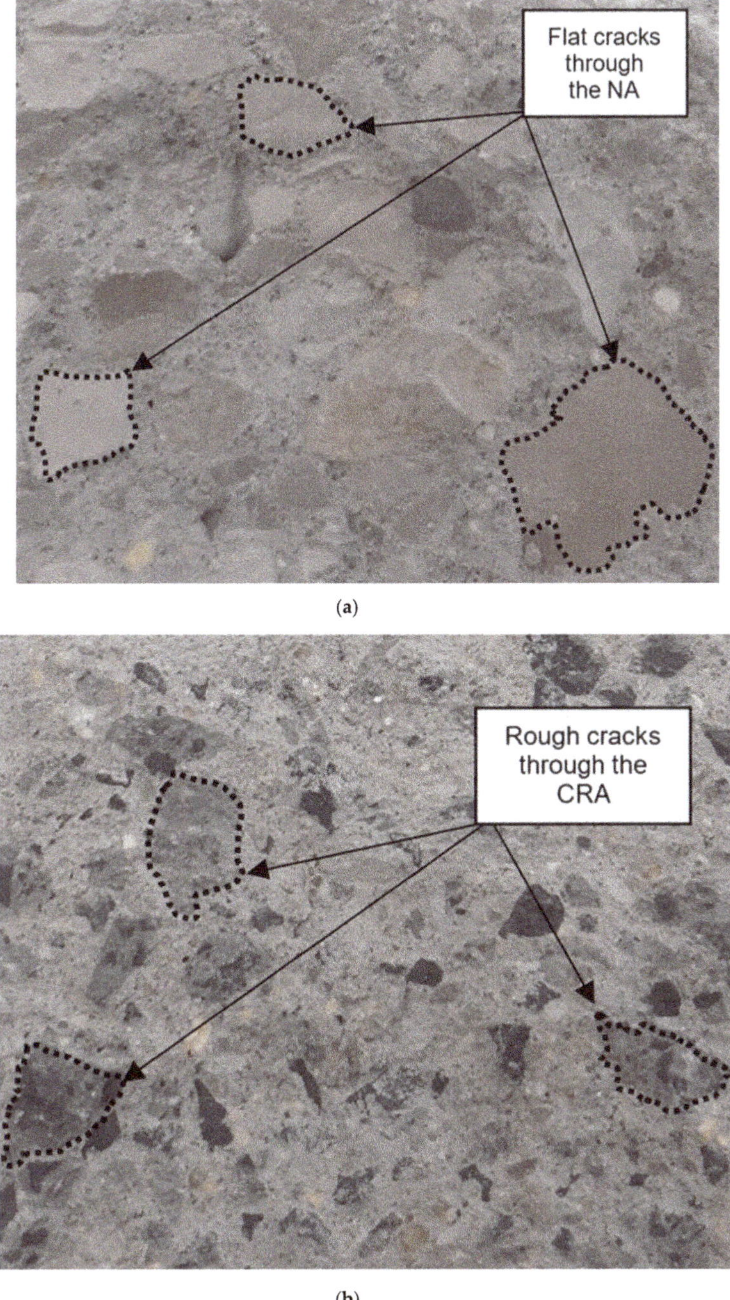

(a)

(b)

Figure 6. Crack patterns of test specimens subjected to tensile strength test for 0% (**a**) and 100% (**b**) replacement ratio.

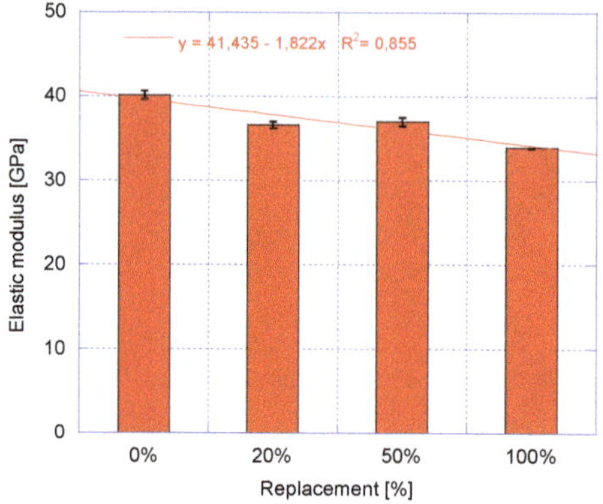

Figure 7. Modulus of elasticity of all mixes at 28 days.

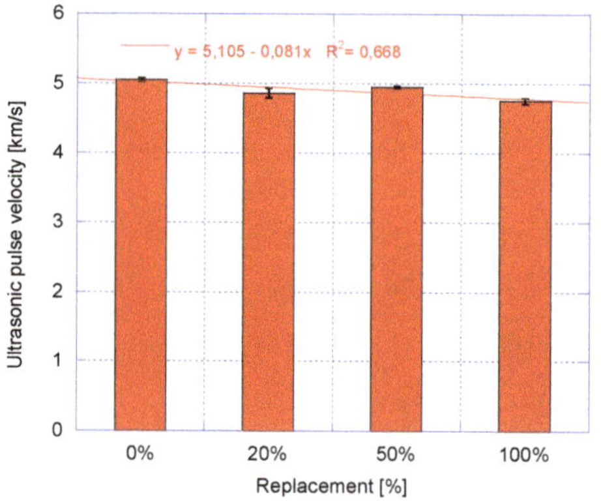

Figure 8. Ultrasonic pulse velocity at 28 days.

3.3. Durability

Carbonation is a phenomenon that alters the alkaline behaviour of concrete. When CO_2 from the atmosphere diffuses into the concrete matrix, it reacts with water forming carbonic acid and the latter reacts with portlandite forming calcium carbonate. This process reduces the pH of concrete, depassivating the reinforcement. Figure 9 shows the evolution of the carbonation depth over time of all mixes. At 7 days, the carbonation depth is very small due to the low w/c ratio of the mixes. The test scatter caused by reading tolerances is relevant but mixes with 100% and 50% replacements appear to be slightly less durable. However, at 28 days, a slight increase in carbonation depth with the increase in CRA incorporation can be observed, which can reach about 10% for total replacement. This trend continues at 91 days, when carbonation becomes more noticeable for higher replacements: 50 and 100%. For total replacement, carbonation depths can be 15% higher than for reference concrete.

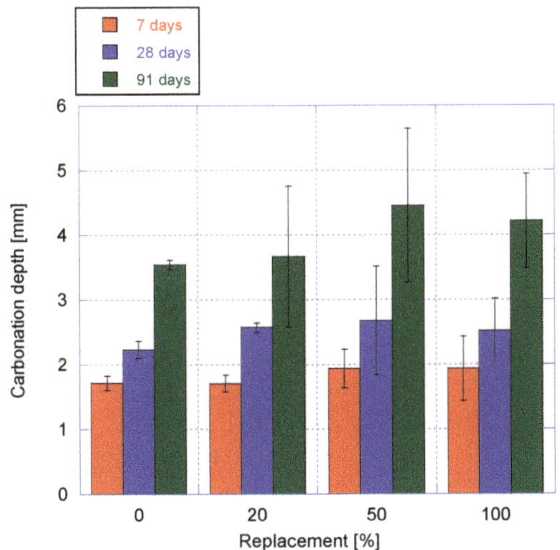

Figure 9. Carbonation depth for all mixes and ages.

The general increase in carbonation depth with replacement ratio and age is due to several causes. CRA is composed of the source concrete mortar, hence part of this cement is already carbonated. Another relevant fact is that the open porosity of RAC is 18% higher than that of NAC, which favours the increase of permeability and hence the diffusion of CO_2 through the cement matrix, since there is easier access to more paste surface. On the other hand, by having a greater volume of total mortar, there is also a greater amount of total portlandite compared to the reference concrete (a greater amount of calcium carbonate can be generated) and hence a greater amount of CO_2 is required for the same carbonation depth.

Other authors have also obtained increases in carbonation depth as the replacement level increases. For similar mixes and 100% replacement, de Brito et al. [10] obtained a marginal increase of the depth of carbonation for rejected high-quality precast concrete at all ages tested. Ryu et al. [62] obtained a slight increase in carbonation depth for similar mixes and similar compressive strengths, namely an increment of 14% for total replacement. Finally, in the review of Silva et al. [45], it is found that, for total replacement, the average carbonation depth may increase by a factor of 2 with respect to the control concrete, establishing the permeability of concrete, which depends directly on the water absorption (and therefore on the open porosity) of RCA, as a key factor.

Chlorides negatively affect concrete, reacting with tricalcium aluminate forming Friedel's salt (bound chlorides) and oxidizing the reinforcement after penetrating concrete (free chlorides) [63]. The resistance to the penetration of chlorides is shown in Figure 10. The chlorides diffusion coefficient increases with replacement ratio and decreases with age. At 28 days, there is a quasi linear increase in this coefficient, reaching 11% for total replacement. At 91 days, the increase is 9.5% for total replacement, due to the extra contribution of mortar from the source concrete that continues to hydrate. On the other hand, there is a large reduction in chlorides diffusion coefficient with age, a reduction greater than 25% for all replacements.

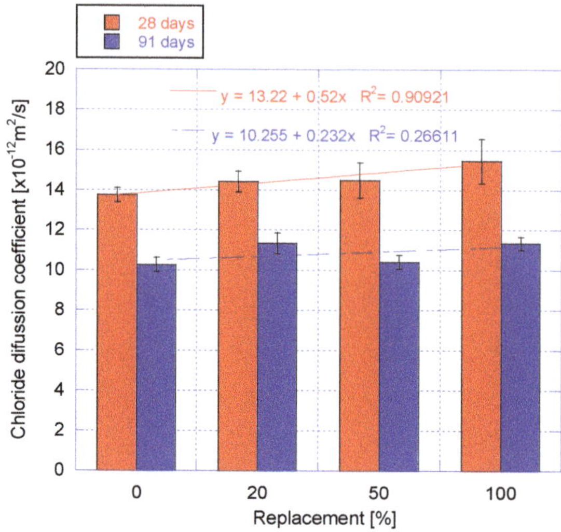

Figure 10. Chloride diffusion coefficient for all mixes and ages.

Chlorides penetration in concrete (an advective-diffusive phenomenon) largely depends on its porosity and the tortuosity of its capillary framework [64]. A strong correlation has been obtained between the chlorides diffusion coefficient and both open porosity ($R^2 = 0.73$) and capillarity absorption ($R^2 = 0.89$). Critical variables such as the content or type of cement remain constant, so it seems consistent that the higher ingress of chlorides in RAC with larger replacement is due to the increase in open porosity, owing to the higher volume of voids in RCA.

The incorporation of EAFS usually improves the chloride diffusion coefficient relative to the reference concrete by about 10% [56], so it is expected that the behaviour of RAC with RCA based on EAFS can show an improvement over mixes with current RCA. Likewise, concrete with current RCA generally have higher chloride diffusion coefficient than concrete produced with NA. Bravo et al. [24] obtained average increases of more than 40% in the chlorides diffusion coefficient for total replacement of CRA produced from unsorted CDW. Other studies, however, argue that this property is unrelated with the RA replacement level [10], stating that it depends mostly on the quality of CRA.

Drying shrinkage is a significant phenomenon during cement hydration. The difference in relative humidity between the environment and the specimen causes the water adsorbed to the hydrated paste to pass into the atmosphere following Fick's law. Drying the specimen involves a reduction in volume that can cause cracking of the paste in severe cases. The evolution of drying shrinkage for the different mixes is shown in Figure 11. The drying shrinkage obtained for RAC with total replacement is about 530 μm/m at 91 days, while for NAC it is 457 μm/m, so there is an increase of 13%.

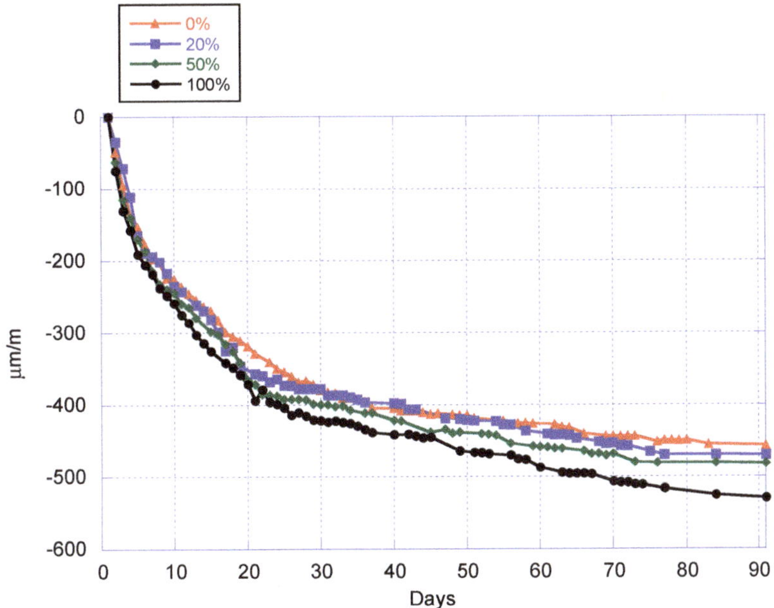

Figure 11. Mean drying shrinkage over time for all mixes.

It is well known that there is a strong relationship between drying shrinkage and modulus of elasticity in concrete [65]. The shape and texture of aggregates are variables that affect the drying shrinkage, although modulus of elasticity (or any of the variables that affect the latter) is the parameter that most affects the former [59]. An increase in modulus of elasticity generates a greater restriction against strain, i.e., stiffer aggregates more efficiently oppose the strain generated by shrinkage than more deformable ones. As stated before, the modulus of elasticity drops with increase of the replacement ratio (17% for full replacement) and drying shrinkage decreases similarly. Another factor to keep in mind is that RCA is still shrinking, especially if the source concrete is young concrete, while NA do not shrink. Recycling concrete that has already reached the end of its service life is not a problem, but in the case of laboratory-produced concrete (the case here) it is something to keep in mind. Shrinkage measured from RAC with laboratory-produced source concrete is thus expected to overestimate the shrinkage of RAC. Losses for total RCA replacement range from 10% [10] to 50% [43], the latter being the value established in the Xuping review as a typical value for RAC.

4. Conclusions

In this study, a new concrete based on the use of concrete waste has been tested. The concrete waste used incorporates a by-product of the steelmaking industry (EAFS), thus waste production and natural aggregate consumption are reduced. The physical, mechanical and durability properties of the new material have been characterized and compared with the results of a reference conventional concrete with the same mix design of the recycled aggregate concrete tested. Moreover, experiments from other authors on the properties of concrete with recycled aggregates from other sources were also compared with those of this experiment. The following conclusions were drawn:

- The use of RCA with a water absorption twice that of NA requires compensation water, but the rounded shape of the RCA produced from concrete with EAFS (shape index 20% lower than NAC) makes it possible to use slightly lower effective w/c ratio to obtain the same workability;

- The replacement level affects the physical properties of RAC. Bulk density increased by 7%, unlike in current RAC (typically density decreases), due to the iron present in this recycled aggregate. Both porosity and capillarity slightly increase (18% and 9% for full replacement respectively) as replacement increases due to the porosity of the source concrete mortar;

- Compressive strength is not affected by the replacement level for ages over 28 days due to the hydration of the unhydrated cement grains of the source concrete and to a good bond strength between RCA and the new mortar. This good bond, favoured by the cavernous form of EAFS in the source concrete, allows tensile strength to be very similar to that of NAC. On the other hand, the modulus of elasticity is the property most affected by the replacement of RCA, showing losses that can reach 17% with respect to NAC. Typical decreases of the modulus of elasticity caused by full recycled aggregate incorporation are higher than this value;

- Durability properties are negatively but slightly affected. Carbonation depths and chlorides diffusion coefficients are 15% and 25% higher respectively for full replacement at 91 days. Both properties are linked to the porosity of RAC and hence with the quality of the RCA used, which is reflected in a low w/c ratio and a good bond between aggregate and paste;

- Drying shrinkage values 13% higher for total replacements have been obtained at 91 days, consistent with the modulus of elasticity obtained;

- The mechanical and durability tests results obtained are satisfactory, usually near the lower limit of the interval of change found in the literature and sometimes showing a behaviour very similar to that of the reference NAC;

- Concrete designed by recycling an EAFS-based concrete has potential applications as a structural material and can also be used in applications where its self-weight is important, as in bridge counterweights, seawalls or radiation-proof structures.

Author Contributions: Data curation, J.d.B.; Formal analysis, P.T., J.P., C.T. and J.d.B.; Funding acquisition, C.T. and J.R.; Investigation, P.T., J.P. and C.T.; Methodology, C.T.; Project administration, C.T.; Resources, J.R.; Writing-original draft, P.T. and J.P.; Writing-review & editing, C.T. and J.d.B. All authors have read and agreed to the published version of the manuscript.

Funding: This research was co-financed by the European Regional Development Fund (ERDF) and the Ministry of Economy, Industry and Competitiveness (MINECO) within the framework of the project RTC-2016-5637-3 and in collaboration with the company INGECID and the department LADICIM (University of Cantabria).

Acknowledgments: The authors gratefully acknowledge the support of the CERIS Research Unit, IST, University of Lisbon and of FCT.

References

1. Tam, V.W.; Soomro, M.; Evangelista, A.C.J. A review of recycled aggregate in concrete applications (2000–2017). *Constr. Build. Mater.* **2018**, *172*, 272–292. [CrossRef]

2. Rao, A.; Jha, K.N.; Misra, S. Use of aggregates from recycled construction and demolition waste in concrete, Resour. *Conserv. Recycl.* **2007**, *50*, 71–81. [CrossRef]

3. Faleschini, F.; De Marzi, P.; Pellegrino, C. Recycled concrete containing EAF slag: Environmental assessment through LCA. *Eur. J. Environ. Civ. Eng.* **2014**, *18*, 1009–1024. [CrossRef]

4. Braymand, S.; Feraille, A.; Serres, N.; Feugeas, F. Benchmarking of materials composition and transportation influences on Life Cycle Assessment of Recycled Aggregate Concrete. In Proceedings of the 5th International Conference on Sustainable Solid Waste Management, Athens, Greece, 21–24 June 2017; pp. 1–3.

5. McNeil, K.; Kang, T.H.-K. Recycled Concrete Aggregates: A Review. *Int. J. Concr. Struct. Mater.* **2013**, *7*, 61–69. [CrossRef]

6. Shayan, A.; Xu, A. Performance and properties of structural concrete made with recycled concrete aggregate. *Mater. J.* **2003**, *100*, 371–380.

7. Ministerio de Fomento de España. EHE-08: Instrucción de Hormigón Estructural. 2008. Available online: http://www.fomento.gob.es/MFOM/LANG_CASTELLANO/ORGANOS_COLEGIADOS/CPH/instrucciones/EHE08INGLES/ (accessed on 22 December 2019).

8. UNE-EN 12620:2003+A1:2009. In *Áridos Para Hormigón*; Normalización Española: Madrid, Spain, 2009.

9. Etxeberria, M.; Vázquez, E.; Marí, A.; Barra, M. Influence of amount of recycled coarse aggregates and production process on properties of recycled aggregate concrete. *Cem. Concr. Res.* **2007**, *37*, 735–742. [CrossRef]

10. De Brito, J.; Ferreira, J.; Pacheco, J.; Soares, D.; Guerreiro, M. Structural, material, mechanical and durability properties and behaviour of recycled aggregates concrete. *J. Build. Eng.* **2016**, *6*, 1–16. [CrossRef]

11. Silva, R.V.; De Brito, J.; Dhir, R.K. Establishing a relationship between modulus of elasticity and compressive strength of recycled aggregate concrete. *J. Clean. Prod.* **2016**, *112*, 2171–2186. [CrossRef]

12. Marinkovic, S.; Radonjanin, V.; Malešev, M.; Ignjatović, I. Comparative environmental assessment of natural and recycled aggregate concrete. *Waste Manag.* **2010**, *30*, 2255–2264. [CrossRef]

13. Anastasiou, E.; Filikas, K.G.; Stefanidou, M. Utilization of fine recycled aggregates in concrete with fly ash and steel slag. *Constr. Build. Mater.* **2014**, *50*, 154–161. [CrossRef]

14. Silva, R.; De Brito, J.; Dhir, R. Properties and composition of recycled aggregates from construction and demolition waste suitable for concrete production. *Constr. Build. Mater.* **2014**, *65*, 201–217. [CrossRef]

15. Thomas, C.; Setién, J.; Polanco, J.A.; Alaejos, P.; Sánchez De Juan, M. Durability of recycled aggregate concrete. *Constr. Build. Mater.* **2013**, *40*, 1054–1065. [CrossRef]

16. Bravo, M.; De Brito, J.; Pontes, J.; Evangelista, L. Mechanical performance of concrete made with aggregates from construction and demolition waste recycling plants. *J. Clean. Prod.* **2015**, *99*, 59–74. [CrossRef]

17. Topcu, I.B.; Sengel, S. Properties of concretes produced with waste concrete aggregate. *Cem. Concr. Res.* **2004**, *34*, 1307–1312. [CrossRef]

18. Fiol, F.; Thomas, C.; Muñoz, C.; Ortega-López, V.; Manso, J.M. The influence of recycled aggregates from precast elements on the mechanical properties of structural self-compacting concrete. *Constr. Build. Mater.* **2018**, *182*, 309–323. [CrossRef]

19. Matias, D.; de Brito, J.; Rosa, A.; Pedro, D. Durability of concrete with recycled coarse aggregates: Influence of superplasticizers. *J. Mater. Civ. Eng.* **2013**, *26*, 6014011. [CrossRef]

20. Thomas, C.; Setién, J.; Polanco, J.; Cimentada, A.; Medina, C. Influence of curing conditions on recycled aggregate concrete. *Constr. Build. Mater.* **2018**, *172*, 618–625. [CrossRef]

21. Thomas, C.; Cimentada, A.; Polanco, J.A.; Setién, J.; Méndez, D.; Rico, J. Influence of recycled aggregates containing sulphur on properties of recycled aggregate mortar and concrete, Compos. *Part B Eng.* **2013**, *45*, 474–485. [CrossRef]

22. Gómez-Soberón, J.M.V. Porosity of recycled concrete with substitution of recycled concrete aggregate: An experimental study. *Cem. Concr. Res.* **2002**, *32*, 1301–1311. [CrossRef]

23. de Brito, J.; Agrela, F. *New Trends in Eco-Efficient and Recycled Concrete*; Woodhead Publishing: Cambridge, UK, 2018.

24. Bravo, M.; De Brito, J.; Pontes, J.; Evangelista, L. Durability performance of concrete with recycled aggregates from construction and demolition waste plants. *Constr. Build. Mater.* **2015**, *77*, 357–369. [CrossRef]

25. Barra, M.; Vazquez, E. Properties of concretes with recycled aggregates: Influence of properties of the aggregates and their interpretation. In *Sustainable Construction: Use of Recycled Concrete Aggregate: Proceedings of the International Symposium organised by the Concrete Technology Unit, University of Dundee and held at the Department of Trade and Industry Conference Centre, London, UK on 11–12 November 1998*; Thomas Telford Publishing: London, UK, 2018; pp. 18–30.

26. Ajdukiewicz, A.; Alina, K. Influence of recycled aggregates on mechanical properties of HS/HPC. *Cem. Concr. Compos.* **2002**, *24*, 269–279. [CrossRef]

27. Tabsh, S.W.; Abdelfatah, A.S. Influence of recycled concrete aggregates on strength properties of concrete. *Constr. Build. Mater.* **2009**, *23*, 1163–1167. [CrossRef]

28. Padmini, A.K.; Ramamurthy, K.; Mathews, M.S. Influence of parent concrete on the properties of recycled aggregate concrete. *Constr. Build. Mater.* **2009**, *23*, 829–836. [CrossRef]

29. Arribas, I.; Vegas, I.; San-José, J.; Manso, J.M. Durability studies on steelmaking slag concretes. *Mater. Des.* **2014**, *63*, 168–176. [CrossRef]

30. Poon, C.; Shui, Z.; Lam, L. Effect of microstructure of ITZ on compressive strength of concrete prepared with recycled aggregates. *Constr. Build. Mater.* **2004**, *18*, 461–468. [CrossRef]
31. Li, W.; Xiao, J.; Sun, Z.; Kawashima, S.; Shah, S.P. Interfacial transition zones in recycled aggregate concrete with different mixing approaches. *Constr. Build. Mater.* **2012**, *35*, 1045–1055. [CrossRef]
32. Rahal, K. Mechanical properties of concrete with recycled coarse aggregate. *Build. Environ.* **2007**, *42*, 407–415. [CrossRef]
33. Alexandridou, C.; Angelopoulos, G.N.; Coutelieris, F.A. Mechanical and durability performance of concrete produced with recycled aggregates from Greek construction and demolition waste plants. *J. Clean. Prod.* **2018**, *176*, 745–757. [CrossRef]
34. Pacheco, J.; De Brito, J.; Chastre, C.; Evangelista, L. Experimental investigation on the variability of the main mechanical properties of concrete produced with coarse recycled concrete aggregates. *Constr. Build. Mater.* **2019**, *201*, 110–120. [CrossRef]
35. Xiao, J.; Li, J.; Zhang, C. Mechanical properties of recycled aggregate concrete under uniaxial loading. *Cem. Concr. Res.* **2005**, *35*, 1187–1194. [CrossRef]
36. Kwan, W.H.; Ramli, M.; Kam, K.J.; Sulieman, M.Z. Influence of the amount of recycled coarse aggregate in concrete design and durability properties. *Constr. Build. Mater.* **2012**, *26*, 565–573. [CrossRef]
37. Xiao, J.; Li, W.; Poon, C. Recent studies on mechanical properties of recycled aggregate concrete in China—A review. *Sci. China Technol. Sci.* **2012**, *55*, 1463–1480. [CrossRef]
38. Pacheco, J.; De Brito, J.; Chastre, C.; Evangelista, L. Scatter of Constitutive Models of the Mechanical Properties of Concrete: Comparison of Major International Codes. *J. Adv. Concr. Technol.* **2019**, *17*, 102–125. [CrossRef]
39. Corinaldesi, V. Mechanical and elastic behaviour of concretes made of recycled-concrete coarse aggregates. *Constr. Build. Mater.* **2010**, *24*, 1616–1620. [CrossRef]
40. Thomas, C.; Carrascal, I.; Setién, J.; Polanco, J.A. Determinación del límite a fatiga en hormigones reciclados de aplicación estructural—Determining the fatigue limit recycled concrete structural application. *An. Mecánica La Fract.* **2009**, *1*, 283–289. (In Spanish)
41. Thomas, C.; Setién, J.; Polanco, J.A.; Lombillo, I.; Cimentada, A.I. Fatigue limit of recycled aggregate concrete. *Constr. Build. Mater.* **2014**, *52*, 146–154. [CrossRef]
42. Thomas, C.; De Brito, J.; Gil, V.; Sainz-Aja, J.; Cimentada, A. Multiple recycled aggregate properties analysed by X-ray microtomography. *Constr. Build. Mater.* **2018**, *166*, 171–180. [CrossRef]
43. Li, X. Recycling and reuse of waste concrete in China: Part I. Material behaviour of recycled aggregate concrete. *Resour. Conserv. Recycl.* **2008**, *53*, 36–44. [CrossRef]
44. Kou, S.; Poon, C. Properties of self-compacting concrete prepared with recycled glass aggregate. *Cem. Concr. Compos.* **2009**, *31*, 107–113. [CrossRef]
45. Silva, R.V.; Neves, R.; De Brito, J.; Dhir, R.K. Carbonation behaviour of recycled aggregate concrete. *Cem. Concr. Compos.* **2015**, *62*, 22–32. [CrossRef]
46. de Juan, M.S.; Gutiérrez, P.A. Study on the influence of attached mortar content on the properties of recycled concrete aggregate. *Constr. Build. Mater.* **2009**, *23*, 872–877. [CrossRef]
47. Wang, L.; Wang, J.; Qian, X.; Chen, P.; Xu, Y.; Guo, J. An environmentally friendly method to improve the quality of recycled concrete aggregates. *Constr. Build. Mater.* **2017**, *144*, 432–441. [CrossRef]
48. Shima, H.; Tateyashiki, H.; Matsuhashi, R.; Yoshida, Y. An Advanced Concrete Recycling Technology and its Applicability Assessment through Input-Output Analysis. *J. Adv. Concr. Technol.* **2005**, *3*, 53–67. [CrossRef]
49. Pawluczuk, E.; Kalinowska-Wichrowska, K.; Bołtryk, M.; Jiménez, J.R.; Fernández, J.M. The Influence of Heat and Mechanical Treatment of Concrete Rubble on the Properties of Recycled Aggregate Concrete. *Materials* **2019**, *12*, 367. [CrossRef] [PubMed]
50. Braymand, S.; Roux, S.; Fares, H.; Déodonne, K.; Feugeas, F. Separation and quantification of attached mortar in recycled concrete aggregates. *Waste Biomass Valorization* **2017**, *8*, 1393–1407. [CrossRef]
51. Lotfi, S.; Eggimann, M.; Wagner, E.; Mróz, R.; Deja, J. Performance of recycled aggregate concrete based on a new concrete recycling technology. *Constr. Build. Mater.* **2015**, *95*, 243–256. [CrossRef]
52. Khatib, J.M. Properties of concrete incorporating fine recycled aggregate. *Cem. Concr. Res.* **2005**, *35*, 763–769. [CrossRef]
53. Papayianni, I.; Anastasiou, E. Utilization of electric arc furnace steel slags in concrete products. In *6th European Slag Conference: Proceedings Ferrous Slag-Resource Development for an Environmentally Sustainable World*; EUROSLAG Publication: Madrid, Spain, 2010; pp. 319–334.

54. Monosi, S.; Ruello, M.L.; Sani, D. Electric arc furnace slag as natural aggregate replacement in concrete production. *Cem. Concr. Compos.* **2016**, *66*, 66–72. [CrossRef]

55. Abu-Eishah, S.I.; El-Dieb, A.S.; Bedir, M.S. Performance of concrete mixtures made with electric arc furnace (EAF) steel slag aggregate produced in the Arabian Gulf region. *Constr. Build. Mater.* **2012**, *34*, 249–256. [CrossRef]

56. Faleschini, F.; Fernández-Ruíz, M.A.; Zanini, M.A.; Brunelli, K.; Pellegrino, C.; Hernández-Montes, E. High performance concrete with electric arc furnace slag as aggregate: Mechanical and durability properties. *Constr. Build. Mater.* **2015**, *101*, 113–121. [CrossRef]

57. Fonseca, N.; De Brito, J.; Evangelista, L. The influence of curing conditions on the mechanical performance of concrete made with recycled concrete waste. *Cem. Concr. Compos.* **2011**, *33*, 637–643. [CrossRef]

58. Rodrigues, F.; Evangelista, L.; De Brito, J. A new method to determine the density and water absorption of fine recycled aggregates. *Mater. Res.* **2013**, *16*, 1045–1051. [CrossRef]

59. Mehta, P.K.; Monteiro, P.J.M. Concrete microstructure, Properties and Materials. 2017. Available online: http://matteopro.com/images/materiali-non-convenzionali/Concrete_microstrutture_-_book.pdf (accessed on 22 December 2019).

60. Silva, R.V.; De Brito, J.; Dhir, R.K. Tensile strength behaviour of recycled aggregate concrete. *Constr. Build. Mater.* **2015**, *83*, 108–118. [CrossRef]

61. Kou, S.-C.; Poon, C.-S.; Wan, H.-W. Properties of concrete prepared with low-grade recycled aggregates. *Constr. Build. Mater.* **2012**, *36*, 881–889. [CrossRef]

62. Ryu, J.S. An experimental study on the effect of recycled aggregate on concrete properties. *Mag. Concr. Res.* **2002**, *54*, 7–12. [CrossRef]

63. Suryavanshi, A.K.; Narayan Swamy, R. Stability of Friedel's salt in carbonated concrete structural elements. *Cem. Concr. Res.* **1996**, *26*, 729–741. [CrossRef]

64. Maekawa, K.; Ishida, T.; Kishi, T. Multi-scale Modeling of Concrete Performance. *J. Adv. Concr. Technol.* **2003**, *1*, 91–126. [CrossRef]

65. Fédération Internationale du Bétom. *Code-Type Models for Structural Behaviour of Concrete*; Fédération Internationale du Béton: Bulletin, Lausanne, Switzerland, 2013.

Article

High-Frequency Fatigue Testing of Recycled Aggregate Concrete

Jose Sainz-Aja, Carlos Thomas *, Juan A. Polanco and Isidro Carrascal

LADICIM (Laboratory of Materials Science and Engineering), University of Cantabria,
E.T.S. de Ingenieros de Caminos, Canales y Puertos, Av./Los Castros 44, 39005 Santander, Spain;
jose.sainz-aja@unican.es (J.S.-A.); polancoa@unican.es (J.A.P.); carrasci@unican.es (I.C.)
* Correspondence: carlos.thomas@unican.es

Received: 11 November 2019; Accepted: 11 December 2019; Published: 18 December 2019

Abstract: Concrete fatigue behaviour has not been extensively studied, in part because of the difficulty and cost. Some concrete elements subjected to this type of load include the railway superstructure of sleepers or slab track, bridges for both road and rail traffic and the foundations of wind turbine towers or offshore structures. In order to address fatigue problems, a methodology was proposed that reduces the lengthy testing time and high cost by increasing the test frequency up to the resonance frequency of the set formed by the specimen and the test machine. After comparing this test method with conventional frequency tests, it was found that tests performed at a high frequency (90 ± 5 Hz) were more conservative than those performed at a moderate frequency (10 Hz); this effect was magnified in those concretes with recycled aggregates coming from crushed concrete (RC-S). In addition, it was found that the resonance frequency of the specimen–test machine set was a parameter capable of identifying whether the specimen was close to failure.

Keywords: high-frequency fatigue test; recycled aggregate; recycled aggregate concrete; fatigue; Locati test

---

1. Introduction

Recycled aggregates have an environmental benefit [1,2] and recycling is a necessity as the rate of waste generation is such that landfills are close to saturation [3]. Additionally, the possibility of obtaining recycled concrete (RAC) with good mechanical [4] and durability [5] properties has been proven. The ability of these recycled aggregates for use in self-compacting concrete [6], even with fine recycled aggregates [7], has also been proven. Even Kareem et al. [8] have used RAC for the manufacture of hot-mix asphalt.

Concrete fatigue behaviour has not been extensively studied, partly because of the difficulty, cost and time required. Concrete elements subjected to this type of load include railway superstructures, sleepers or slab tracks [9], rail and road bridges [10], offshore structures subject to variable wind and tidal loads [10,11] and/or wind generators [12]. Khosravani et al. [13] have defined a procedure for analyzing ultra-high performance concrete's response to impacts. As Skarżyński et al. state, knowledge about the effect of cyclic loads on concrete is currently very limited [12]. Several authors have analysed the responses of concrete. Xiao et al. [11] analysed the behavior of RAC to both compression and bending fatigue. Li et al. [14] analysed the influence of compressive fatigue on a fiber-reinforced cementitious material. Thomas et al. analysed the concrete fatigue behavior using two different methods: the staircase method [15] and the Locati method [16]. Innovative techniques, such as micro computed tomography (micro-CT) have also been used to analyze the behavior of concrete at fatigue, both in compressive fatigue [12] and bending fatigue [17]. Moreover, some micro-CT studies, which is a technique able to analyse the pores and cracks on concrete [18,19], had been developed to understand the concrete damage fatigue micromechanism [20,21]. In general, fatigue is known to lead to microcracks in concrete growing at lower loads than in static tests, which can lead to concrete failure earlier than expected [22].

It is well known that the fatigue limit of concrete depends on different factors. At higher stresses, the fatigue strength decreases with decreasing frequency [12,23,24]. The fatigue strength is also affected by the water/cement ratio, cement content, concrete type, rest periods, curing conditions and age during loading [25]. It is assumed that damage linearly increases with the number of cycles applied at a certain stress level [24]. The strain at the concrete failure during fatigue tests approximately corresponds to that at the peak load during quasi-static tests [26]. The failure meso-mechanism in concrete under fatigue compressive tests is almost the same as in monotonic compressive tests [27].

In order to reduce the characterization time as much as possible, the influence of increasing the test frequency up to the resonance frequency of the set of specimens and test machine [28] has been analyzed. In response to this proposal, several authors [23,29,30] indicate that the range of frequencies to be tested should be between 1 and 15 Hz, since they state that within this range, the effect of frequency is limited. It is justifiable to set the minimum at 1 Hz in order to avoid an excessive increase in the effect of creep in very long duration tests. However, there is no justification for setting the maximum at 15 Hz.

Fatigue tests are proposed for classification according to the frequency of the test, distinguishing between low frequency tests, moderate frequency tests and high frequency tests. Low frequency tests are carried out at less than 1 Hz, moderate frequency tests between 1 and 15 Hz and high frequency tests at more than 15 Hz.

Three types of recycled self-compacting concrete were characterized in this investigation of compression fatigue tests. This characterization was developed both at a moderate frequency (10 Hz) and a high frequency (90 Hz). First, the results of the Locati tests were compared with 2×10^5 cycles per step at moderate and high frequencies, where it was possible to show that the tests carried out at a high frequency were notably more conservative than those at a moderate frequency. Second, the influence of the number of cycles per step in the high frequency Locati tests was analysed, where similar results were obtained using both test methodologies.

2. Materials and Methods

2.1. Aggregates

Recycled aggregates used for the manufacture of concretes were obtained from the crushing of ballast (RA-B) and out-of-use sleepers (RA-S). After crushing these materials, three granulometric fractions were obtained from each of these wastes (see Table 1). The grading of the six fractions of the two crushed products can be seen in Figure 1. Additionally, Table 1 shows the results of the relative density of the coarse aggregate and the real densities of the sands. The flakiness index of the two coarse aggregates was also determined, obtaining a value of 14% for RA-B-CA and 5% for RA-S-CA.

Table 1. Aggregate properties.

Code	Description	Min.–Max. Size (mm)	Density (g/cm^3)
RA-B-CA	Ballast coarse aggregates	5–12	2.5
RA-S-CA	Sleeper coarse aggregate	5–12	2.3
RA-B-LS	Ballast coarse sand	2–5	2.7
RA-S-LS	Sleeper coarse sand	2–5	2.4
RA-B-FS	Ballast fine sand	0–2	2.8
RA-S-FS	Sleeper fine sand	0–2	2.5

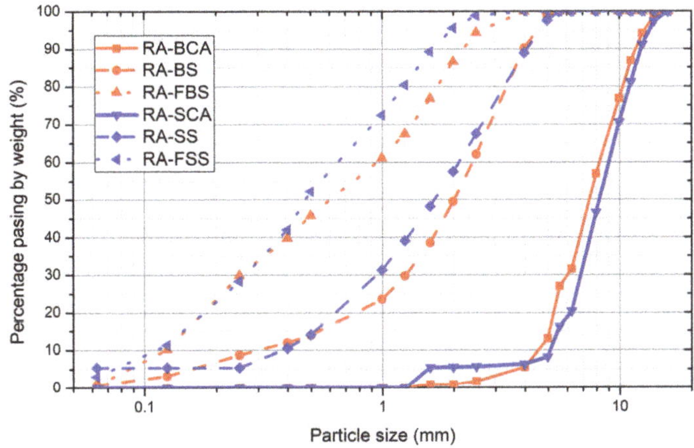

Figure 1. Aggregate grading curves.

2.2. Cement

A CEM IV (V) 32.5 N type cement according to EN 197-1 [31] was used, provided by Alpha cements [32], with a density of 2.85 g/cm^3 determined according to UNE 80103 [33] and a Blaine specific surface of 3885 cm^2/g obtained according to EN 196-6 [34]. The chemical composition of the cement is given in Table 2.

Table 2. Cement chemical composition.

	\multicolumn{8}{c}{Composition (wt.%)}							
	CaO	SiO_2	Al_2O_3	Fe_2O_3	MgO	K_2O	SO_3	Ignition Loss
CEM IV	35.5	41.2	13.3	4.4	1.2	1.4	1.3	1.7

2.3. Mix Proportions

Three self-compacting recycled concretes were mixed with recycled aggregates from out-of-use track elements: the first, exclusively with RA-B, called RC-B; the second, exclusively with RA-S, called RC-S; and finally, a third concrete with both types of aggregates in the proportion in which these wastes were found in a ballast track, called RC-M. Table 3 shows the three mix proportions.

Table 3. Mix proportions.

Material	RC-B	RC-S	RC-M
Water	225	200	221
Cement	500	500	500
Superplasticizer additive	10	10	10
RA-FBS	790	-	677
RA-BS	320	-	274
RA-BCA	522	-	447
RA-FSS	-	690	98
RA-SS	-	283	40
RA-SCA	-	587	83
Water/cement ratio	0.45	0.40	0.44
% sand (0–2 mm) from the total sand	70	70	70
% coarse aggregate from the total aggregates	35	40	36
% superplasticizer additive/cement	2.00	2.00	2.00

It should be noted that the mix proportions were determined using the criteria that the three self-compacting concretes had a similar workability; therefore, as the RA-B-CA had a notably sharper geometry than the RA-S-CA, it was necessary to increase the amount of water to achieve the same workability [35].

2.4. Mechanical Properties

The compressive strength tests were performed according to the EN 12390-3 and EN 13290-3/AC [36,37] standards, using cubes of 100 mm side at ages of 1, 2, 3, 5, 7, 28, 90 and 180 days, testing 5 specimens for each age and material. The stabilized secant moduli of elasticity was determined according to EN 12390-13 [38] using cylindrical specimens of 200 mm in height and 100 mm in diameter at the ages of 7, 28, 90 and 180 days, testing 2 specimens for each age and material.

2.5. Fatigue Tests

The influence of conducting trials at a very high frequency was analyzed. For this purpose, the reference was tests carried out at moderate frequency, i.e., usual test frequencies, namely 10 Hz. For very high frequency tests, the highest possible test frequency was used, namely, tests were performed on a resonant fatigue machine. This machine carried out the tests at the resonance frequency of the set test machine and the specimen. In the specific case of the resonance machine used and the specimens used, these resonance frequencies were in the range of 90 ± 5 Hz.

An explicative scheme of the fatigue tests that were carried out can be found in Figure 2a. Fatigue tests at both a moderate frequency (10 Hz) and a very high frequency (90 Hz) were performed using cylindrical specimens of 200 mm in height and 100 mm in diameter in all tests when these specimens were older than 90 days; this was done to ensure that the properties of the concrete had reached a stationary state. All the tests carried out recorded both the load applied to the specimen and the evolution of the strain suffered by the specimen. The strain was recorded by means of two strain gauges fixed to the specimens in two diametrically opposed generatrixes by considering the mean of the values provided by both gauges as the strain value of the specimen. In the high-frequency tests, the frequency at which the system was located was recorded, noting that the resonant frequency of the system will evolve with the variations in its stiffness. For each of the fatigue test types, two specimens were tested, ensuring that similar results were obtained in both.

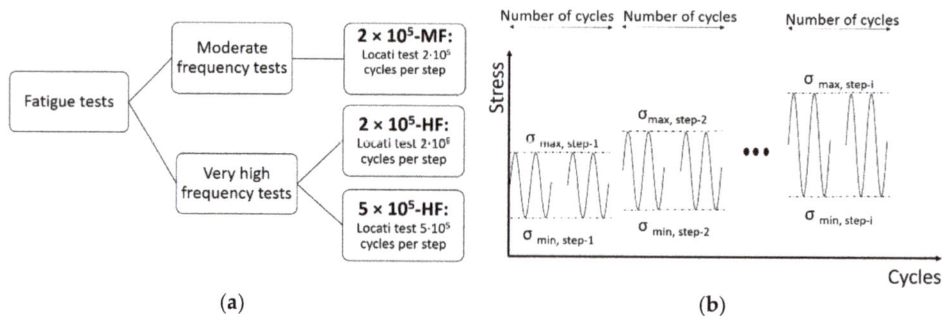

Figure 2. (**a**) Experimental fatigue campaign. (**b**) Locati methodology description. MF: moderate frequency, HF: high frequency.

Fatigue characterization at a moderate frequency was performed using the Locati method [39,40]. The Locati methodology consists of applying increasing steps of sinusoidal loads, maintaining a constant ratio $\sigma_{max}/\sigma_{min} = 0.1$ over a constant number of cycles, in this case 2×10^5 cycles (Figure 2b). These tests were performed on a servo-hydraulic machine with a maximum capacity of 1000 kN (Figure 2b). The criterion used to determine the fatigue limit using this type of test was that followed by Thomas in his PhD thesis [41], which obtained the fatigue limit stress range to be 80% of the stress range of the step in which the breaking occurs. The stress range was the difference between the maximum and minimum stress, i.e., $\Delta\sigma = \sigma_{max} - \sigma_{min}$. This method is called method-1.

In order to analyse whether an increase in frequency had an influence on the concrete fatigue limit, tests were performed under identical loads at a moderate frequency and a very high frequency. A 400-kN capacity machine was used to perform fatigue tests at the resonance frequency of the test set. This resonance frequency was found in all cases to be in the range 90 ± 5 Hz (see Figure 3b). In this case, in addition to analysing the results following the procedure followed by Thomas in his PhD thesis [41], an additional failure criterion was introduced, namely, to define the stress range of the step prior to a resonance frequency drop as the stress range of the endurance limit. This method was denominated as method-2.

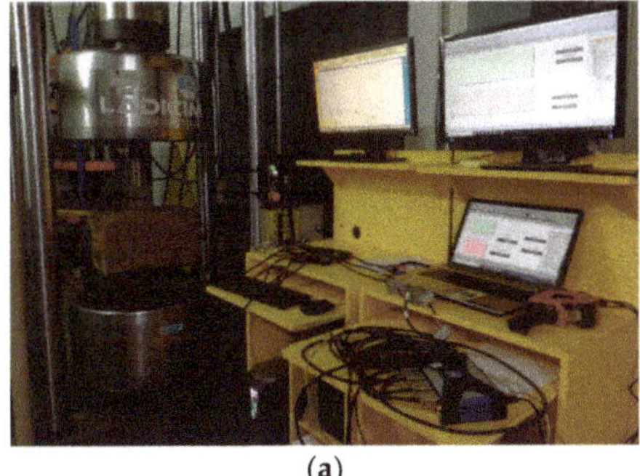
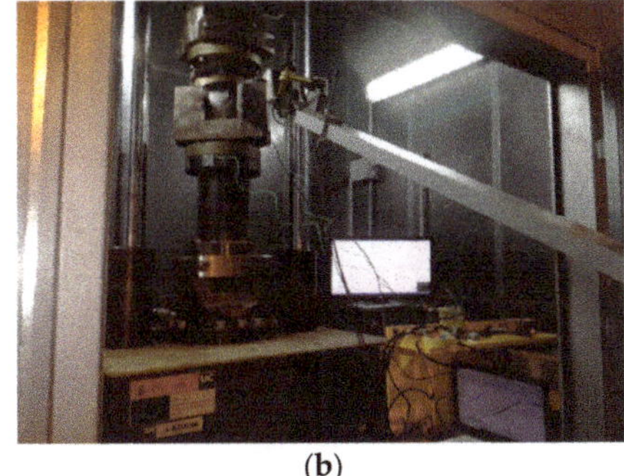

(a)   (b)

Figure 3. Fatigue tests. (**a**) Low-frequency fatigue testing and (**b**) high-frequency fatigue testing.

In order to analyse the influence of the number of cycles in the Locati tests, very high frequency tests were performed with the same stress values but applying 5×10^5 cycles per step instead of 2×10^5 cycles. For these tests, the same testing machine was used, and the same criteria were used to define the endurance tension range.

The common objective of all these studies was to determine the step associated with the fatigue limit of the concrete in each case. In order to be able to compare the results as directly as possible, it was decided to fix the load steps based on the compressive strength of each material. For this purpose, the value of the maximum stress of each step was defined as a coefficient (k) multiplied by the compressive strength at the time the fatigue tests began, that is, 90 days after manufacture of each of the concretes; furthermore, the tensional ratio $\sigma_{max}/\sigma_{min}$ was set to 0.1. Table 4 shows a summary of the tension values used in each of the load steps applied.

Table 4. Fatigue test stress values.

N	k	RC-B			RC-S			RC-M		
		σ_{max} (MPa)	σ_{min} (MPa)	Range (MPa)	σ_{max} (MPa)	σ_{min} (MPa)	Range (MPa)	σ_{max} (MPa)	σ_{min} (MPa)	Range (MPa)
1	0.30	17.8	1.8	16.0	23.3	2.3	21.0	18.9	1.9	17.0
2	0.35	20.8	2.1	18.7	27.2	2.7	24.5	22.0	2.2	19.8
3	0.40	23.8	2.4	21.4	31.1	3.1	28.0	25.1	2.5	22.6
4	0.45	26.7	2.7	24.0	35.0	3.5	31.5	28.3	2.8	25.5
5	0.50	29.7	3.0	26.7	38.9	3.9	35.0	31.4	3.1	28.3
6	0.55	32.7	3.3	29.4	42.8	4.3	38.5	34.6	3.5	31.1
7	0.60	35.7	3.6	32.1	46.7	4.7	42.0	37.7	3.8	33.9
8	0.65	38.6	3.9	34.7	50.6	5.1	45.5	40.8	4.1	36.7
9	0.70	41.6	4.2	37.4	54.5	5.4	49.1	44.0	4.4	39.6

In order to compare the degree of influence of repeated loads on each of the materials, the influence coefficient (IC) was defined as the ratio of the stress range corresponding to the fatigue limit of the concrete to its compressive strength.

3. Results and Discussions

3.1. Compressive Strength and Young's Modulus

Figure 4 shows the evolution of the compressive strength as a function of time, while Figure 5 shows the evolution of Young's modulus as a function of the age of the different concretes.

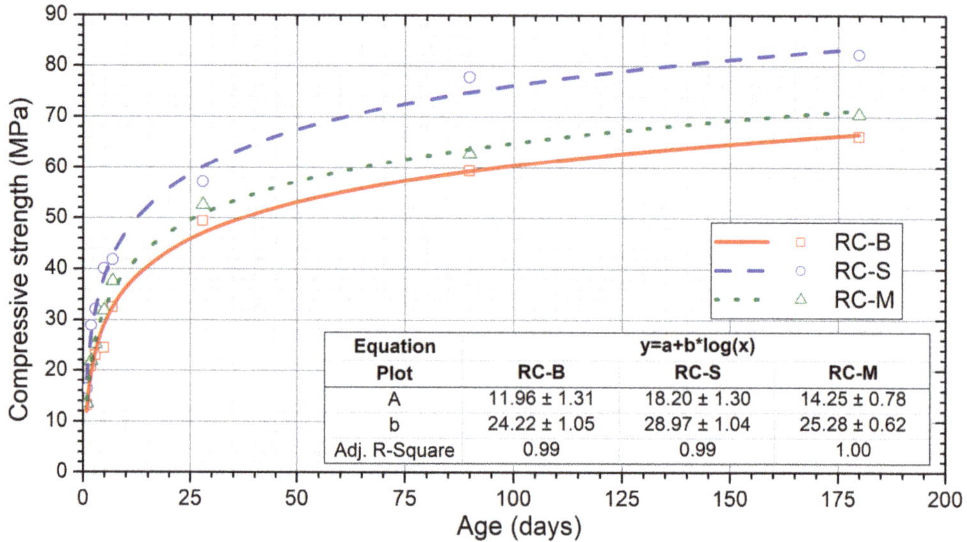

Figure 4. Evolution of the compressive strength.

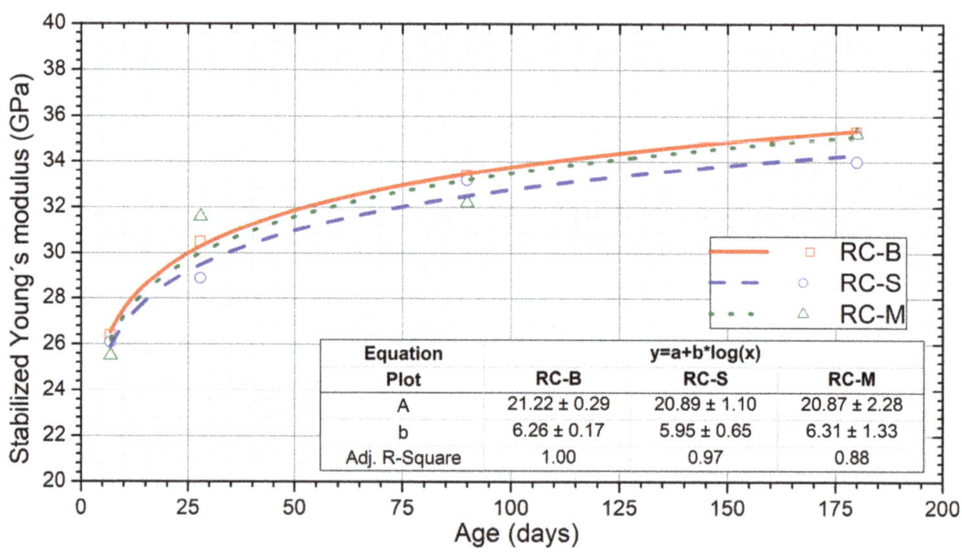

Figure 5. Evolution of the Young's modulus.

The great influence of the water/cement ratio on the compressive strength of the RC-S can be observed. On the other hand, RC-B had the lowest compressive strengths and RC-M had intermediate compressive strengths between the other two concretes.

In the case of Young's modulus, although the RC-S paste was of a better quality than that of RC-B, the noticeably lower stiffness of the mortar, adhered to the natural aggregates that made up the RA-S, meant that the elastic modulus of RC-S was lower than that of RC-B. As in the case of the compressive strength, RC-M was found between RC-B and RC-S concretes in all cases.

3.2. Influence of the Frequency on Fatigue

Figure 6 shows an example of the maximum deformation envelope recorded during the Locati tests at a frequency of 10 Hz.

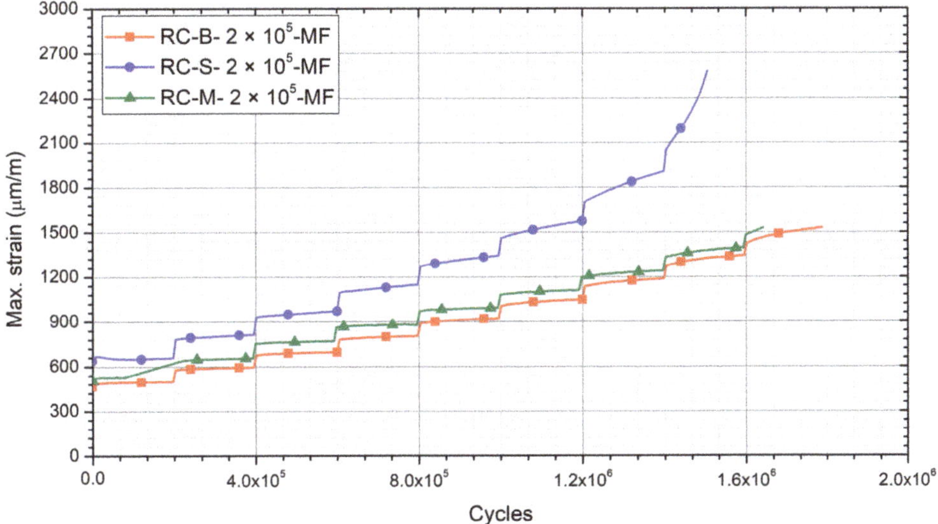

Figure 6. Maximum strain of low-frequency Locati fatigue tests with 2×10^5 cycles per step.

Figure 6 shows, first of all, that RC-B was able to resist the most steps, which meant that it was the material with the highest IC, with RC-S being able to resist the least and RC-M was in an intermediate situation between the other two materials. These results agree with the results of other authors who state that the presence of adhered mortar in the RA reduces this coefficient [15,16,42]. It can also be seen that the deformation values suffered by the specimens was lower in the case of RC-B. It should be noted that, as each of the steps is a fixed percentage compared to the compressive strength of each material, they are not directly comparable to each other as they are different stress values.

Table 5 shows a brief summary of the results obtained from the moderate-frequency tests carried out using method-1.

Table 5. Low-frequency fatigue limit (f_L) obtained using the Locati method with 2×10^5 cycles per step. IC: influence coefficient.

Material	$\Delta\sigma_{max}$	f_L	IC
	(MPa)	**(MPa)**	**(%)**
RC-B	38.75	31.0	52.2
RC-S	45.5	36.4	46.8
RC-M	39.6	31.7	50.4

Analysis of these results shows that the material with the highest stress range was RC-S, the least was RC-B and the RC-M had an intermediate value. Likewise, it is possible to determine that, although the RC-S had the highest stress range values, it had the lowest IC. This loss of compressive strength was due to the presence of mortar adhered to the aggregate. In any case, these values of IC are within the usual range for concretes and agree with values found in the literature [15,16].

Figure 7 shows an example of the maximum deformation suffered by a specimen of each material during Locati tests at the resonance frequency of the machine test set.

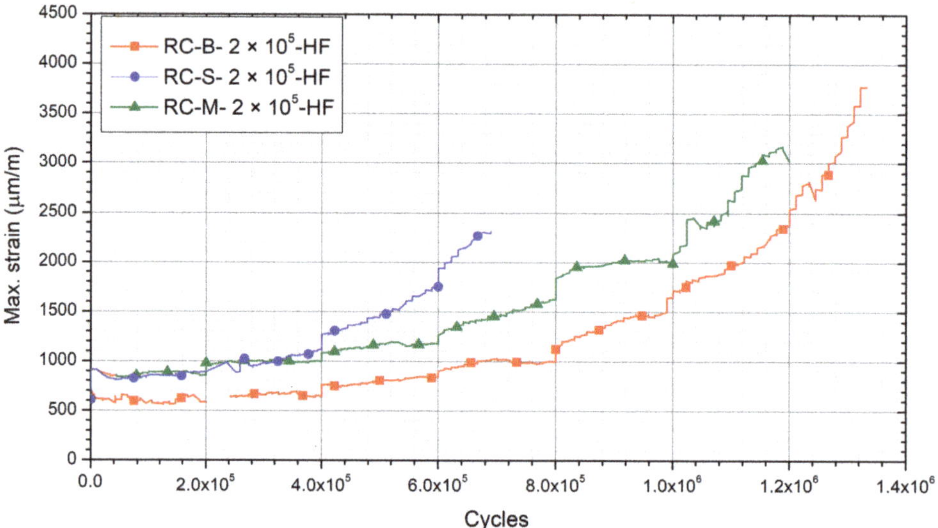

Figure 7. Maximum strain of high-frequency Locati fatigue tests with 2×10^5 cycles per step.

Figure 7 shows that RC-S was the material that had the lowest IC. RC-M was generally between RC-B and RC-S.

Figure 8 shows that the resonance frequency evolved throughout the test. For the first steps, there was an increase in the frequency with the cycles throughout each step, as well as a punctual increase when there was a change of step. At the end of each test, a drop in the resonance frequency of the system was seen. This resonance frequency depended on the stiffness of the system; an increase in the stiffness of the system resulted in an increase in the resonance frequency, while a reduction in the stiffness of the system reduced it. For this reason, it was interpreted that both point and distributed frequency increases occurred as a consequence of a stiffening of the system, while the fall that occurred in the final part of the test was a consequence of the damage suffered by the specimen, where the cracks had reached such a size that they produced a flexibilization of the specimen, which indicated that it was close to breaking.

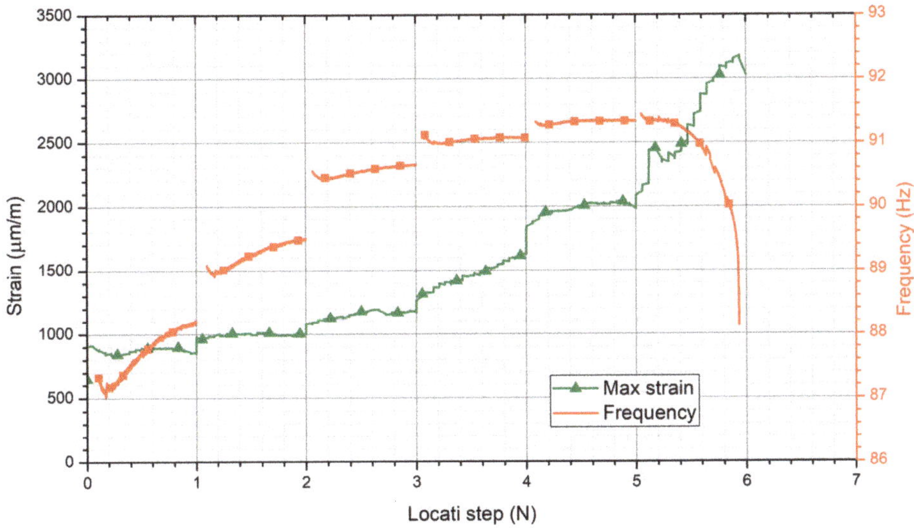

Figure 8. Maximum strain and resonance frequency comparison during high-frequency Locati fatigue tests with 2×10^5 cycles per step.

Table 6 shows the results obtained from the Locati tests with 2×10^5 cycles per step at a very high frequency using the two criteria previously established.

Table 6. High-frequency fatigue limit obtained using the Locati method with 2×10^5 cycles per step.

Material	$\Delta\sigma_{max}$	Method-1 f_L	IC	Method-2 f_L	IC
	(MPa)	(MPa]	(%)	(MPa)	(%)
RC-B	32.09	25.67	43.19	26.74	44.93
RC-S	31.51	25.21	32.40	28.01	35.99
RC-M	31.10	24.88	39.60	28.27	45.04

Analysing the results of Table 6, it can be determined that, as had been deduced from the evolution of the maximum strain, RC-B was the material with the highest IC, although the compression strength of RC-S was the highest of all. The stress ranges corresponding to the fatigue limit of the three concretes were similar. It can also be appreciated that, although the values of both the fatigue limit and IC, obtained using the two analysis criteria were similar, a 5% difference was found relative to the limit provided by Thomas [41], which was usually more conservative than the one obtained using the resonance frequency.

3.3. Influence of the Number of Cycles Per Step during a Locati Test

In order to determine the influence of increasing the number of cycles in each step of the Locati test, identical tests were carried out to those described above, but the number of cycles per step was set at 5×10^5 instead of 2×10^5. Figure 9 shows the evolution of the maximum deformation as a function of the number of cycles throughout the Locati test for 5×10^5 cycles per step.

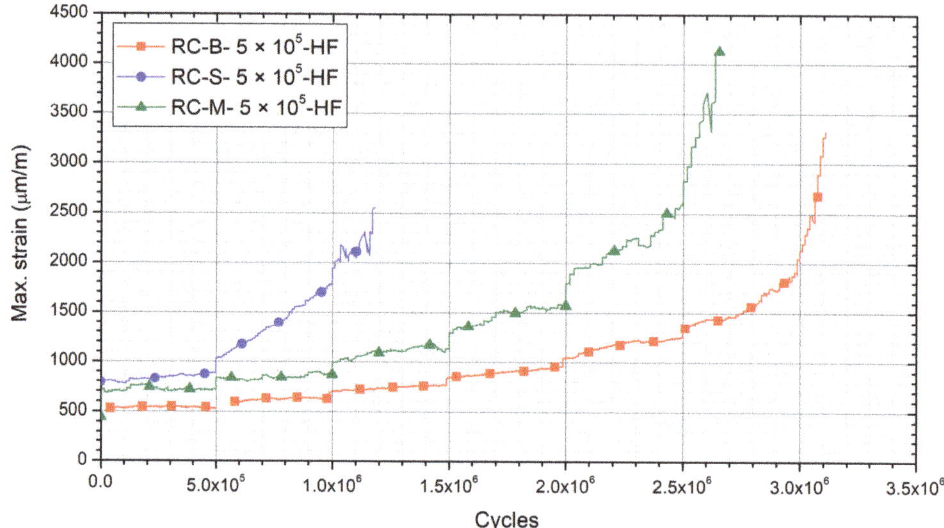

Figure 9. Maximum strain of high-frequency Locati fatigue tests with 5×10^5 cycles per step.

Figure 9 shows that RC-S was the material that had the lowest IC. RC-M was between RC-B and RC-S. In this case, as in the previous case, the two analysis criteria were used to determine the stress range corresponding to the fatigue limit. These results are shown in Table 7.

Table 7. High-frequency fatigue limit obtained using the Locati method with 5×10^5 cycles per step.

Material	$\Delta\sigma_{max}$	Method-1		Method-2	
		f_L	f_L/f_c	f_L	f_L/f_c
	(MPa)	(MPa)	(%)	(MPa)	(%)
RC-B	32.09	24.6	41.39	26.74	44.93
RC-S	29.76	23.81	30.60	24.51	31.49
RC-M	29.685	23.75	37.80	25.45	40.59

In order to compare the evolution of the maximum strain throughout the test, it was decided to divide the number of cycles performed by the number of cycles per step, in this way it was be possible to compare the results of the three variants of the test. An example of the comparison between the Locati test types for each material is given in Figure 10; Figure 12.

Figure 10 shows the influence of increasing the frequency on the fatigue behaviour of the RC-B by comparing the RC-B-HF-2×10^5 with the RC-B-MF-2×10^5. It is possible to conclude that, in the first phase, increasing the test frequency had no influence on the deformation suffered by the specimens. After step 4, the two curves separated, increasing more rapidly in the case of RC-B-HF-2×10^5. This behaviour can be justified by arguing that, in the first phase, a phase in which the concrete was not damaged, the effect of the frequency seemed irrelevant, whereas when the cracks reached a critical size, an increase in frequency produced a reduction in the number of cycles that RC-B could withstand. It was observed that the specimens tested at very high frequencies showed a markedly higher temperature increase than in the case of tests performed at moderate frequency. This fact is believed to be related to the reduction in fatigue life of the material.

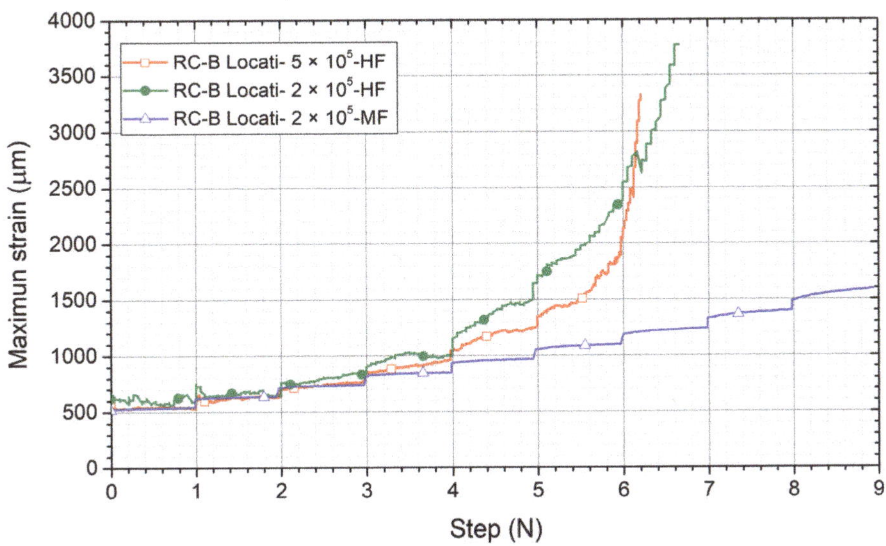

Figure 10. Comparison of the maximum strain of the RC-B for a Locati test with 2×10^5 cycles per step at a low frequency (RC-B-MF-2×10^5), a Locati test with 2×10^5 cycles per step at a high frequency (RC-B-HF-2×10^5) and a Locati test with 5×10^5 cycles per step at a high frequency (RC-B-HF-5×10^5).

From this same Figure 10, it is also possible to analyse the influence of increasing the number of cycles per step from 2×10^5 to 5×10^5. Comparing the two curves, it is observed that the effect of each step was similar until the specimen was close to breaking. For this reason, it can be assumed that the influence of increasing the number of cycles per step during a Locati test was small given that the difference between steps had a greater influence than increasing the number of cycles per step by 150%.

Figure 11 shows the influence of increasing the frequency on the fatigue behaviour of the RC-S by comparing RC-S-HF-2×10^5 with RC-S-MF-2×10^5. As in RC-B, it can be concluded that, in the first phase, increasing the test frequency had no impact on the deformation suffered by the specimens. From step 2, it can be seen how the two curves split, increasing more rapidly in the case of the RC-S-HF-2×10^5. This effect was more severe in RC-S than in RC-B, which was reflected by a greater variation in the IC of RC-S as the test frequency increased.

Figure 11 shows the influence of increasing the number of cycles per step from 2×10^5 to 5×10^5 for RC-S. A similar behaviour to that of RC-B was observed, but in this case, the effect of increasing the number of cycles per step was greater than in RC-B, which was reflected in the accelerated separation of the curves. As in the case of RC-B, in this case, the greatest difference in fatigue limit was found between RC-B-HF-2×10^5 and RC-B-HF-5×10^5 and was one step; for this reason, it can be assumed that the influence of increasing the number of cycles per step during a Locati test was small, given that the jump between steps had a greater influence than increasing the number of cycles per step by 150%.

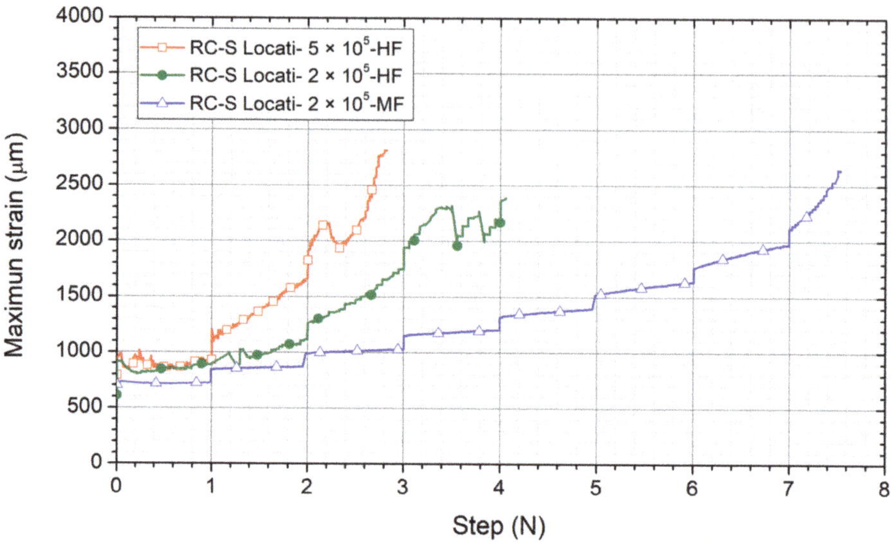

Figure 11. Comparison of the maximum strain of the RC-S for a Locati test with 2×10^5 cycles per step at a low frequency (RC-S-MF-2×10^5), a Locati test with 2×10^5 cycles per step at a high frequency (RC-S-HF-2×10^5) and a Locati test with 5×10^5 cycles per step at high frequency (RC-S-HF-5×10^5).

In Figure 12, both the frequency and number of cycles affected the fatigue life during the Locati RC-M test. The behaviour of this material was an intermediate situation between RC-B and RC-S, being in all cases, more similar to the behaviour of RC-B.

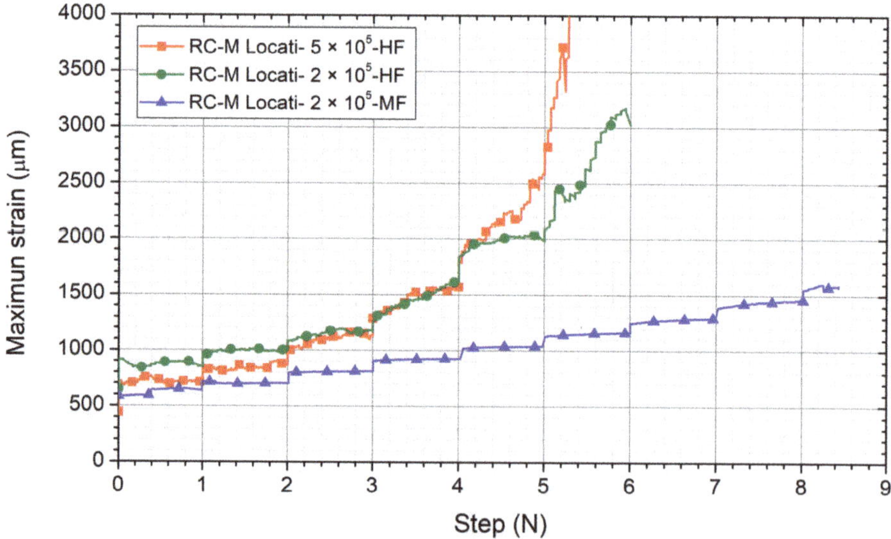

Figure 12. Comparison of the maximum strain of the RC-M for a Locati test with 2×10^5 cycles per step at a low frequency (RC-M-MF-2×10^5), a Locati test with 2×10^5 cycles per step at a high frequency (RC-M-HF-2×10^5) and a Locati test with 5×10^5 cycles per step at a high frequency (RC-M-HF-5×10^5).

To summarise, it can be stated that performing tests at a very high frequency reduced the number of cycles that a specimen was capable of resisting compared to performing the test at a moderate frequency. On the other hand, it was also demonstrated that 2×10^5 cycles per step was enough to characterize a fatigued concrete.

3.4. Comparison of the Different Methodologies

In this section the endurance values for each of the three concretes tested was compared using the five proposed analysis criteria. Figure 13 shows the values of the stress range corresponding to the fatigue limit of the three concretes analysed using the five previously defined procedures.

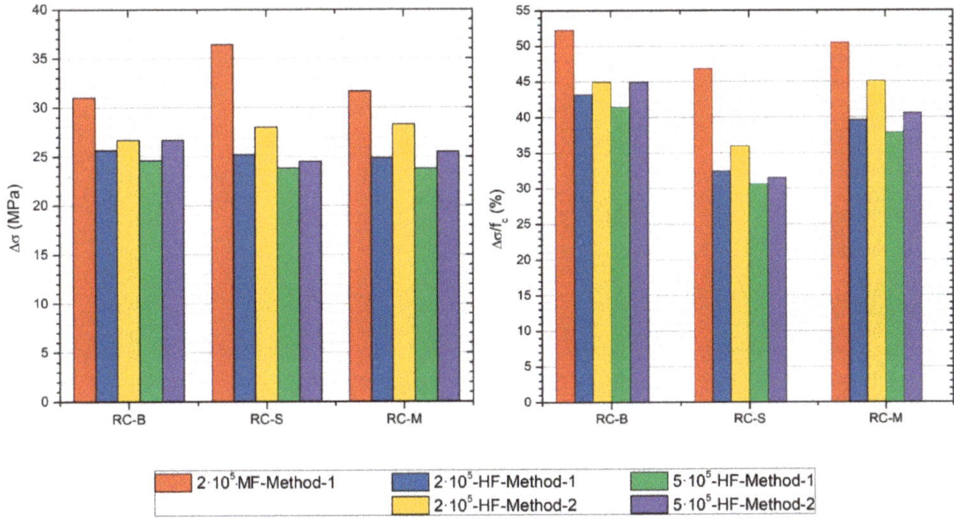

Figure 13. Comparison of the high-frequency fatigue limit obtained using five different procedures and the low-frequency fatigue limit.

It can be observed that, in all cases, the fatigue limits obtained using moderate frequency tests gave higher values than in very high frequency tests, regardless of the characterization method.

On the other hand, the two methods used to determine the fatigue limit at a very high frequency produced similar values in all cases, where the one that defines the fatigue limit as 80% of the tension range of the breaking step being more conservative in all cases. This was due to the weakness introduced by the mortar adhered to the aggregates, as well as RC-S being the material most affected by creep.

4. Conclusions

Although concrete is a material that is not usually characterized under fatigue, there is no doubt that there are certain elements typically made from concrete that are subjected to variable loads, and therefore, which need to be characterized under fatigue. In this work, three types of concrete were characterised under fatigue using the Locati accelerated fatigue method. Fatigue tests were carried out to analyse two fundamental variables in the duration of the tests, namely the frequency of testing and the number of cycles per step of the Locati method. The following conclusions can be observed from the results obtained:

- In all cases, the material with the lowest fatigue limit/compression resistance ratio was RC-S, which was due to the weakness introduced by the mortar adhered to the aggregates.

- The Locati method was validated as a method to determine the fatigue limit, showing that 2×10^5 cycles per step was enough to determine the fatigue limit.

- It was found that the resonance frequency of the system was a parameter that could enable the identification of sensitive variations in the stiffness of the whole, and a symptom that the specimen was close to breaking. For this reason, the stress range of the step prior to the step in which a drop in the resonance frequency of the system occurs was defined as the criterion for determining the fatigue limit by means of the Locati tests.

- It is proposed that during very high frequency tests there is an increase in temperature that may reduce the fatigue life of the concrete. This study opens the door to the analysis of this hypothesis, which may explain why an increase in frequency reduces the fatigue life of the elements.

Author Contributions: J.S.-A.: investigation, methodology, formal analysis, writing original draft. C.T.: formal analysis, surpervision, writing original draft, review and editing. I.C.: founding, methodology, formal analysis, surpervision, review and editing. J.A.P.: methodology, formal analysis, surpervision, review and editing. All authors have read and agreed to the published version of the manuscript.

Funding: This research was funded Ministerio de Economía y Competitividad grant number MAT2014-57544-R.

Acknowledgments: The authors would like to thank The LADICIM, Laboratory of Materials Science and Engineering of the University of Cantabria and Instituto Superior Técnico of the University of Lisbon for making the facilities used in this research available to the authors.

Conflicts of Interest: The authors declare no conflict of interest.

References

1. Alnahhal, M.F.; Alengaram, U.J.; Jumaat, M.Z.; Abutaha, F.; Alqedra, M.A.; Nayaka, R.R. Assessment on engineering properties and CO_2 emissions of recycled aggregate concrete incorporating waste products as supplements to Portland cement. *J. Clean. Prod.* **2018**, *203*, 822–835. [CrossRef]
2. Meyer, C. The greening of the concrete industry. *Cem. Concr. Compos.* **2009**, *31*, 601–605. [CrossRef]
3. Poon, C.-S.; Chan, D. The use of recycled aggregate in concrete in Hong Kong, Resour. *Conserv. Recycl.* **2007**, *50*, 293–305. [CrossRef]
4. De Brito, J.; Ferreira, J.; Pacheco, J.; Soares, D.; Guerreiro, M. Structural, material, mechanical and durability properties and behaviour of recycled aggregates concrete. *J. Build. Eng.* **2016**, *6*, 1–16. [CrossRef]
5. Thomas, C.; Setién, J.; Polanco, J.A.J.; Alaejos, P.; de Juan, M.S.; de Juan, M.S. Durability of recycled aggregate concrete. *Constr. Build. Mater.* **2013**, *40*, 1054–1065. [CrossRef]
6. De Schutter, G.; Audenaert, K. *Report 38: Durability of Self-Compacting Concrete-State-of-the-Art Report of RILEM Technical Committee 205-DSC*; RILEM Publications: Paris, France, 2007.
7. Kou, S.C.C.; Poon, C.S.S. Properties of self-compacting concrete prepared with coarse and fine recycled concrete aggregates. *Cem. Concr. Compos.* **2009**, *31*, 622–627. [CrossRef]
8. Kareem, A.I.; Nikraz, H.; Asadi, H. Performance of hot-mix asphalt produced with double coated recycled concrete aggregates. *Constr. Build. Mater.* **2019**, *205*, 425–433. [CrossRef]
9. Sainz-Aja, J.A.; Pombo, J.; Tholken, D.; Carrascal, I.; Polanco, J.; Ferreno, D.; Casado, J.; Diego, S.; Pérez, A.; Filho, J.A.; et al. Dynamic Calibration of Slab Track Models for Railway Applications using Full-Scale Testing. *Comput. Struct.* **2020**, *228*, 106180. [CrossRef]
10. Alliche, A. Damage model for fatigue loading of concrete. *Int. J. Fatigue* **2004**, *26*, 915–921. [CrossRef]
11. Xiao, J.; Li, H.; Yang, Z. Fatigue behavior of recycled aggregate concrete under compression and bending cyclic loadings. *Constr. Build. Mater.* **2013**, *38*, 681–688. [CrossRef]
12. Skarżyński, Ł.; Marzec, I.; Tejchman, J. Fracture evolution in concrete compressive fatigue experiments based on X-ray micro-CT images. *Int. J. Fatigue* **2019**, *122*, 256–272. [CrossRef]
13. Khosravani, M.R.; Wagner, P.; Fröhlich, D.; Weinberg, K. Dynamic fracture investigations of ultra-high performance concrete by spalling tests. *Eng. Struct.* **2019**, *201*, 109844. [CrossRef]
14. Li, Q.; Huang, B.; Xu, S.; Zhou, B.; Yu, R.C. Compressive fatigue damage and failure mechanism of fiber reinforced cementitious material with high ductility. *Cem. Concr. Res.* **2016**, *90*, 174–183. [CrossRef]
15. Thomas, C.; Setién, J.; Polanco, J.A.A.; Lombillo, I.; Cimentada, A. Fatigue limit of recycled aggregate concrete. *Constr. Build. Mater.* **2014**, *52*, 146–154. [CrossRef]

16. Thomas, C.; Sosa, I.; Setién, J.; Polanco, J.A.; Cimentada, A.I. Evaluation of the fatigue behavior of recycled aggregate concrete. *J. Clean. Prod.* **2014**, *65*, 397–405. [CrossRef]

17. Vicente, M.A.; Mínguez, J.; González, D.C. Computed tomography scanning of the internal microstructure, crack mechanisms, and structural behavior of fiber-reinforced concrete under static and cyclic bending tests. *Int. J. Fatigue* **2019**, *121*, 9–19. [CrossRef]

18. Thomas, C.; de Brito, J.; Gil, V.; Sainz-Aja, J.A.; Cimentada, A. Multiple recycled aggregate properties analysed by X-ray microtomography. *Constr. Build. Mater.* **2018**, *166*, 171–180. [CrossRef]

19. Thomas, C.; de Brito, J.; Cimentada, A.; Sainz-Aja, J.A. Macro-and micro-properties of multi-recycled aggregate concrete. *J. Clean. Prod.* **2019**, 118843. [CrossRef]

20. Sainz-Aja, J.A.; Carrascal, I.A.; Polanco, J.; Sosa, I.; Thomas, C. Análisis del comportamiento a fatiga de morteros con áridos reciclados provenientes de vía de ferrocarril. In Proceedings of the REHABEND 2018, Caceres, Spain, 15–18 May 2018; pp. 1313–1322.

21. Sainz-Aja, J.; Carrascal, I.; Polanco, J.A.; Thomas, C. Fatigue failure micromechanisms in ed aggregate mortar by μCT analysis. *J. Build. Eng.* **2019**, 101027. [CrossRef]

22. Carloni, C.; Subramaniam, K.V. Investigation of sub-critical fatigue crack growth in FRP/concrete cohesive interface using digital image analysis. *Compos. Part B Eng.* **2013**, *51*, 35–43. [CrossRef]

23. Rep, C.; Hanson, J.M.; Ballinger, C.A.; Linger, D. Considerations for Design of Concrete Structures Subjected to Fatigue Loading. *J Am. Concr. Inst.* **1974**, *71*, 97–121.

24. European Concrete Committee, Ceb Committee GTG 15. *Fatigue of Concrete Structures*; CEB Bulletin d'information, N° 188; CEB-FIP: Lausanne, Switzerland, 1988.

25. Lantsoght, E.O.L.; van der Veen, C.; de Boer, A. Proposal for the fatigue strength of concrete under cycles of compression. *Constr. Build. Mater.* **2016**, *107*, 138–156. [CrossRef]

26. Balázs, G.L. Fatigue of bond. *ACI Mater. J.* **1991**, *88*, 620–629.

27. Vicente, M.A.; Ruiz, G.; González, D.C.; Mínguez, J.; Tarifa, M.; Zhang, X. CT-Scan study of crack patterns of fiber-reinforced concrete loaded monotonically and under low-cycle fatigue. *Int. J. Fatigue* **2018**, *114*, 138–147. [CrossRef]

28. Yoris, A.; Thomas, C.; Medina, C.; Polanco, J.A.; de Rojas, M.I.S.; Frias, M.; Cantero, B. Comportamiento a fatiga resonante en compresión de hormigones reciclados para uso estructural. In Proceedings of the 7th Euro-American Congress on Construction Pathology, Rehabilitation Technology and Heritage Management, REHABEND 2018, Caceres, Spain, 15–18 May 2018; pp. 1460–1467.

29. Medeiros, A.; Zhang, X.; Ruiz, G.; Yu, R.C.; Velasco, M.d.L. Effect of the loading frequency on the compressive fatigue behavior of plain and fiber reinforced concrete. *Int. J. Fatigue* **2015**, *70*, 342–350. [CrossRef]

30. Murdock, J.W. *A Critical Review of Research on Fatigue of Plain Concrete*; University of Illinois at Urbana Champaign, College of Engineering: Urbana, IL, USA, 1965.

31. EN 197-1, Cement Part 1: Composition, Specifications and Conformity Criteria for Common Cements, Br. Stand. 2011. Available online: https://doi.org/10.1103/PhysRevLett.64.88 (accessed on 18 December 2019).

32. Alfa, C. Ficha técnica EN 197-1 CEM IV/B (V) 32,5 N. in press.

33. Test Methods of Cements. Physical Analysis. Actual Density Determination, U.N.E. 80103:2013. 2013. Available online: https://infostore.saiglobal.com/en-us/Standards/UNE-80103-2013-22841_SAIG_AENOR_AENOR_50269/ (accessed on 18 December 2019).

34. Methods of Testing Cement—Part 6: Determination of Fineness, E.N. 196-6:2010. 2010. Available online: https://webstore.ansi.org/standards/din/dinen1962010 (accessed on 18 December 2019).

35. Sainz-Aja, J.; Carrascal, I.; Polanco, J.A.; Thomas, C.; Sosa, I.; Casado, J.; Diego, S. Self-compacting recycled aggregate concrete using out-of-service railway superstructure wastes. *J. Clean. Prod.* **2019**, *4*, 386. [CrossRef]

36. Testing Hardened Concrete—Part 3: Compressive Strength of Test Specimens, U.-E. 12390-3:2009. 2009. Available online: https://www.scirp.org/(S(vtj3fa45qm1ean45vvffcz55))/reference/ReferencesPapers.aspx?ReferenceID=1884402 (accessed on 18 December 2019).

37. Testing Hardened Concrete—Part 3: Compressive Strength of Test Specimens, E.N. 12390-3:2099/AC:2011. 2011.

38. Testing Hardened Concrete—Part 13: Determination of Secant Modulus of Elasticity in Compression, E.N. 12390-13:2013. 2014. Available online: http://78.100.132.106/External%20Documents/Intenational%20Specifications/British%20Standards/BS%20EN/BS%20EN%2012390-13-2013.pdf (accessed on 18 December 2019).

39. Locati, L. *La Fatica dei Materiali Metallici*; Ulrico Hoepli: Milan, Italy, 1950.

40. Locati, L. Programmed fatigue test, variable amplitude rotat. *Metall. Ital.* **1952**, *44*, 135–144.

41. García, C.T. *Hormigón Reciclado de Aplicación Estructural*; Universidad de Cantabria: Santander, Spain, 2012.
42. Thomas, C.; Carrascal, I.; Setién, J.; Polanco, J.A. Determinación del límite a fatiga en hormigones reciclados de aplicación estructural—Determining the fatigue limit recycled concrete structural application. *An. Mec. Fract.* **2009**, *1*, 283–289. (In Spanish)

MDPI

St. Alban-Anlage 66

4052 Basel

Switzerland

Tel. +41 61 683 77 34

Fax +41 61 302 89 18

www.mdpi.com

Applied Sciences Editorial Office

E-mail: applsci@mdpi.com

www.mdpi.com/journal/applsci